Color Tab Index to Bi

With Photos by

Barth Schorre,

Vernon E. Grove Jr.,

David F. Parmelee,

Kevin Winker,

and VIREO

(Academy of Natural

Sciences)

Birds of the Mid-Atlantic Region

and where to find them

John H. Rappole

The Johns Hopkins University Press • Baltimore and London

Printed in China on acid-free paper
9 8 7 6 5 4 3 2 1

The Johns Hopkins University Press
2715 North Charles Street
Baltimore, Maryland 21218-4363
www.press.jhu.edu

Library of Congress Cataloging-in-Publication Data
Rappole, John H.
Birds of the Mid-Atlantic region and where to find them /
John H. Rappole.
p. cm.
Includes bibliographical references (p.) and index.
ISBN 0-8018-7075-5 (hardcover : alk. paper) —
ISBN 0-8018-7077-1 (pbk. : alk. paper)
1. Birds—Middle Atlantic States—Identification. I. Title.
QL683.M54 R36 2002
598′.0975—dc21 2001008637

A catalog record for this book is available from the British Library.

To Bonnie:
Oh, you were the best of all my days.
—Frank O'Hara, *Animals*

Contents

Preface and Acknowledgments

Every region of the country has its share of spectacular places to visit and find birds, but the Mid-Atlantic is among the best in the nation. There are obvious sites like Chincoteague on Virginia's Eastern Shore; Hawk Mountain, Pennsylvania; or New Jersey's Cape May and Brigantine Marsh, four of the crown jewels of American birding. Some of the less obvious gems are Cranberry Glade, West Virginia; Swallow Falls, Maryland; and Trap Pond, Delaware. I have spent much of the past decade studying the region's birdlife and visiting these and similar sites throughout the six states and the District of Columbia, which compose the Mid-Atlantic region, so that those who visit, plan to visit, or just want to take a vicarious vacation will have the best and most complete information in the form of this book. Wherever your travels take you in the region, whether actual or imaginary, this book will help you make the most of that experience—guiding you to the sites, locating and identifying the birds, and bringing back the memories once you have gone. That is what the regional guide is all about.

People often like to purchase books when visiting a new part of the country as introduction and souvenir for that particular location: a volume of Rand-McNally maps, an AAA tour guide to places to stay and see, a U.S. Park Service pamphlet describing scenic spots. For those who travel and enjoy the outdoors, a regional bird guide can satisfy the same kind of need. Whether the destination is an exotic place like Canaan Valley, West Virginia, or an urban oasis like Philadelphia's Tinicum Refuge, a beautiful guide to the birds and habitats of the new area can serve as a handy reference while you are there and a delightful reminder of the trip when you get home.

Until now, there have been no good *regional* field guides available to American birders—guides that provide accounts, photos, and maps for every species, along with photographs of exciting birding spots and instructions on how to find them. There are many excellent and more general national guides, such as the Peterson and Audubon series, National Geographic, Golden Guides, Masters guides, and Stokes guides. But national guides don't have the space to tell you whether you are likely to hear a Hermit Thrush on Virginia's Mount Rogers in June or scare up a Short-eared Owl on abandoned strip mines in Pennsylvania's Clarion County. In addition, while most regions of interest to birders now have detailed handbooks explaining precisely which county road to take to find a particular rarity, these books, too, fail to fill the important niche of the regional guide, lacking detailed descriptions, maps, and identification photos. In short, there is a need that neither national guides nor regional handbooks can satisfy. If you want to develop some sense of what a particular bird is about within a relatively confined geographical area, the regional field guide, with its unique combination of information on the bird *and* its regional status, is the best answer.

I thank Nate Rappole, who drew the maps, and Jeff Diez, who provided help with computer graphics. David Johnston was extremely helpful in providing a thorough review of the manuscript. Several of the most outstanding wildlife photographers in North America donated slides for this book, including Barth Schorre, Vernon E. Grove

Jr., and David F. Parmelee. Doug Wechsler of the Academy of Natural Sciences's Bird Photo Collection (VIREO) was extremely helpful in providing additional photos. Certain sections of this guide (e.g., some species and range descriptions) are taken from *Birds of Texas: A Field Guide*, published by Texas A&M University Press, copyright 1994 by John H. Rappole and Gene W. Blacklock. Material from that work reprinted herein is used with the kind permission of my coauthor, Gene W. Blacklock. Finally, I thank my wife, Bonnie Rappole, for her invaluable encouragement and support throughout the long evolutionary history of the project.

Birds of the Mid-Atlantic Region

and where to find them

Using This Guide

Each account begins with the species' common name followed by the scientific name in italics. Nomenclature and taxonomic organization follow the American Ornithologists' Union's *Check-list of North American Birds,* seventh edition (1998), as amended by the 42nd supplement (American Ornithologists' Union 2000). The size of the bird is given in parentheses: L = length from tip of the bill to tip of the tail in inches; W = wingspread in inches. A description of the adult male in breeding plumage follows. Other plumages are described where necessary; for example, immature (*First Basic,* i.e., first winter after hatching) and female plumages are given when they differ markedly from that of the breeding male. The plumage description necessarily involves the use of a few arcane morphological terms. These parts of the bird are shown in figure 1. One photograph depicts each species. Although photos can be misleading in terms of plumage coloration, they capture an essence of how the bird carries itself that is peculiar to that particular species, a quality that is difficult to capture in a painting. In addition, photos often place the bird in a fairly typical habitat. A distribution map is provided for every species found regularly in the Mid-Atlantic, showing where the bird can be expected to occur by season (summer, winter, and fall/spring transient).

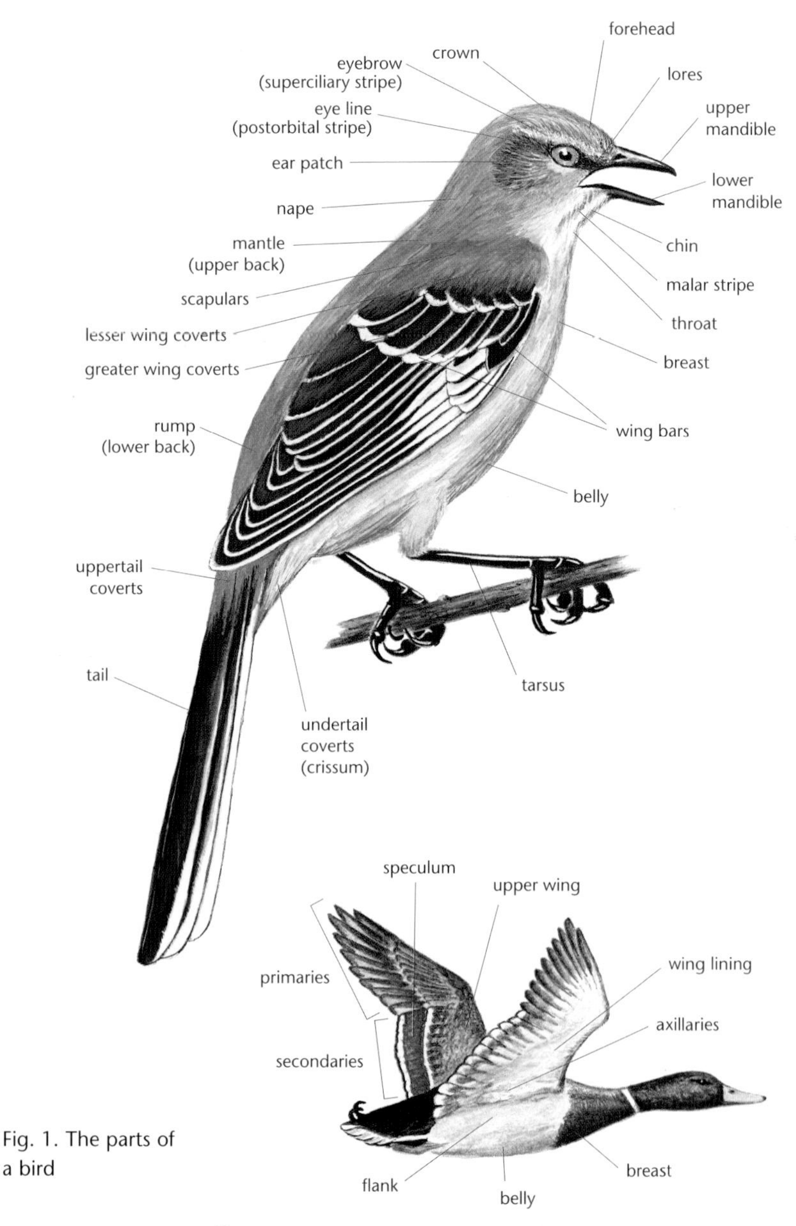

Fig. 1. The parts of a bird

Habits — This category is included only when some peculiarity of the bird's behavior can be useful in identifying the species (e.g., "hangs upside down while foraging" or "flicks tail").

Voice — This section may include both (1) a description of the song, which is usually given only by the male during the breeding season, and (2) the call, which is normally given by both sexes at any time during the year. For the song description, a more or less "typical" song is described. Songs are described based on field notes, recorded songs, or, where necessary, descriptions in the literature (author credited). Songs may vary not only regionally but individually, and one person's interpretation of what a song sounds like will differ from another's. To some people "chip" is "ship" or "tschip" or "slip."

Similar Species — This information is given only if another species is similar in size, pattern, and coloration.

Habitat — Those habitats most often used by the bird in the Mid-Atlantic region and on its North American breeding grounds are given here. No attempt is made to describe tropical habitats used by birds when they are away from the Mid-Atlantic. Transients are found in a variety of habitats in which they would not normally forage, and no attempt is made to catalog all such habitats.

Abundance and Distribution — The abundance, principal time of occurrence, and range for the bird in the Mid-Atlantic are presented. Normally only regular occurrences (common, uncommon, and rare) of the bird within the region are reported. All species that have been recorded as breeding in the region are marked with an asterisk (*) following a statement of their residency status. The abundance categories used are

Common: Ubiquitous in specified habitat; high probability of finding several individuals (>5) in a day.
Uncommon: Present in specified habitat; high probability of finding a few individuals (<5) in a day.
Rare: Scarce in specified habitat, with only a few records per season; low probability of finding the bird.
Casual: A few records per decade.
Accidental: Not expected to recur.
Hypothetical: Recorded for the region but not accepted by the state records committees.

Where to Find

Some species are ubiquitous, and for these it is almost silly to provide a specific locality. There is an old story about a bird-watching newspaper columnist who, when called and asked, "What's this bird in my yard?" answered, "House Sparrow." Good guess. Some birds are everywhere, but you will have a much better chance of finding most species if you are directed to a particular place. In this section I try to give at least three localities from different parts of the region. Directions to all localities are provided in the locator section in the front of the book. When considering sites for listing, I attempted, when possible, to choose those that are well-known, public-use areas requiring relatively little familiarity with the region to locate. Nevertheless, some listed localities will require fairly adventurous navigating. Another caveat is that there is no guarantee that the bird will actually be there when you get there. The locality information is derived principally from published accounts in the many "bird-finding" guides on various parts of the region. In most cases, I have tried to pinpoint the best places and time of year in which to find the bird. But many species simply are not found anywhere with any degree of regularity, and even in well-known regions there are places that have escaped the notice of bird watchers. To obtain further information on the best places to find birds, the reader should consult bird-finding guides, which may include several widely distributed localities where a bird may be found. The principal sources for bird-finding sites are listed below and in the references: Johnston's *A Birder's Guide to Virginia* (1997), Ford's *Birder's Guide to Pennsylvania* (1995), Boyle's *A Guide to Bird Finding in New Jersey* (1986), Sibley's *The Birds of Cape May* (1993), Wilds's *Finding Birds in the National Capital Area* (1992), and Simpson's *Birds of the Blue Ridge Mountains* (1992). Not all of the locality information used in this book was derived from these sources; breeding bird atlases, checklists from specific sites, and personal experience were also used. However, all of the birding guides have particulars on exactly where and when to find most of the species that occur in the regions they cover—and how to get there. Hunting down specialties and rarities using these guides can be challenging, exciting, and fun.

Range The total world range of the bird is provided in abbreviated form, based on information from the American Ornithologists' Union *Check-list*, 7th edition (1998).

Maps For every species found regularly in the Mid-Atlantic, a distribution map shows where the bird can be expected to occur by season (summer—tan, winter—blue, fall/spring transient—gray, year-round—purple).

Quick Guide to the Most Common Birds

Thirty-three of the most common species in the Mid-Atlantic region are shown here as a quick reference. Of course, what is "common" depends on where you are and when you are there. This group is broken down according to major human summertime haunts.

Downtown

European Starling

House Sparrow

Rock Dove

Backyard

American Robin

Northern Cardinal

Downy Woodpecker

Carolina Chickadee

Baltimore Oriole

At the Beach

Brown Pelican

Laughing Gull

Great Black-backed Gull

Ruddy Turnstone

American Oystercatcher

The Open Road

Red-tailed Hawk

Turkey Vulture

American Crow

Indigo Bunting

American Kestrel

On the Farm

Eastern Meadowlark

Barn Swallow

Eastern Bluebird

Red-winged Blackbird

Mourning Dove

On the River

Great Blue Heron

Belted Kingfisher

Mallard

Canada Goose

Northern Rough-winged Swallow

In the Woods

Wood Thrush

Scarlet Tanager

American Redstart

Acadian Flycatcher

Pileated Woodpecker

General Description of the Mid-Atlantic Region

Thirty-five million people live in the Mid-Atlantic region, and most of us know little about the immense natural richness of our surroundings. We move through it as though in a dream, focused on our jobs, love life, kids, or the day-to-day frustrations of living at the beginning of a new millenium, unaware of the abundance of life around us. We wake in a start when confronted by an intrusion—usually unwelcome—from the natural world, by, for example, wind, heat, cold, rain, snow, biting insects, confused squirrels, hungry rodents, or overabundant pigeons. To many of us, the changing seasons are marked mainly by what sports are on television and what discomforts and inconveniences might be encountered when we leave home for the day. To understate the message of Henry David Thoreau, Edward Abbey, and John Muir: we are missing a lot.

Looking at birds is one way to stop missing so much. You begin to understand something about your surroundings when you are able to recognize and name some of its more obvious parts, and birds are among the easiest, most accessible, and most satisfying features of the natural

world to identify and enjoy. In the Mid-Atlantic, we are blessed with a rich variety: a total of about 472 species of birds has been recorded, of which 346 occur regularly during one or more seasons each year, only 60 fewer than are found throughout all of Britain and Europe. Complete descriptions and accounts for each of those birds occurring *regularly* in the region (likely to be found at least once a year) are contained in this book, and the 126 species that are considered *casual* (occur a few times per decade), *accidental* (not expected to recur), *hypothetical* (not documented by specimen or photo), or *extinct* are listed in the appendix.
In the Mid Atlantic region we are privileged to have spectacular places to view birds that are just a short drive from wherever we are, places that people from around the country and the world come thousands of miles to see. Cape May, Hawk Mountain, Skyline Drive, Assateague Island, Great Dismal Swamp, Brigantine, Presque Isle, Bombay Hook, and Canaan Valley are among the most famous birding destinations in the United States, and no one of them is more than an easy day's drive from almost anywhere in the Mid-Atlantic states.

A History of the Region from an Ornithological Perspective

Knowledge of the birds in the Mid-Atlantic grew very slowly, and most of what we now know was learned within the past 150 years. The first permanent European settlement was established in Jamestown, Virginia, on May 14, 1607. Captain John Smith, John Rolfe, and other colonial leaders wrote little in their journals about the avifauna of the New World and, indeed, not much was written by any of the early pioneers about birds except, perhaps, to comment on their flavor. A German explorer, John Lederer, who made three trips to Virginia's Blue Ridge mountains in 1669 and 1670, did a respectable job of describing the basic fauna, a feat for any visitor of the day, considering the effort required to avoid starvation, freezing, dying of thirst, or being killed by Indians or other members of the local megafauna. Lederer listed bears, wolves, elk, mountain lions, raccoons, foxes, bobcats, rattlesnakes, squirrels, beaver, otter, swans, geese, ducks, turkeys, pigeons, grouse, and a recluse spider (which bit him) but provided no great detail on the birdlife. Surveyors like William Byrd II, Thomas Lewis, and George Washington (*the* George Washington) often kept meticulous notes on what they encountered on their surveys. Lewis, for instance, writing in 1746, mentioned many of the common tree species observed and used as pole markers on his traverse of the Fairfax Line (running along the present-day boundary between Shenandoah and Rockingham Counties in Virginia and defining the property inherited by Lord Fairfax from his mother, daughter of a former Virginia governor), although, like Lederer, the only birds he remarked upon were those he ate. Thomas Jefferson made an early list of the bird species of Virginia, as did the great botanist and chronicler of natural history, William Bartram, who lived and worked in eastern Pennsylvania, but he produced nothing from this area to rival his classic work, *Travels of William Bartram* (1791), which provides remarkable detail on the animal and plant life of the southeastern United States (Bartram 1983 in reprint).

The first thorough treatment of the birds of the region was written by Alexander Wilson nearly two hundred years after settlement of the first colony. Wilson's *American Ornithology* is a multivolume work; the first installment was published in 1808 and the last in 1833 with help from George

Ord and Charles Lucien Bonaparte (Napoleon's nephew). John James Audubon was the first to present precise and comprehensive depictions of the birds of the Mid-Atlantic in *The Birds of America* (1827–39), in which paintings of 435 species, mostly from eastern North America, were represented. Mark Catesby's (1731–48) *Natural History of Carolina, Florida, and the Bahama Islands* predated Audubon's work by almost a century. His paintings include more than one hundred species that occur in the Mid-Atlantic.

Compilation of detailed lists of the birds from specific parts of the region began in the late 1800s with the work of Coues and Prentiss (*Avifauna Columbiana,* 1883) for Washington, D.C., and surrounding areas of Virginia and Maryland, Rives's *Catalogue of the Birds of the Virginias* (1889–90), Warren's *Report on the Birds of Pennsylvania* (1890), Shriner's *The Birds of New Jersey* (1896), and Maynard's *Birds of Washington and Vicinity* (1902). More thorough accounts providing aspects of life history for each species treated came later: Bailey's *The Birds of Virginia* (1913), Johnston's *Birds of West Virginia* (1923), Stone's *Bird Studies at Old Cape May* (1937), Todd's *Birds of Western Pennsylvania* (1940), and Stewart and Robbins's *Birds of Maryland and the District of Columbia* (1958). Excellent recent treatments have included: Wood's *Birds of Pennsylvania* (1979), Leck's *The Status and Distribution of New Jersey's Birds* (1984), and Hall's *West Virginia Birds* (1983).

Perhaps the most exciting recent development for the science of ornithology in the Mid-Atlantic has been the institution of Breeding Bird Atlas programs. Staffed by hard-working volunteers, these programs produce meticulous documentation of exact localities for every bird species found breeding in a state. To date, such atlases have been completed for Pennsylvania (Brauning 1992), West Virginia (Buckelew and Hall 1994), and Maryland (Robbins 1996).

Landforms, Climate, and Habitat

The topography, the lengths and characteristics of the seasons, and the plant life in the Mid-Atlantic determine the rich variety of birds found here. Thus, some knowledge of these factors is helpful in developing an understanding of its birdlife and an appreciation for the region's wealth of natural beauty.

Though it is hard to believe now, when people crisscross the Appalachians at 70 mph, hardly noticing them in their passage, the mountain chain was a major barrier to westward settlement at least until the early 1800s. The aforementioned John Lederer and his contemporaries certainly did not see the Appalachians as insignificant in the late 1600s. In fact, half of Lederer's party, led by his military escort, Major Harris, deserted at the mere prospect of crossing the upper James River and approaching the mighty "Apalatan." This mountain range forms the dominant physiographic feature of the Mid-Atlantic. Six major landforms (geological provinces) are found in the region. From east to west, these are the (1) Coastal Plain, (2) Piedmont, (3) Blue Ridge, (4) Ridge and Valley, (5) Appalachian Plateau, and (6) Central Lowlands (fig. 2).

Characterized by its flatness, the Coastal Plain stretches twenty-two hundred miles along the edge of the continent from Massachusetts to the Mexican border, forming the shifting boundary between land and ocean. In the Mid-Atlantic, the southern end of this province is about 125 miles wide at the Virginia–North Carolina border, but it narrows sharply and disappears at its northern end in central New Jersey, where the Piedmont extends to the ocean shore. Several of the great cities of the Mid-

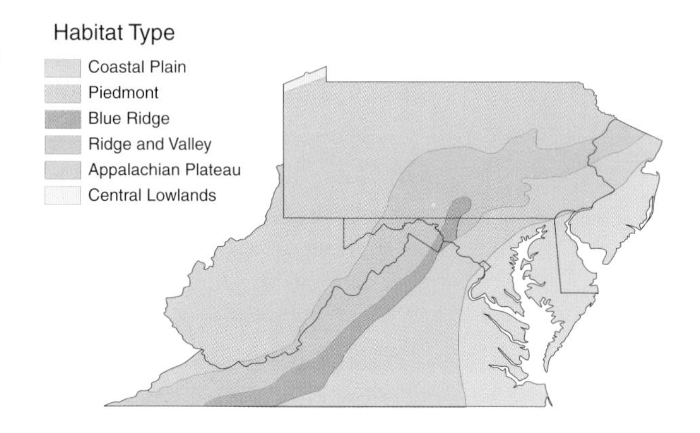

Fig. 2. Map of the Mid-Atlantic states showing principal landforms

Atlantic, such as Washington, Baltimore, Philadelphia, and Richmond, are located at the boundary between Coastal Plain and Piedmont known as the Fall Line. Here rivers like the James, Delaware, and Potomac tumble out of the hard bedrock of the rolling Piedmont onto the sediments of the level Coastal Plain, and falls mark the head of navigation. The Piedmont is the easternmost of the four provinces that form subcategories of a single geological feature—the Appalachian Highlands (Piedmont, Blue Ridge, Ridge and Valley, and Appalachian Plateau). The hills of the Piedmont extend westward from the Fall Line to the base of the Blue Ridge Mountains in Virginia, Maryland, and southern Pennsylvania and to the mountains of the Ridge and Valley Province of Pennsylvania and New Jersey. The Blue Ridge is a long, narrow province consisting of a single ridge or series of parallel ridges and valleys running generally from southwest to northeast 550 miles from north Georgia to Carlisle, Pennsylvania. North and west of the Blue Ridge is a wider series (up to 80 miles) of higher ridges and deeper valleys that also run southwest to northeast; these constitute the Ridge and Valley Province, which extends twelve hundred miles from the Saint Lawrence River at the Canadian border to central Alabama. West and north of this province are the steep hills of the Appalachian Plateau, covering most of West Virginia and western Pennsylvania. A sixth province, the Central Lowlands, occurs in only a small portion of our region, that part of Pennsylvania bordering Lake Erie, where Presque Isle and the city of Erie are located.

Four major climatic zones occur in the Mid-Atlantic. To categorize them, I use the "life zone" classification system developed by C. Hart Merriam, based principally on climatic characteristics, as follows: (1) the Austral or Austroriparian is characterized by hot summers, mild winters, and moderate precipitation and is found only in southeastern Virginia; (2) the Carolinian is characterized by hot summers, cool winters, and moderate precipitation and covers the Piedmont region and valleys of the southern Appalachians; (3) the Alleghenian is characterized by warm summers, cold winters, and moderate precipitation and covers most of the Appalachian Highlands; and (4) the Canadian is characterized by cool summers, harsh winters, and moderate precipitation and is found only at the highest elevations in our region.

The varied landforms and climate of the Mid-Atlantic create the basis for the wonderful diversity of plant and animal communities, or habitats, that occur here. Specific associations or communities of plants and animals constitute a habitat for a given species, and, with experience, one can learn to recognize these communities and know which birds to expect in them. However, there are differences among the classifications provided by biologists. For instance, C. Hart Merriam recognized four major life zones, based principally on climate, for the Mid-Atlantic, whereas the plant ecologist Küchler (1975) has described nine principal habitat types: Northern Cordgrass Prairie, Northeastern Spruce-Fir Forest, Beech Maple Forest, Mixed Mesophytic Forest, Appalachian Oak Forest, Northern Hardwoods, Northeastern Oak-Pine Forest, Oak Hickory Pine Forest, and Southern Floodplain Forest. The difference in the number of groupings selected derives largely from considerations of scale. At one end of the spectrum, one could place all creatures on the planet in a single habitat called "Earth"; at the other, we could assign the space occupied by each individual organism as its own particular habitat. In the eight habitat categories presented below, I have attempted to define plant associations that are easily recognizable and possess characteristic bird communities.

Marine, Coastal Waters, and Shoreline

The coastal marine habitat can be broken into three major subdivisions, each with its own characteristic group of avian species: (1) Pelagic (open ocean) and Bays, (2) Beaches and Dunes; and (3) Estuaries, Salt Marshes, and Tidal Flats. Typical pelagic species include loons, gulls, terns, grebes, and sea ducks (inshore and bays) and petrels, shearwaters, gannets, dovekies, kittiwakes, and jaegers (offshore). Some of the beach and dune species are American Oystercatcher, Wilson's Plover, Least Tern, Willet, Royal Tern, Common Tern, and Great Black-backed Gull. Salt marsh and tidal flat species include Little Blue Heron, Snowy Egret, Great Egret, Tricolored Heron, Clapper Rail, Black Skimmer, and Gull-billed Tern.

Freshwater Wetlands

Freshwater wetlands include lakes, ponds, impoundments, rivers, and marshes. The defining characteristic is the presence of fresh water, which stimulates the growth of such plants as cattails *(Typha)*, sedges *(Carex)*, and bulrushes *(Scirpus)*.

Marine, coastal waters and shoreline habitat at Assateague National Seashore, Maryland

Freshwater wetlands are among the most endangered of habitats in our region, perhaps because they tend to limit human economic activities and so are dammed, dredged, drained, channeled, and filled out of existence. Those that remain serve mainly as conduits for waste. Protected wetlands have rich breeding communities with such birds as American Bittern, Common Yellowthroat, Red-winged Blackbird, Green Heron, Pied-billed Grebe, Virginia Rail, American Black Duck, Great Blue Heron, and Common Snipe.

Freshwater wetland habitat at Sandstone Falls, New River, West Virginia

Grassland and savanna habitat at Brigantine National Wildlife Refuge, New Jersey

Grassland

Most grasslands that now occur in our region have been created by human activity, such as the grasslands now covering reclaimed strip mines in Pennsylvania's Clarion County, the broad pasture lands of Virginia's Piedmont, or Delaware's hayfields, although natural coastal cordgrass prairies still occur along the immediate coast, especially in places like Brigantine and Barnegat marshes in New Jersey. The primeval forests of the eastern United States are legendary; supposedly, during the time of Captain John Smith, a squirrel could travel from the coast to the Mississippi and never touch the ground. Nevertheless, that there were extensive areas of grassland has been well documented. Wayland (1989), for instance, reported that much of the broad Shenandoah Valley was grassland at the time of the arrival of the first European settlers, perhaps maintained by Indians with fire. At least one bird was native to grasslands, the Heath Hen, an eastern subspecies of the Greater Prairie-Chicken *(Tympanuchus cupido)*, now extinct. Breeding birds of native, eastern grasslands include Henslow's Sparrow, Northern Harrier, Eastern Meadowlark, Bobolink, Savannah Sparrow, Northern Bobwhite, Short-eared Owl, and American

Broadleaf deciduous and mixed forest habitat, Kittatinny Ridge, Pennsylvania

Kestrel. Unfortunately, some of these species are now scarce or absent from our grassland habitats.

Broadleaf Deciduous and Mixed Forest

Several of Küchler's (1975) forest communities are lumped within this designation, including the Beech *(Fagus)*–Maple *(Acer)* Forest, as found near Sharon and New Castle in western Pennsylvania; the Mixed Mesophytic Forest of maple, beech, oak *(Quercus),* tulip poplar *(Liriodendron),* and horse chestnut *(Aesculus),* found in much of the Appalachian Plateau of West Virginia; Appalachian Oak Forest (and formerly American Chestnut), which covers the lower slopes of the Appalachian Highlands of Virginia, Maryland, Pennsylvania, and New Jersey; Northeastern Oak-Pine *(Pinus)* Forest, otherwise known as the pine barrens of central and southern New Jersey; and the Oak-Hickory *(Carya)*-Pine Forest found throughout most of the Piedmont. The avian communities of these different plant associations share 80–90 percent of their species, which are the typical birds of the eastern deciduous forest—Wood Thrush,

Northern mixed hardwood forest habitat on Spruce Knob, West Virginia

Eastern Towhee, Ovenbird, Downy Woodpecker, Red-shouldered Hawk, and Red-eyed Vireo.

Northern Mixed Hardwoods

The Northern Mixed Hardwood is a maple, birch *(Betula)*, beech, hemlock *(Tsuga)*, and white pine forest. This forest type is often referred to as Transitional Forest, as in transition from the primarily deciduous forests of the temperate portions of the continent to the mainly coniferous forests of the boreal regions. The bird community of this habitat shares many species with the eastern deciduous forest but also has birds generally not found as breeding populations in that forest type, such as Magnolia Warbler, Blue-headed Vireo, Black-capped Chickadee, and Red-breasted Nuthatch.

Highland Coniferous Forest

This forest type is referred to as Northeastern Spruce *(Picea)*–Fir *(Abies)* Forest by Küchler (1975) and actually represents an outlier of the

Highland coniferous forest habitat on Mount Rogers, Virginia

Southern floodplain forest habitat in Suffolk County, Virginia

vast Holarctic boreal forests that cover much of Canada, Siberia, and northern Europe. In our region, the Highland Coniferous Forest is found as isolated patches at high elevations (above four thousand feet in the southern mountains of our region, lower in more northern latitudes) and has a very distinctive breeding bird community including, among others, the Northern Goshawk, Northern Saw-whet Owl, Hermit Thrush, Blackburnian Warbler, Golden-crowned Kinglet, Olive-sided Flycatcher, and Yellow-rumped Warbler.

Southern Floodplain Forest

Like the Highland Coniferous Forest, the Southern Floodplain Forest is an outlier in our region, repre-

Agriculture/residential habitat (corn fields) in Maryland

senting the northernmost extent of the great southeastern bottomlands that once choked every river with tangled swamps and bayous along the Coastal Plain from southeastern Virginia all the way to east Texas but now persist only as remnants. It also has a distinctive breeding bird community, which includes Swainson's Warbler, Yellow-throated Warbler, Carolina Wren, Chuck-will's-widow, Anhinga, and Yellow-crowned Night Heron. The upland long-leaf pine/palmetto savanna habitats of the Austroriparian are now even more rare than the bottomlands. Bachman's Sparrows and Red-cockaded Woodpeckers are famous inhabitants of this endangered habitat type, which can be found in Virginia's Suffolk County.

Agriculture/Residential

Unfortunately, many of the beautiful and unique habitats of the Mid-Atlantic states have been converted to these universal types: a cornfield looks pretty much the same in coastal Maryland as it does in Iowa or Ecuador, for that matter, and plowed dirt is plowed dirt, no matter where you find it. These habitats share common species in many parts of the world, such as European Starlings, House Sparrows, and Rock Doves. Still, some species native to the region's aboriginal grasslands and woodlands can be found in "improved" pastures, orchards, suburban lawns and gardens, and similarly altered environments, for example, Northern Mockingbirds, Ruby-throated Hummingbirds, Yellow Warblers, Song Sparrows, and American Crows.

Mid-Atlantic Bird Migrations

In the United States, we take knowledge of the massive, seasonal movements of birds as an obvious harbinger, as clear as fall's changing leaf colors or first frost or spring's blooming redbuds or first robin's song. These movements are much less obvious in Europe, where those birds that do migrate seem to be there one moment and be gone the next, leading to centuries of speculation into the late 1700s among even well-known and respected scientists that the swallows and other migrants hibernated in the mud like frogs. Indeed, it was observations made by a man who lived for several years in Virginia, where the progress of migration is relatively easy to see, especially along the coast, that helped to solve the mystery. Mark Catesby, the famous artist of southeastern birdlife, presented a paper describing bird migration in the Mid-Atlantic region of the American colonies at a meeting of Britain's Royal Society in 1747.

More than 70 percent of the species that occur in the Mid-Atlantic are migrants, present for a single season—breeding, transient, or wintering. Most of our forest birds spend only four or five months with us and then depart for winter homes in the

tropics. Typical representatives include the flycatchers, thrushes, tanagers, warblers, and vireos. For many of these birds, the Gulf of Mexico exerts a major influence on their migratory route, a fact known at least since the early 1900s (Cooke 1915). The coast and mountains probably exert a similar influence on the routes across the region taken by migrants, though relatively little has been done to determine the shapes of continental migration routes. Many migrants that traverse the Mid-Atlantic states during migration may follow different routes in fall and spring to take advantage of prevailing winds, avoid harsh mountain barriers, or exploit seasonally available resources, but documentation of such seasonal differences in migration route has yet to be made for most species.

The Atlantic Ocean, naturally, also plays a major role in shaping migration routes. It serves as a barrier to many species, causing large numbers to collect along the shore and follow the coastline in both fall and spring. As a result, these areas are excellent for viewing of herons, egrets, waterfowl, shorebirds, and hawks as they parallel the coastline on their migratory flights. Often hundreds or thousands of birds will pass along the beaches or just offshore in the course of a single day during peak migratory periods. Songbirds also back up in coastal habitats during migration at places like Cape May, Island Beach, Bombay Hook, and Kiptopeke; most of these are young birds lacking experience in preparation and navigation. For other species, the western Atlantic is an important pathway. Major seabird migrations brush the Atlantic coast and can be viewed from such places as Chincoteague, Cape May, and the Chesapeake Bay Bridge-Tunnel. The paths and process of migration are among the most fascinating and readily observable of natural history phenomena, and we here in the Mid-Atlantic region have a front-row seat.

Perhaps the most renowned migration phenomena that can be observed in the Mid-Atlantic are the hawk migrations, when tens of thousands of individuals of at least sixteen species travel along the eastern ridges of the Appalachians, especially on crisp, fall days in late September or early October. Hawk Mountain Sanctuary, located on the Kittatinny Ridge in eastern Pennsylvania, is the most famous of many places in the Mid-Atlantic where spectacular concentrations of raptors can be observed with relative ease.

My personal favorite among migration experiences, though, is the songbird migration of late April and early May, especially in the mountains or northern parts of the region when the trees are not quite fully leafed out. During its peak, it is not unusual to see ten or fifteen species of brightly colored warblers in a single tree or grove, presenting a delightful challenge for the novice and the familiar joy of encountering old friends under the best of circumstances for the experienced birder.

Mid-Atlantic Birding Sites: A Locator

This section includes directions to many of the Mid-Atlantic's best birding sites. Each is mentioned in one or more of the species accounts as being one of the best or, in some cases, the only place where a particular bird occurs in the region. I have tried to provide at least three such sites for each species. In compiling the list, I have leaned heavily on existing bird-finding guides—although responsibility for any errors is entirely mine, and, unfortunately, there are errors. Nevertheless, if you use this list with "an open mind and a sense of humor" (as my dissertation chairman, Dwain Warner, used to say), you should have some fun prospecting for birds.

When using the list, please keep in mind that mileages are estimates, for the most part, designed to put you in the right vicinity. Also note that interstate exit numbers are subject to change. Consider these little problems to be a challenge to your sense of adventure, and you will probably be happier. It is also helpful to have detailed maps. In addition to state highway maps, a gazetteer for each of the states can be extremely useful. DeLorme (207-865-4171) makes gazetteers for Virginia, West Virginia, Pennsylvania, Maryland, and

Delaware. In addition, you can find detailed maps for nearly any spot in the country on the internet (mapquest.com).

In most cases, the bird-finding guides have much more detailed information than I provide on the routes to take and what you are likely to find. The sources consulted for many of the sites in this list are *A Birder's Guide to Virginia,* edited by David W. Johnston (1997); *Finding Birds in the National Capital Area* by the late Claudia Wilds (1992); *Birder's Guide to Pennsylvania* by Paula Ford (1995); *The Status and Distribution of New Jersey's Birds* by Charles Leck (1984); *Birds of the Blue Ridge Mountains* by Marcus Simpson (1992); *The Mountain and the Migration: A Guide to Hawk Mountain* by James Brett (1991); *A Guide to Bird Finding in New Jersey* by William Boyle Jr. (1986), and *The Birds of Cape May* by David Sibley (1993).

Abbreviations used in the descriptions are as follows:

CR	County Road
DE	Delaware State Highway
FR	Forest Service Road
hq	headquarters
I	Interstate Highway
jct	junction
LR	Local Route (a designation for rural Pennsylvania highways)
MD	Maryland State Highway
mi	miles
NJ	New Jersey State Highway
PA	Pennsylvania State Highway
RD	Rural Delivery (a designation for rural Delaware roads)
SR	State Route

TR	Township Road
US	U.S. Highway
VA	Virginia State Highway
WV	West Virginia State Highway

Delaware

Assawoman Wildlife Area

From the town of Williamsville just north of the Delaware-Maryland state line, proceed northeast on DE 54 a little over a mile to CR 381. Turn left on CR 381 and go about 2 mi to jct with CR 384. Bear right on CR 384 and "follow signs for Camp Barnes until they are joined by the ones for Assawoman State Wildlife Area" (Wilds 1992, 185). Follow signs to the entrance. Habitats include bottomlands, marshes, swamps, and bay.

Blackbird State Forest

From the city of Smyrna, go north on US 13 about 5.7 mi to RD 471 (Blackbird Forest Road). Turn left (southwest) on RD 471 and proceed 1.9 mi. One of five sections of Blackbird State Forest is on your right (north) along this road, which parallels Blackbird Creek. This site is good for upland forest and field birds of the Coastal Plain.

Blackiston Wildlife Area

From the city of Smyrna, go west from jct of US 13 and DE 6. Go west on DE 6/DE 300 about 1.1 mi to where DE 6 branches off to the right. Turn right (west) on DE 6 for about 7.2 mi. Turn right (north) on Holletts Corner Rd (RD 126). The wildlife area is on the left (west) side of the road. Other sections of the area are located north and south of this site. This site is good for upland forest and field birds of the Coastal Plain.

Bombay Hook National Wildlife Refuge

From jct of US 13 and DE 300/DE 6 in the city of Smyrna, go south on US 13 about 0.9 mi to jct with RD 12 (Smyrna-Leipsic Rd). Turn left (east)

Bombay Hook National Wildlife Refuge, Delaware. This land was purchased by settlers from Kahansink Indian Chief Machacksett in 1679 for "one gun, fower hands full of powder, three Mats coats, one anckor of Liquors and one Kittle," which seems like a pretty good deal now. The tour loops, towers, and trails provide excellent viewing opportunities for a wide range of waterfowl and shorebirds.

on RD 12 and proceed about 5.3 mi to jct with RD 85 (Whitehall Neck Rd). Turn left (east) on RD 85 and go about 2.5 mi to the refuge. Bombay Hook is famous for its wading birds, shorebirds, and waterfowl.

Brandywine Creek State Park

From the jct of I-95 and US 202 in Wilmington, go north on US 220 about 1.4 mi to jct with DE 141. Turn left on DE 141 and proceed to jct with DE 100. Turn right on DE 100 and go to jct with Adams Dam Rd. Turn right on Adams Dam Rd and follow signs to the park entrance.

Cape Henlopen State Park

In the city of Lewes, go north on US 9 to jct with Cape Henlopen Drive (RD 19) at the beach. Turn right (east) on Cape Henlopen Drive and go roughly 1.2 mi to the entrance to the park on the

right. Cape Henlopen has nesting Piping Plovers, American Oystercatchers, Saltmarsh Sharp-tailed Sparrows, Least Terns, and Ospreys, along with many other species of beach, dune, and saltmarsh habitats. The Ipswich Sparrow, a pale race of the Savannah Sparrow, can be found in grassy back-beach habitats in winter.

Cypress Swamp Conservation Area (North Pocomoke Swamp)

From jct of US 50 and US 13 in Salisbury, Maryland, go north on US 13 about 6.9 mi to jct with State Line Rd (RD 419/DE 54). Turn right (east) on RD 419 and continue 11.2 mi until the road forks at Bethel. Take the right fork (Bethel Rd) and go southeast 0.6 mi to jct with Sheppards Crossing Rd. Turn left (east) on Sheppards Crossing Rd and go 1.1 mi to jct with Nelson Rd. Turn left (north) on Nelson Rd, which enters the swamp. Yellow-throated Vireos; Black-and-White, Prothonotary, and Yellow-throated warblers, Louisiana Waterthrushes, and Barred Owls are among the attractions.

Delaware Seashore State Park

From jct of DE 26 and DE 1 in the southeastern coastal town of Bethany Beach, go north 4.7 mi on DE 1 to the park entrance (on the left). The park has typical coastal habitat attracting Sanderlings and plovers (Semipalmated, Black-bellied, American Golden), as well as such birds of coastal waters as Eared Grebes.

Indian River Inlet

From jct of DE 26 and DE 1 in the southeastern coastal town of Bethany Beach, go north 4.7 mi on DE 1 to the bridge over the inlet (parking lot available). The inlet is a good site for winter waterfowl, including Brant, Long-tailed Duck, Ruddy Duck, American Goldeneye, and Greater Scaup. Brown Pelicans, American Oystercatchers, and Piping Plovers are present during the warmer months.

Killens Pond State Park

At jct of DE 14 and US 13 in Harrington, go north 4.0 mi on US 13 to jct with Paradise Alley Rd (RD 426). Turn right (east) on RD 426 and go 2.1 mi to jct with Killens Pond Rd (RD 384). Turn left (north) on RD 384 and proceed 0.3 mi to the pond. Killens is a good site to visit for birds typical of the freshwater wetlands of the Delaware Coastal Plain, including Great Blue Heron, Green Heron, Wood Duck, and Canada Goose.

Little Creek Wildlife Area

In Dover at jct of US 13 and DE 8 (N Little Creek Rd), go east 3.8 mi on DE 8 to the T with Bayside Dr. Turn right (south) on Bayside Dr and go 0.4 mi to jct with Port Mahon Rd (RD 89). Turn left (east) on RD 89. The Little Creek Wildlife Area is

Stands of mixed loblolly and oak alternate with clearcuts at Little Creek Wildlife Area, Delaware, providing excellent habitat for Red-headed Woodpeckers, American Kestrels, Wild Turkeys, and Northern Bobwhite.

on both sides of the road. Marshes here harbor Short-eared Owls, Rough-legged Hawks, and Northern Harriers in winter; Saltmarsh Sharp-tailed Sparrows, Seaside Sparrows, and Marsh and Sedge wrens in summer. In addition, there are excellent tidal flats for migrating shorebirds.

Lums Pond State Park

From I-95 in western Delaware, take the exit for DE 896/US 301 South. Go about 5.4 mi on DE 896 to jct with Howell School Rd (RD 54). Turn left (east) on Howell School Rd to enter the park. Look here for species typical of the woods, fields, and freshwater wetlands of western Delaware.

Milford Neck Wildlife Area (7 tracts)

From jct of DE 1 and DE 36 east of the town of Milford, go east on DE 36 about 5.6 mi to the beach. Turn right (south) on Bay Ave. (RD 204) to pass through the wildlife area. Beach and coastal waters attract a variety of waterbirds.

New Castle County Airport

South of the city of Wilmington at jct of I-495 and US 13, take US 13 south 4.4 mi to jct with Old Churchmans Rd. Turn right (northwest) on Old Churchmans Rd. US 13 and Old Churchmans Rd border the airport. Upland Sandpipers, American

Pipits, American Kestrels, Vesper Sparrows, and Eastern Meadowlarks can be found in the open grasslands of the airport.

Norman G. Wilder Wildlife Area (2 tracts)

From jct of DE 14 and US 13 in Harrington, go north about 7.8 mi on US 13 to jct with Plymouth Rd (RD 239). Turn left (west) on RD 239 and go 2.4 mi (RD 239 becomes RD 108). The wildlife area is on your right (north). Look for common species of the Delaware uplands—Turkey Vulture, Carolina Chickadee, and White-eyed Vireo, among others.

North Pocomoke Swamp

See Cypress Swamp Conservation Area.

Port Mahon Impoundment

In Dover from jct of US 13 and DE 8 (N Little Creek Rd), go east 3.8 mi on DE 8 to the T with Bayside Dr. Turn right (south) on Bayside Dr and go 0.4 mi to jct with Port Mahon Rd (RD 89). Turn left (east) on RD 89 and proceed about 1.0 mi. The impoundment is on your right (south side of the road). The coastal marshes, ponds, and tidal flats of the impoundment attract a rich variety of wetland species, including such occasional rarities as Eurasian Wigeon, Red-necked Stint, Yellow-headed Blackbird, Black-headed Gull, and Little Gull.

Prime Hook National Wildlife Refuge

From jct of DE 1 and DE 16 east of the town of Milton, go east on DE 16 (Broadkill Beach Rd) about 1.3 mi. Turn left to enter the refuge. The refuge is outstanding for shorebirds, waterfowl, marsh birds, and transient raptors.

Redden State Forest

From jct of US 113 and US 9 just southeast of Georgetown, go north on US 113 about 4.4 mi to jct with RD 565 (Redden Crossroads). Turn left on RD 565 and proceed 1.6 mi. State forest lands are

"Prime Hook National Wildlife Refuge is the best-kept secret in Delaware." So a gentleman told me at the refuge boat launch when I visited in September 1998. Unfortunately, the secret was out as far as the mosquitoes were concerned. They were abundant and probably sorry that more hardy visitors did not come their way. Still, mosquitoes go with good tidal marsh, and Prime Hook offers some of the best.

on both sides of the road, providing habitat for such woodland birds as Summer Tanager, Wood Thrush, and Acadian Flycatcher.

Robert L. Graham (Nanticoke) Wildlife Area

From jct of US 13 and DE 24 in Laurel, go west on DE 24 about 5 mi to jct with CR 493. Turn right (north) on CR 493 and go about 3 mi to where CR 496 goes to the left in Portsville (CR 493 goes to the right). Follow CR 496 west for 2 mi or so through the wildlife area.

Trap Pond State Park

From the east side of the town of Laurel at jct of US 13 and DE 24, go east 5.6 mi to jct with RD 449. Turn right (south) on RD 449 and go about 1 mi to the pond. Ducks, geese, mergansers, and grebes frequent the pond, while such birds as the

The Prothonotary Warblers, Yellow-throated Vireos, Brown-headed Nuthatches, and Red-bellied Woodpeckers complete the southern "bayou" atmosphere created by cypress bays at delightful Trap Pond State Park, Delaware.

Pine Warbler, Gray Catbird, Eastern Towhee, and Blue-gray Gnatcatcher frequent the wet woodlands.

Woodland Beach Wildlife Area (2 tracts)

From the east side of the town of Smyrna at jct of DE 1 and DE 6 (Woodland Beach Rd), go east on Woodland Beach Rd. After crossing DE 9 (Hay Point Landing Rd) at about 5 mi, continue east on Woodland Beach Rd. The wildlife area on the left (north) side of the road is good for a variety of transient waterbirds.

Maryland

Assateague Island National Seashore

From jct of US 50 and US 113 just north of Berlin, go south about 1.5 mi on US 113 to jct with Bay St in Berlin (MD 376, which becomes Assateague Rd). Turn left (east) on Bay St and go about 4.1 mi to jct with MD 611 (Stephen Decatur Hwy). Turn right (south) on MD 611 and go 2.5 mi to where MD 611 bends (east). Continue on MD 611 2 mi to the seashore. With all the waterbirds typical of the Eastern Shore, Assateague is also famous for raptor and landbird migrations, especially in the fall.

Wild horses seem to be the main natural history attraction at Assateague Island National Seashore, Maryland, but the island's marshes, dunes, tidal flats, and beaches host a rich assortment of birdlife, including Seaside Sparrows, Ospreys, Peregrine Falcons (migration), and Piping Plovers.

Back River Waste Water Treatment Plant

In south Baltimore on I-695 take the exit for MD 150 (Eastern Ave.) east. Proceed on MD 150 to the first light (Willis Ave). Turn right on Willis Ave to enter the plant's grounds, where a rich diversity of gull species has been found in the appropriate seasons.

Baltimore Harbor

From I-95 in downtown Baltimore, take the exit for MD 2 (Hanover St.) south. Turn left at Cromwell St and right on Light St. Go to the end of Light St and park at Ferry Bar Park overlooking the harbor. Loons, grebes, cormorants, bay ducks, shorebirds, and gulls can be found in the harbor waters and along the shoreline at different times of the year.

Blackwater National Wildlife Refuge

From jct of US 50 and MD 16 just east of Cambridge, go south and west on MD 16 (Church Creek Rd) about 2.3 mi to jct with Egypt Rd. Turn left (south) on Egypt Rd. Go south on Egypt Rd about 7.5 mi to the T with Key Wallace Dr. Turn right on Key Wallace Dr and go 1.3 mi to the

The Chesapeake & Ohio (C&O) Canal was once a busy arm of commerce linking coastal shipping lanes with interior land routes via a 184-mile artificial waterway paralleling the Potomac River from Washington, D.C., to Maryland's Cumberland Plateau. Bicyclists and hikers have replaced mules along the tow paths, while canoeists ply those sections that still have water. Birders come to the canal to visit outstanding riparian habitats for Carolina Wrens, Louisiana Waterthrushes, and Kentucky Warblers.

refuge. In addition to being a prime site for wintering waterfowl, Blackwater has one of the largest East Coast nesting populations of Bald Eagles.

Catoctin Mountain Park

From jct of US 15 and MD 77 at Thurmont, go west on MD 77 (Foxville Rd) 3.3 mi to jct with Park Central Rd. Turn right on Park Central Rd to enter the park. Barred Owls, Cerulean Warblers, and other woodland species can be found in the park.

Cedarville State Forest

From jct of MD 5 (Blue Star Memorial Hwy) and Cedarville Rd just north of the town of Mattawoman, go east on Cedarville Rd about 4.2 mi to

jct with Cedar Forest Rd. Turn right (south) on Cedar Forest Rd and go about 2 mi to enter the state forest, a good site for Red-shouldered Hawk and other forest birds.

Chesapeake and Ohio (C&O) Canal

From I-270 south of Frederick, exit west on MD 109 at Hyattstown. Proceed southwest on MD 109 (Old Hundred Rd) about 5 mi to jct with Barnesville Rd. Turn right and then take an immediate left to continue on MD 109 (now Beallsville Rd). Proceed about 6 mi to jct with MD 107 (Fisher Rd) at Poolesville. Turn right and then take an immediate left on west Willard Rd and go about 4.5 mi to a T at River Rd. Turn left on River Rd, and take the next right (at 0.2 mi) on Sycamore Landing Rd. Follow Sycamore Landing Rd 0.8 mi to parking for the old canal, towpath, and river shore, with excellent riparian habitat for Carolina Wrens, Great Crested Flycatchers, Yellow-billed Cuckoos, and Baltimore Orioles.

Deal Island Wildlife Management Area

"Take United States 13 south from Salisbury to Princess Anne and turn right on Md. 363. Drive 9.5 miles and fork left at the Deal Island WMA sign" (Wilds 1992, 191). Deal Island is an especially productive marsh for waterbirds and marsh birds like Least Bitterns, Common Moorhens, Black-necked Stilts, and Sedge Wrens.

Deep Creek Lake State Park

From Oakland, go north on US 219 about 8 mi to jct with Braddock Miller Rd. Turn right on Braddock Miller Rd, then left at 0.1 mi on Glendale Rd. Bear left at the next three road junctions to follow the lake shore and enter the state park. The lake is an excellent site in western Maryland for loons, grebes, and ducks.

Eastern Neck National Wildlife Refuge

From jct of MD 20 and MD 514 about 1 mi northwest of Chestertown, go west 5 mi on MD 20 to Fairlee, where MD 20 bends south. Continue on MD 20 another 7 mi to jct with MD 445 at Rock Hall. Take MD 445 south 8 mi to the refuge. Waters of the Chester River and Chesapeake Bay at the refuge attract an exceptional variety of waterfowl.

Elliott Island

From the village of Vienna on the Vienna–Henrys Crossroads Rd, go 6 mi to Henrys Crossroads (jct with Lewis Wharf Rd). Continue south on the Vienna–Henrys Crossroads Rd (now Elliot Island Rd) another 12 mi to Elliott Island. Breeding Black Rails are the principal attraction, but American

The park at Fort McHenry allows visitors to combine a taste for history with broad vistas of busy Baltimore Harbor, where sea ducks, gulls, terns, grebes, and loons can often be seen in winter.

Woodcock, Chuck-will's-widow, and Blue Grosbeak are also present.

Fort McHenry

Take Exit 55 from I-95 on the south side of Baltimore. Follow signs for Fort McHenry. There is parking and a visitor center at the fort, along with a nice view of the harbor, sea wall, and some marsh. Loons, grebes, bay ducks, gulls, and shorebirds frequent the waters bordering the fort, while its broad lawns can attract transient Upland Sandpipers and American Pipits.

Francis Scott Key Bridge, Baltimore

I-695 crosses the harbor via this bridge southeast of Baltimore. Take Exit 44 on the east side of the bridge onto Broening Hwy, and then take the first left to reach Sollers Point below the bridge, an excellent site for viewing scaup, Canvasbacks, Redheads, and Ruddy Ducks in winter.

Green Ridge State Forest

From I-68 about 11 mi west of Hancock, exit south onto Orleans Rd South. Proceed about 5.7 mi to jct with Oldtown Rd in Little Orleans. Turn right on Oldtown Rd and follow this as it winds into the state forest. Bear left at junctions to follow the river, C&O Canal, and towpath. Wild Turkeys, grouse, Golden-winged Warblers, and Brown Creepers all breed at Green Ridge.

Gunpowder Falls State Park

This is a linear park that extends from the bay northeast of Baltimore several miles west inland along the Gunpowder River. These instructions lead to the entrance into the east (bay) end of the park. From I-95 east of Baltimore, exit east onto Whitemarsh Blvd (MD 43). Go 1.4 east on MD 43 to jct with US 40. Turn left (north) on US 40 and go 0.6 mi to jct with Ebenezer Rd. Turn right on Ebenezer Rd and go east 4.2 mi to the park (Ebenezer Rd becomes Graces Dr in the town of Chase). Virginia Rails, Least Bitterns, and Bald Eagles frequent the marshes, while Orchard Orioles, Whip-poor-wills, and Yellow-breasted Chats can be found in the uplands.

Hughes Hollow

See McKee-Beshers Wildlife Management Area.

Inner Harbor, Baltimore

Take Exit 54 from I-95 on the south side of Baltimore onto MD 2 (Hanover St). Proceed north on MD 2, which turns right on Sharp St and left on Light St. Turn right on Pratt St. The piers for the Inner Harbor are on your right. Park where you can, and walk out on the piers to the World Trade Center (Pier 2), National Aquarium (Pier 3), Christopher Columbus Center (Pier 5), etc. Despite the urban surroundings, the Inner Harbor attracts a variety of waterfowl.

Jug Bay Natural Area

From jct of MD 4 and US 301 in the town of Upper Marlboro east of Washington, D.C., go south on US 301 about 3 mi to jct with MD 382 (Croom Rd). Turn left (east) on MD 382 (Croom Rd) and go about 3.5 mi to jct with Croom Airport Rd. Turn left (east) on Croom Airport Rd and proceed a little over a mile to a T with Duvall Rd.

Bear right at the T, remaining on Croom Airport Rd and following it all the way to the end. The park has trails, a headquarters building, and an observation tower. Habitats include ponds, swamps, fields, bottomland forest, and the west shore of Jug Bay on the Patuxent River, excellent for Sedge Wren, Grasshopper Sparrow, Prothonotary Warbler, Pileated Woodpecker, Wood Duck, and many other grassland, marsh, and riparian species.

Lily Pons Water Gardens

From I-270 about 30 mi northwest of Washington, D.C., exit onto MD 80 west (Fingerboard Rd). Go 1.7 mi on MD 80 to jct with Park Mills Rd. Turn left on Park Mills Rd and go 5.3 mi to jct with Lily Pons Rd. Turn right on Lily Pons Rd to enter the gardens. King Rails, Belted Kingfishers, Tree Swallows, ibises, egrets, and Ospreys can be found at the ponds.

Martin National Wildlife Refuge

Accessible by ferry. At jct of US 13 and MD 413 about 17 mi south of Salisbury, bear right on MD 413 and go 14 mi to the bay shore at Crisfield. From there, the refuge is about a 7-mi boat trip. Black- and Yellow-crowned night-herons, Great and Snowy egrets, cormorants, bay ducks, and swans, along with many other members of the rich waterbird fauna of the Chesapeake Bay, can be sampled at the refuge.

McKee-Beshers Wildlife Management Area (Hughes Hollow)

From I-270 south of Frederick, exit west on MD 109 at Hyattstown. Proceed southwest on MD 109 (Old Hundred Rd) about 5 mi to jct with Barnesville Rd. Turn right and then take an immediate left to continue on MD 109 (now Beallsville Rd). Proceed about 6 mi to jct with MD 107 (Fisher Rd) at Poolesville. Turn right and then take an immediate left on west Willard Rd and go about 4.5 mi to a T at River Rd. Turn left on River Rd, and take the next right (0.2 mi) on Sycamore Landing Rd. Parking sites for the management area are found along Sycamore Landing Rd and also on the right of River Rd if you continue on this route southeast past the Sycamore Landing turnoff. Also, continuing south past the Sycamore Landing turnoff, you can take the next left on Hughes Rd to enter another part of this site. Look here for birds of old field and forest edge, including Blue Grosbeak, Willow Flycatcher, Red-headed Woodpecker, and Prairie Warbler.

Meadowside Nature Center

About 10 mi northwest of Washington, D.C., in Rockville, exit from I-270 onto MD 28 east. At 2.5 mi, turn left to continue on MD 28 (Norbeck Rd) another 1.4 mi to jct with Avery Rd. Turn left on Avery Rd and go 1.3 mi to a T with Muncaster Hill Rd (MD 115). Turn right on Muncaster Hill Rd and go 0.7 mi to Meadowside Ln. Turn right on Meadowside Ln to enter the nature center. The old fields, lakeside, and woodlands support such grassland and edge birds as Common Yellowthroat, Indigo Bunting, Brown Thrasher, and Yellow-breasted Chat.

Ocean City

From jct of US 13 and US 50 in Salisbury, go east on US 50 about 30 mi to Ocean City. The beachfront is a good place to scan for such winter visitors as cormorants, loons, grebes, several gull species, ducks, and mergansers.

Ocean City Inlet

From jct of US 13 and US 50 in Salisbury, go east on US 50 about 30 mi to Ocean City. Turn right on Philadelphia Ave and go 0.4 mi to the inlet. The rocky jetty bordering the inlet is a good place for Ruddy Turnstones and Purple Sandpipers in winter.

Patuxent Research Refuge	From I-95 east of Washington, D.C., exit northeast onto the Baltimore-Washington Parkway. Go about 5.6 mi on the parkway to the exit for MD 197 (Laurel-Bowie Rd). Take MD 197 southeast about 2 mi to the entrance to the refuge on the left. The refuge provides a good site for viewing turkeys, quail, and pheasants in addition to other common species of forest and field habitats.
Pelagic Birding Trips	Contact Brian Patteson, P.O. Box 1135, Amherst, VA 24521, Tel. (804) 933-8687; also the *OC Princess* offers charter trips Aug–Oct; Tel. (800) 457-6650. For additional information, consult Patteson's section in Johnston (1997, 197–203). These trips open possibilities for viewing an entirely new avifauna—oceanic birds including fulmars, petrels, storm-petrels, jaegers, and shearwaters.
Piscataway National Park	From I-495 south of Washington, D.C., exit onto MD 210 south (Indian Head Hwy) and proceed about 9.7 mi to jct with Bryan Point Rd. Turn right (west) on Bryan Point Rd, and go about 5.4 mi to the entrance to the park on the right. The park borders the Potomac River and has an exciting diversity of winter waterfowl.
Point Lookout State Park	From the jct of MD 246 and MD 235 in the town of Lexington Park on the Chesapeake Bay, proceed south on MD 235 about 12 mi to jct with MD 5 in the village of Ridge. Turn left on MD 5 and go 9 mi to the park.
Rocky Gap State Park	About 7 mi east of Cumberland, exit north from I-68 for Pleasant Valley and follow signs to the park. The lake at Rocky Gap attracts transient and wintering loons, grebes, and ducks.

Three-hundred-year old hemlocks line the Youghiogenny River at Swallow Falls State Park, Maryland, providing breeding habitat for such northern forest species as Golden-crowned Kinglets, Blue-headed Vireos, Magnolia Warblers, and Northern Waterthrushes.

Sandy Point State Park

From jct of I-495 and US 50 east of Washington, D.C., follow US 50 east about 27 mi to Skidmore and take Exit 32 north to the park. Located on the Chesapeake Bay, the park offers excellent viewing of inshore seabirds.

Savage River State Forest

From I-68 about 20 mi west of Cumberland, exit south onto Chestnut Ridge Rd. Follow Chestnut Ridge Rd about 2.5 mi to T with New Germany Rd. Turn left on New Germany Rd to enter the state forest. Follow New Germany Rd about 1.4 mi to New Germany State Park. Continue another 4.4 mi to jct with Big Run Rd. Turn left on Big Run Rd to reach Big Run State Park. Forests and meadows here harbor Veeries, Black-throated Blue Warblers, and Chestnut-sided and Golden-winged warblers.

Seneca Creek State Park

From I-270 about 4 mi northwest of Gaithersburg, take exit west on MD 118 (Germantown Rd) about 2.2 mi to jct with Clopper Rd (MD 117). Turn left on MD 117 and go about 2.1 mi to the park entrance. Seneca Creek is good for grassland species, such as Grasshopper Sparrow and Eastern Meadowlark, and old field birds like Orchard Oriole, Yellow-breasted Chat, and Prairie Warbler.

Swallow Falls State Park

From Oakland, take US 219 about 6.8 mi north to jct with Mayhew Run Rd. Turn left on Mayhew Run and go about 4.4 mi to jct with the Oakland–Sang Run Rd. Turn left on the Oakland–Sang Run Rd, and go 1.3 mi to the park entrance. Look here for birds typical of Appalachian highland coniferous and mixed forest: Least Flycatcher, Northern Waterthrush, Blue-headed Vireo, and Magnolia and Black-throated Green warblers.

Violette's Lock

From I-495 northwest of Washington, D.C., take the River Rd (MD 190) exit west. Continue on River Rd about 11 mi to where it makes a sharp bend north. Turn left after the bend onto Violette's Lock Rd. If you come to a T at Seneca Rd (MD 112), you have gone about a half-mile too far.

Washington Monument Knob

West of Frederick, exit onto US 40 Alternate west. Go about 10 mi on US 40 Alt to signs for the park at Zittlestown Rd. Turn right and go 1 mi to the parking lot. A trail goes from here to the hawk-watching site on the knob.

West Ocean City Pond

From jct of US 13 and US 50 in Salisbury, go east on US 50 about 29 mi to West Ocean City. Turn left at the light at jct of US 50 and Golf Course Rd, and go 0.4 mi north to the pond (left side of road), which offers good winter viewing of swans, ducks, and gulls.

New Jersey

Allemuchy Mountain State Park

From jct of I-80 and CR 517 (Exit 19), go south on CR 517 a little over a mile to the park entrance on the left (east) side of CR 517. The park has a variety of woodland and old field habitats, and some of its deciduous forest harbors breeding Cerulean Warblers, especially around the town of Waterloo.

Assunpink Wildlife Management Area

From I-195 about 13 mi east of Trenton, exit north onto CR 43 for about 0.2 mi to jct with CR 524. Turn right (east) on CR 524 and continue about 3..5 mi to jct with CR 571 just north of Clarksburg. Turn left (north) on CR 571 and go about 3 mi to Roosevelt. The management area is on your left for most of the distance between Exit 11 and Roosevelt. The lake attracts waterfowl during migration and winter, summering Veeries have been heard in the refuge, and a pair of Canada Warblers nested in 1980.

Avalon

From the Garden State Parkway about 16 mi north of Cape May, exit east onto CR 601. Proceed about 3 mi on CR 601 to the town of Avalon. The best viewing point for seabirds is "a small parking lot at the E end of 8th Street" (Sibley 1993, 10).

Barnegat National Wildlife Refuge

From the Garden State Parkway, take Exit 67 east onto CR 554 and go about to 2 mi to jct of CR 554 and US 9 in Barnegat. Turn right (south) on US 9. For the next 13 mi or so (to Tuckerton), the Barnegat Division of the Edwin B. Forsythe National Wildlife Refuge borders the east (left) side of US 9. There are no public-use facilities, though the birdlife is similar to that found at Barnegat's famous sister refuge, Brigantine.

Less than half an hour from the boardwalk, slots, and casinos of Atlantic City, Brigantine National Wildlife Refuge, New Jersey, offers visitors quite a different experience. Bitterns, egrets, ibises, geese, ducks, plovers, sandpipers, and a variety of other water birds frequent the marshes and waterways of the refuge, making this site one of the top birding destinations in North America.

Beaver Swamp Wildlife Management Area, Cape May

From the Garden State Parkway 13 mi north of Cape May, take Exit 13 west to the village of Swainton on CR 601. At jct of CR 601 and US 9 in Swainton, turn right (north) on US 9 and go about 2 mi to jct of US 9 and NJ 83 at Clermont. Turn left (west) in Clermont onto NJ 83 toward South Dennis. Beaver Swamp is along the left (south) side of the road for the next 4 mi or so. Species typical of southern swamps and riparian habitat can be found here, including Prothonotary Warbler, Carolina Wren, and Yellow-throated Warbler.

Brigantine National Wildlife Refuge

From the Atlantic City Expressway, 5 mi west of Atlantic City, exit onto US 9/NJ 157 north and proceed about 9 mi to the town of Oceanville. Turn right at the sign for the refuge in Oceanville

Cape May Point, New Jersey, is one of those places that every keen bird enthusiast must visit. The tradition of outstanding birding in a setting of extraordinary natural beauty is irresistible. In addition, the site has its own ornithological literature to enrich the visitor's experience. Witmer Stone's *Bird Studies at Old Cape May* (1937) is a classic, while David Sibley's more recent *The Birds of Cape May* (1993) serves as an excellent guide to what birds can be found in the area and when. Visit the bookstore at the Cape May Bird Observatory, located on Lake Drive in Cape May Point, for these and other publications on Cape May and also for friendly chat about special birds seen recently. Purple Sandpipers are a winter specialty on the rocky sea walls.

on Great Creek Rd and go 0.7 mi to the refuge entrance. Brigantine is justly famous as one of the top five birding localities in the eastern United States. Waterbirds, such as herons, bitterns, egrets, ibises, ducks, geese, swans, and shorebirds, are the principal attraction.

Bull's Island Section, Delaware and Raritan Canal State Park

From jct of NJ 29 and US 202 just north of Lambertville, proceed north on NJ 29 about 6 mi to the park entrance on the left (west).

Cape May Migratory Bird Refuge

See South Cape May Meadows.

Cape May Point

Follow the Garden State Parkway to its southern end, where it becomes NJ 109. Continue on this road south all the way into downtown Cape May and jct with CR 606 (about 2 mi). Turn right (west) on CR 606 (Sunset Blvd) and go about 2 mi to a left turn for the town of Cape May Point. Turn left and enter town. Cape May State Park (at the lighthouse) and numerous boardwalks provide access to the beach and sea walls. Perhaps the most renowned birding destination on the east coast, Cape May offers superb birding opportunities in a breathtaking setting, especially during migration. Unfortunately, it's no secret, and you are likely to share your experience with a large number of other visitors.

Delaware Bay

At jct of CR 553 and CR 649 in downtown Port Norris, go west on CR 553 0.4 mi to jct with CR 631. Turn left (south) on CR 631 toward the village of Bivalve. Go 0.9 mi on CR 631 until you come to the parking lot for Schooner Sails on your right. Turn right at the far end of this lot and drive to the boardwalk and parking area for the Commercial Township Wetlands Restoration Site. The boardwalk and towers provide a good view of the marsh and tidal flats, where vast numbers of shorebirds congregate during migration.

Dennis Creek Wildlife Management Area

From jct of CR 610 and NJ 47 in the southern New Jersey village of Dennisville, go northwest on NJ 47 about 1.4 mi to jct with Jakes Landing Rd. Turn left on Jakes Landing Rd, which runs through the wildlife management area. The site is good for

Shorebird lovers and conservationists around the world know of Delaware Bay, New Jersey (see p. 53). Its tidal flats and marshes host large segments of eastern North America's shorebird populations each migratory period as the birds stop over to refuel on their way to breeding grounds in the high Arctic and wintering grounds in Central and South America. In particular, look for Red Knots (spring), Hudsonian Godwits (fall), White-rumped Sandpipers (spring), and Stilt Sandpipers (fall).

birds of the marsh and shorebirds, wintering raptors, and Yellow-throated Warblers in summer.

Edwin B. Forsythe National Wildlife Refuge

See Brigantine National Wildlife Refuge and Barnegat National Wildlife Refuge.

Great Swamp National Wildlife Refuge

From the town of Basking Ridge, located about 5 mi southwest of Morristown, follow CR 657 south and east about 5.5 mi to jct with CR 604 in Meyersville. Turn left (north) on CR 604 to cross the refuge. With its combination of marsh, forest, hardwood swamp, pasture, and cropland, the Great Swamp presents excellent viewing opportunities for a variety of birds, including Swamp

Sparrow, Bobolink, Cedar Waxwing, Black Vulture, American Bittern, and Green-winged Teal.

Hereford Inlet

From the Garden State Parkway about 9 mi north of Cape May, exit east onto NJ 147. Follow NJ 147 about 2 mi to jct with CR 619 (north) and CR 621 (east). Continue east on CR 621 about 2 mi to North Wildwood. The inlet is on your left along this section of CR 621. Sibley states that the inlet is best viewed "from Angelsea Drive in North Wildwood. Sandbars in the inlet are constantly shifting and the best viewing changes accordingly" (1993, 9). The sandbars serve as nesting and\or roosting sites for Black Skimmer, Common Tern, Piping Plover, and American Oystercatcher, among other species.

Higbee Beach Wildlife Management Area

Follow the Garden State Parkway to its southern end, where it becomes NJ 109. Continue on this road south all the way into downtown Cape May and jct with CR 606 (about 2 mi). Turn right (west) on CR 606 (Sunset Blvd) and proceed about 1 mi to jct with Bayshore Rd. Turn right (north) on Bayshore Rd and go a little over a mile to the T with New England Rd. Turn left (west) on New England Rd and go about 1 mi to the parking lot for the beach. This site is good for lots of transients (raptors, Western Kingbird, Yellow-bellied Flycatcher, Dickcissel, Gray-cheeked Thrush) and also has interesting summer birds (Yellow-breasted Chat, Blue Grosbeak).

High Point State Park

From the town of Colesville in extreme northern New Jersey, go north on NJ 23 about 2 mi to enter the park. Follow signs to park hq. The highest point in the state is located in the park (1,803 feet). The highland coniferous forest and bog habitats found in the park are rare in New Jersey and attract several interesting species, such as Northern Waterthrush, Blackburnian Warbler, Magnolia Warbler, Blue-headed Vireo, and Alder Flycatcher.

Hoffman Sanctuary

From the north, take Interstate 287 south to Exit 30B (Bernardsville). From the exit ramp go a short distance to a traffic light at Rt. 202, then proceed straight ahead through the light onto Childs Rd. Go 0.2 mile, and bear right onto Hardscrabble Rd. After about 0.8 mile you will cross a bridge; the second highway (uphill) on the right after the bridge is the entrance to Hoffman Sanctuary" (Boyle 1986, 88). The sanctuary is operated by the New Jersey Audubon Society and has a bookstore, rest rooms, and trails.

Holgate

From the Garden State Parkway, take Exit 63A onto NJ 72 East. Follow NJ 72 about 8 mi to a T at Ship Bottom. Turn right (south), and go about 10 mi to Holgate at the southern tip of Long Beach Island. Barrier island sites like Holgate are excellent for observing beach-nesting shorebirds, migrants in transit, and inshore seabirds.

Hopatcong State Park

Exit I-80 at Landing and go north from the exit about 1 mi until the road comes to a Y. Bear left at the Y. The park will be on your right in about 0.3 mi. The lake attracts a nice variety of waterfowl in migration and winter.

Island Beach State Park

From the Garden State Parkway at Toms River, take Exit 82 onto NJ 37 East. Follow NJ 37 about 8 mi to a T at Seaside Heights. Turn right, and go about 1.5 mi to enter the park. Famous for its banding studies, this barrier island park is inundated with transients during migration.

Lebanon State Forest

From the Garden State Parkway near Barnegat, take Exit 67 west onto CR 554. Follow CR 554 about 4 mi until it joins NJ 72. Continue west on NJ 72 about 11 mi to the forest. For the next 8.5 mi along NJ 72, the state forest lines the north (right) side of NJ 72. Roads branching to the right along this section will take you into the state forestlands, covered mostly by New Jersey's unique "pine barren" habitat. Look and listen here for Whip-poor-wills, Red-headed Woodpeckers, and Pine Warblers.

MacNamara Wildlife Management Area (also known as Tuckahoe–Corbin City Fish & Wildlife Management Area)

From the Garden State Parkway about 17 mi south of Atlantic City, take Exit 20 north onto NJ 50. Go about 4 mi on NJ 50 to the town of Petersburg. Continuing north on NJ 50, the management area is on your right for the next 8 mi or so. Bordering the Tuckahoe River and with several artificial impoundments, the refuge attracts a rich variety of waterfowl.

Raccoon Ridge

"Take Interstate 80 west to Exit 12 (Rt. 521, Blairstown). Follow Rt. 521 north for 5 miles to Rt. 94 and turn left. Go about 3.8 miles to Walnut Valley Rd. and turn right (there is a Dairy Queen on the left just before the turn and a sign on the left directing you to turn right for the Yards Creek Pumped Storage Power Plant). Stop at the gate and tell the guard that you are going hawk-watching. Continue on the road staying right at the first fork, left at the second fork, and right at the third fork (0.4 mile from the gate), following the signs for the Boy Scout Camp. After 1.1 miles (from the entrance), you will enter the camp and see a parking area on the right; park here after checking with the ranger at the house ahead on the left. Just ahead on the left, an old road starts up the mountain. Hike up this road for about three-quarters of a mile (staying right when it forks just past an old stone house) until it intersects with the Appalachian Trail at the top of the ridge. Turn left onto the Appalachian Trail. The first lookout is a short distance up the trail, but the main one is an exposed outcropping of rocks about one-quarter mile southwest along the trail" (Boyle 1986, 439–40). This site is one of the best for hawk watching during fall migration in New Jersey.

Rosedale Park, Princeton

From I-95/I-295 north of Trenton, exit north on US 206 and go about 10 mi to the town. A Northern Shrike has wintered at the park in years past.

South Cape May Meadows

Follow the Garden State Parkway to its southern end, where it becomes NJ 109. Continue on this road south all the way into downtown Cape May and jct with CR 606 (about 2 mi). Turn right (west) on CR 606 (Sunset Blvd) and go about 1 mi. This Nature Conservancy site is on the left, marked by a small sign; it is also known as the Cape May Migratory Bird Refuge. The Meadows is home to birds of freshwater marshes and ponds and has been the locality for sightings of several scarce or rare species, including King Rail, Sedge Wren, Virginia Rail, Black-necked Stilt, Black Rail, Gull-billed Tern, and Wilson's Phalarope.

Stokes State Forest

From jct of US 206 and NJ 15 in Ross Corner, go north on NJ 15 about 5.5 mi to Culvers Inlet. Continuing north on NJ 15, the next 2.5 mi of this road pass through the state forest. Stokes is part of the New Jersey highlands, like neigboring High Point State Park, and shares interesting species with High Point. Four species of *Empidonax* flycatchers have bred here (Alder, Willow, Least, and Acadian), as have Hermit Thrushes; Nashville, Magnolia, and Black-throated Blue warblers; and Long-eared Owls.

Wharton State Forest provides extensive access to backcountry birding, hiking, and camping in New Jersey's unique Pine Barrens. Leck (1975) notes that this habitat is especially good for Pine, Prairie, and Black-and-white warblers, Eastern Towhees, Northern Mockingbirds, and Whip-poor-wills.

Stone Harbor

From the Garden State Parkway about 9 mi north of Cape May, take Exit 6 east onto NJ 147. Follow NJ 147 about 2 mi to jct with CR 619 (north) and CR 621 (east). Turn left on CR 619 to cross Hereford Inlet and proceed about 3.5 mi to the town of Stone Harbor. The sanctuary at the south end of town was once home to a nesting colony of several heron and egret species. The colony has since moved to neighboring Sedge Island.

Wawayanda State Park

From jct of NJ 23 and CR 515 in the town of Stockholm in northern New Jersey, go north on CR 515 about 9 mi to the park. Birds typical of eastern deciduous and mixed forest can be found here, such as Downy Woodpecker, Rose-breasted Grosbeak, White-breasted Nuthatch, Pileated Woodpecker, Scarlet Tanager, and Worm-eating Warbler.

Wharton State Forest

From the Garden State Parkway about 12 mi north of Atlantic City, take Exit 50 onto US 9. Follow

In Allegheny National Forest, a huge natural area in northwestern Pennsylvania, you can see mature northern mixed forest in which boreal and temperate bird communities overlap. Look for Hermit and Swainson's thrushes, Canada and Mourning warblers, Golden-crowned Kinglets, and Winter Wrens in summer; Pine and Evening grosbeaks, Red Crossbills, and Pine Siskins in winter.

US 9 about 1 mi to jct with CR 542. Turn left (west) on CR 542 and go 2.7 mi to jct with CR 653. Turn left, continuing on CR 542 across the bridge, and proceed 2.6 mi to enter the state forest. Roads off to the right enter the forest, and CR 563 traverses a large section.

Pennsylvania

Allegheny National Forest

This forest extends over a large area in northwestern Pennsylvania, covering significant portions of five counties: Warren, McKean, Elk, Jefferson, and Forest. A nice sampling of the habitats of the region can be observed by following PA 59 east from its jct with US 6 just south of Warren past Kinzua Dam, across an arm of the Allegheny Reservoir, to jct with PA 321 (20 mi). Turn right (south) on PA 321, and follow this to Kane. This route takes you past Jakes Rocks, Kinzua Beach, and Red Bridge National Forest picnic areas—all suitable for stopping and exploring. Several boreal species breed in Allegheny, including Winter Wren, Purple Finch, Common Raven, Swainson's Thrush, Pine Siskin, and Blackburnian Warbler.

Beaver Creek Wetlands

From jct of US 322 and PA 208 in Shippenville, take PA 208 south (bends west) about 7 mi to where Beaver Creek crosses the road. Wetlands are visible on both sides of the road, and a parking lot is on the right. The ponds attract a variety of waterfowl, and the marshes, old fields, and woodlands harbor at least three species of owls (Eastern Screech, Great Horned, and Barred).

Big Pocono State Park

At the town of Tannersville near Stroudsburg in eastern Pennsylvania, exit I-80 onto PA 715 west and follow signs to the park.

Bowman's Hill Wildflower Preserve

In the town of New Hope on the Delaware River, go south from jct of US 202 on PA 32 about 2.3 mi. The entrance to the preserve is on the right. Old field, deciduous woodland, and bog habitats host breeding populations of Ruby-throated Hummingbirds, Wood Thrushes, American Redstarts, Veeries, and Baltimore Orioles.

Brucker Great Blue Heron Sanctuary

From jct of US 62 and PA 18 just east of Sharon, go north on PA 18 about 10 mi to the sanctuary on the left (west) side of the road.

Buzzard Swamp Wildlife Management Area

"From Marienville, take FR 128 [Loleta Rd] south to FR 157 [Buzzard Swamp Rd] and turn left. Just ahead on the left is the Songbird Sojourn Trail, a 1.6-mile, self-guided interpretive trail that includes

26 stops that focus on birds' use of vegetation and habitats" (Ford 1995, 69–70). Open water at the swamp lures transient waterfowl and shorebirds, such as Hooded Merganser, Double-crested Cormorant, American Wigeon, Least Sandpiper, and Common Moorhen.

Caledonia State Park

From the southeastern Pennsylvania town of Gettysburg, proceed west on US 30 about 13 mi to signs for the park entrance. Habitats include deciduous and mixed forests and grasslands.

Colonel Denning State Park

From I-81 about 10 mi southwest of Carlisle, exit north onto PA 233. Follow PA 233 about 12 mi to the park entrance on your right. The park woods and lake provide breeding habitat for Belted Kingfisher, Red-headed Woodpecker, Canada Warbler, and Scarlet Tanager, among others.

Cook Forest State Park

From jct of I-80 and PA 36 at Brookville, go north on PA 36 about 15 mi to the park entrance. Old-growth mixed forest and bogs harbor northern species including Yellow-bellied Flycatcher, Red-breasted Nuthatch, Blackburnian Warbler, Blue-headed Vireo, and Canada Warbler.

Cowan's Gap State Park

From I-76 (Pennsylvania Turnpike) about 40 mi west of Carlisle, exit south onto PA 75. Go south on PA 75 about 10 mi to Richmond Rd and follow signs to park. Summer residents include Blue-gray Gnatcatcher, Acadian Flycatcher, Red-bellied Woodpecker, Yellow-throated Vireo, and Worm-eating Warbler.

Curllsville Strip Mines

From jct of US 322 and PA 68 in Clarion, go south on PA 68, "about 8 miles to the intersection of Route 2007. Turn left and drive east on Route 2007 1.2 miles, toward Curllsville. At the intersection of Routes 2007 and 2011, in Curllsville, turn right on 2011 across the creek. Go 2.5 miles and turn left on Up Church Road (TR 442). Continue 0.4 mile farther to a dirt road that goes straight while the township road curves right. This dirt road loops through the grassland habitat and returns to SR 2011 in 1.7 miles" (Ford 1995, 60). These open spaces attract Northern Harrier, Short-eared Owl, Rough-legged Hawk, Vesper Sparrow, American Pipit, Bobolink, and other grassland species.

Delaware Water Gap National Recreation Area

At jct of I-80 and US 209 near the city of Stroudsburg, go north on US 209 about 26 mi to signs at Dingman's Ferry for the visitors' center.

Erie National Wildlife Refuge

From Meadville, go east on PA 27 to jct with SR 2032. Continue straight east on SR 2032 (PA 27 curves to the right) another 3 mi to jct with SR 2015 in Guys Mills. Proceed straight on SR 2015 (PA 198) 0.8 mi to the refuge hq on the right. The refuge has beaver swamps, impoundments, grasslands, and cropland to bring in large numbers of transient waterfowl and shorebirds.

Fairmount Park

From I-476 northwest of Philadelphia, exit south onto Ridge Pike (SR 3009). Follow Ridge Pike south into Philadelphia about 4 mi to where the road splits in a Y. Take the left fork of the Y (Henry Ave, SR 4001). The park borders the left (east) side of Henry Ave for about 3 mi, providing several access points. With extensive riparian habitat, woodlands, and ponds, this large city park (almost 9,000 acres) attracts a variety of species during migration.

Fowler's Hollow State Park

From I-76 (Pennsylvania Turnpike) about 40 mi west of Carlisle, exit north onto PA 75. Go north on PA 75 about 8.6 mi to where PA 274 branches off to the right (east). Follow PA 274 about 9 mi to Upper Buck Ridge Rd. Turn right (south) on Upper Buck Ridge Rd and go about 2 mi to jct with Fowler Hollow Rd (SR 3004). Turn right on Fowler Hollow Rd to reach the park. The mixture of coniferous and deciduous forest and second growth provides habitat for Whip-poor-wills, Black-throated Blue Warblers, Ovenbirds, Wood Thrushes, and Yellow-breasted Chats.

French Creek State Park

South of Reading on I-176, exit east onto PA 724. Go east on PA 724 about 7.3 mi to jct with LR 2083 in Monocacy. Turn right (south) onto LR 2083 and go about 5 mi to the park. The lake and woods are home to Osprey, both vulture species, Tree Swallow, Belted Kingfisher, Green Heron, and Pileated Woodpecker.

Friendship Hill National Historic Site

At jct of PA 21 and PA 166 in Masontown, take PA 166 south about 6 mi along the Monongahela River to Friendship Hill on the left of the road. Riparian forest and wetlands are the principal habitats, hosting four thrush species in summer (Wood, Hermit, Swainson's, and Veery) along with Northern Bobwhite, Brown Thrasher, and many other species.

Gifford Pinchot State Park

From I-83 about 6 mi south of Harrisburg, exit south onto PA 177 and go south about 7 mi to the park entrance on the left (east) side of the road. Conewango Lake is the park's main attraction for waterfowl and shorebird species.

Hawk Mountain Sanctuary

From jct of US 22/I-78 and PA 61 about 17 mi north of Reading, go north on PA 61 about 4.1 mi to jct with CR 895 East. Turn right on CR 895 East and go 1.7 mi to the Hawk Mountain turnoff,

Hawk Mountain Sanctuary, Pennsylvania, is justly among the most famous raptor-watch localities in the world. Established in the 1930s, thanks in large part to the efforts of Rosalie Barrow Edge, the sanctuary on Kittatinny Ridge draws thousands of visitors each year to one of the few places where you can actually see the phenomenon of bird migration in progress. Like other great birding localities, Hawk Mountain has its own lore and literature, including Maurice Broun's *Hawks Aloft* (1948) and Jim Brett's *The Mountain and the Migration* (1991). Stop at the sanctuary's bookstore for these and for the latest information on the status of the migration.

which is well marked by a sign. Turn right at the turnoff, and proceed 1.9 mi to the Visitor Center and Parking Area. Hawk Mountain is recognized as one of the premier birding destinations in the country, providing outstanding views of the spring and fall raptor migrations (16 species) along Kittatinny Ridge.

Jennings Environmental Education Center and Nature Reserve

From jct of US 422 and PA 528 in Prospect, go north on PA 528 about 8 mi to the entrance to the center, which includes forest, prairie, and riparian habitats. Black-billed Cuckoo, American Woodcock, Blue-winged Warbler, Swamp Sparrow, and Hooded Warbler can be found. Common Redpoll,

Red Crossbill, Evening Grosbeak, and White-winged Crossbill have been reported in winter.

John Heinz National Wildlife Refuge	See Tinicum National Wildlife Refuge.
Kahle Lake	From I-80 2 mi east of Emlenton, exit north onto PA 208. Turn right (east) on PA 208 and go about 2.7 mi to jct with T324. Turn left (north) on T324 and go about 1.4 mi to the lake, which attracts grebes, loons, swans, ducks, gulls, terns, and other waterbirds during migration.
Kettle Creek State Park	From the town of Westport on the Susquehanna River at jct of PA 120 and SR 4001 (Kettle Creek Rd), go north on Kettle Creek Rd about 7 mi to the park. Kettle Creek Lake and reservoir host summering Pied-billed Grebes, Buffleheads, and Common Mergansers, while neighboring forest has Broad-winged Hawks and Black-and-white Warblers.
Keystone State Park	From I-76 southeast of Pittsburgh, take exit onto US 22 East. Follow US 22 east about 19 mi to jct with PA 981 at New Alexandria. Turn right (south) on PA 981 and go about 4 mi to jct with Derry Rd. Turn left on Derry Rd and go about 2 mi to the park. Marsh, lake, old field, and forest habitats are home to Great Blue Herons, Northern Rough-winged Swallows, Great Crested Flycatchers, and Orchard Orioles.
Marsh Creek State Park	About 25 mi west of Philadelphia, go north from I-76 (Pennsylvania Turnpike) onto PA 100 about 0.6 mi to jct with Park Rd. Turn left on Park Rd, and go about 1.5 mi to the park. The 525-acre lake is a magnet for migrating waterbirds.

Moraine State Park

From jct of I-79 and US 422 about 25 mi north of Pittsburgh, go east on US 422 about 2 mi to the park entrance. Habitats include old fields, lake, lake shore, and marsh. Cliff Swallows have nested here, and habitat is present for Sedge Wrens.

Mount Zion Strip Mines

From the town of Clarion, go west on I-80 to the exit for SR 3007. Go south on SR 3007 "across the Clarion River at Canoe Ripple Bridge, at 2.1 miles. As SR 3007 ascends the hill, take the first dirt road (TR 425) to the left, at 1.1 miles. . . . At the next intersection, 1.1 miles, a Y, turn left onto another dirt road. . . . At 0.4 mile farther on this road, locked gates on both sides of the road prohibit driving through the reclaimed surface mines owned by C & K Coal Co., but walking is permitted" (Ford 1995, 61). Abandoned strip mines provide grassland habitat for Barn Owl, Northern Harrier, Short-eared Owl, Rough-legged Hawk, Vesper Sparrow, American Pipit, Bobolink, and other open-country species.

Mud Level Road, Shippensburg

From I-81 at Shippensburg, exit onto PA 696 North. Follow PA 696 about 4.8 mi (through Shippensburg) to jct with Mud Level Rd (TR 305). Turn right (east) on TR 305 (becomes LR 4002) and proceed 2–3 mi along the road (Ford 1995, 175). The broad expanse of cropland, grassland, and fallow fields attract open country birds like Upland Sandpiper, Northern Harrier, Barn Owl, Grasshopper Sparrow, Bobolink, and Savannah Sparrow among others.

Nockamixon State Park

From I-476 about 25 mi north of Philadelphia, exit onto PA 663 East. Follow PA 663 about 3 mi to Quakerstown where it joins PA 313. Continue east on PA 313 to jct with PA 563. Turn left (northeast) on PA 563 and go about 4 mi to the park. The lake, old fields, marsh, and deciduous forest of the park are especially attractive to raptors (Northern Harrier; Sharp-shinned, Cooper's, Red-tailed, Rough-legged, Broad-winged, and Red-shouldered hawks; Northern Goshawk; Osprey; and American Kestrel).

Ohiopyle State Park

The town of Ohiopyle is located about 50 mi southeast of Pittsburgh. To get to the state park hq from the town, follow SR 2012. The old fields and coniferous, deciduous, and mixed forest of the park host a remarkable variety of breeding birds, including Eastern Kingbird; Parula, Chestnut-sided, Blackburnian, Black throated Blue, and Golden-winged warblers; White-eyed and Blue-headed vireos; and Savannah, Field, and Chipping sparrows.

Presque Isle State Park

Just south of Erie, take Exit 18 from I-90 onto PA 832 North. Follow PA 832 about 6 mi to Peninsula Dr and continue north on this road another 2 mi to the park. This site on the shores of Lake Erie is outstanding for transient and wintering waterbirds, including gulls, ducks, swans, grebes, loons, terns, and geese.

Prince Gallitzin State Park

About 10 mi northwest of Altoona, take SR 1026 west from jct of PA 53 and SR 1026 in the village of Frugality. Follow SR 1026 about 3 mi to the park. Several habitat types (including lake, streamside, marsh, cropland, old field, mixed woodland, and mowed grassland) attract a diverse assemblage of summer birds including Grasshopper, Vesper, and Henslow's sparrows, Spotted Sandpipers, Sora and Virginia rails, and Acadian, Willow, and Least flycatchers.

Located on Lake Erie's eastern shore, Presque Isle State Park, Pennsylvania, provides "pelagic" viewing opportunities for people living far from the ocean. Waters off the park are often the winter home for scoters, mergansers, goldeneyes, scaup, swans, loons, grebes, and many other northern-nesting waterbirds.

Public Docks, Erie

From I-90 just south of Erie, exit onto PA 97 North (Perry Hwy). Follow PA 97 north for about 0.9 mi until it splits in a Y. PA 97 goes to the right, and PA 505 goes left. Follow PA 505 about 2.5 mi to State St, and then follow State St north to its end at the lake and docks. Like Presque Isle, this site on the shores of Lake Erie is outstanding for transient and wintering waterbirds including gulls, ducks, swans, grebes, loons, terns, and geese.

Pymatuning Wildlife Management Area

From I-79 at Meadville, exit onto US 6 West. Follow US 6 about 6 mi to Conneaut Lake. From the Y where PA 285 continues straight west and US 6 bears right to the northwest, follow US 6 for another 7 mi to jct with SR 3011 at Linesville. Turn left (south) on SR 3011, and proceed about 1.2 mi to the Visitor Center and picnic area. Pymatuning is a superb site for transient waterbirds and is outstanding for transient songbirds as well.

Pymatuning can be translated (rather freely) as "liar's home." Still, there is no exaggerating the remarkable flights of migrant waterfowl and songbirds that pass through the region. Twenty-five species of ducks, geese, and swans and forty species of warblers and vireos, many in spectacular numbers, occur regularly in Pymatuning Wildlife Management Area, Pennsylvania (see p. 69).

Winter State Park	From jct of US 15 and PA 192 in Lewisburg, go west on PA 192 about 16 mi to the park entrance.
Reed's Gap State Park	About 15 mi southeast of State College at jct of US 322 and SR 1002 in Reedsville, go east on SR 1002 about 8 mi to the park. Birds typical of mixed forest occur here, including Ruffed Grouse, Northern Flicker, Red-shouldered Hawk, and Pileated Woodpecker.
Ridley Creek State Park	At jct of I-476 and PA 3 just west of Philadelphia, go west on PA 3 (West Chester Pike) about 6.5 mi to the park entrance on the left (south) side of the road. Riparian and old field habitats attract several interesting breeding species, including Warbling, Yellow-throated, and White-eyed vireos, American

Tinicum National Wildlife Refuge, Pennsylvania, is located smack in the middle of the great megalopolis, only one mile from the Philadelphia International Airport. In fact, I-95 runs right through the refuge. Nevertheless, 288 species of birds have been recorded for this twelve-hundred-acre site, which represents the last remnant of freshwater tidal marsh in the state of Pennsylvania. Even if you can't appreciate the wealth of waterbirds on display, the quiet channels, pools, and ponds provide a welcome respite from the city.

Woodcock, Swamp Sparrow, Blue-gray Gnatcatcher, and Blue-winged Warbler.

Shawnee State Park — From jct of US 220 and US 30 in Bedford, go west on US 30 about 10 mi to the park entrance on the left (south) side of the road. Shawnee Lake attracts a good variety of transient waterbirds.

Tinicum National Wildlife Refuge — Heading south on I-95 from Philadelphia, look for the Bartram Ave and Essington Ave exit shortly after crossing the Walt Whitman Bridge. The exit ramp becomes Bartram Ave. Proceed straight about 1.0 mi to jct with 84th St. Turn right on 84th St and go 0.7 mi to Lindbergh Blvd. Turn left on Lindbergh Blvd and go 0.3 mi to the inconspicuous entrance to the 1,200-acre refuge on your right. Don't try these directions coming from south of Philadelphia on I-95. There is no Bartram Ave exit coming from the south. Tinicum is a shining pearl in a morass of urban, industrial waste. The checklist for this last piece of freshwater tidal marsh remaining in Pennsylvania has nearly 300 species of birds, including Little Blue Heron, Tricolored Heron, Cattle Egret, Glossy Ibis, and many others that are rare or absent elsewhere in the state.

Tioga/Hammond Lakes National Recreation Area	From jct of US 6 and US 15 in Mansfield, go north on US 15 about 10 mi to jct with PA 287. Turn left (west) on PA 287 and go about 4 mi to the lakes. Continue on PA 287, which parallels the shore of Hammond Lake for about 2 mi. The recreation area offers several habitats, including open water, marsh, swamp, riparian and upland forest, and old fields. Some birds found here are Cliff, Bank, and Barn swallows, Bobolinks, Bald Eagles, Eastern Phoebes, and Yellow Warblers.
Tobyhanna State Park	At jct of I-380 and PA 423 southwest of Tobyhanna, exit north onto PA 423 and go about 2 mi to park entrance.
Woodbourne Sanctuary	Eighteen miles northwest of Scranton at jct of US 6 and PA 29 in Tunkahannock, go north on PA 29 about 16 mi to the sanctuary. Birds typical of northern, transitional mixed forest breed here, including Northern Waterthrush, Yellow-bellied Sapsucker, Winter Wren, Hermit Thrush, Dark-eyed Junco, and Canada and Blackburnian warblers.
World's End State Park	From jct of US 220 and PA 154 just north of Laporte, go west on PA 154 about 7 mi to the park entrance. Habitat is northern mixed forest, home to breeding populations of Black-throated Green Warbler, Red-breasted Nuthatch, Blackburnian Warbler, Yellow-rumped Warbler, Canada Warbler, and Northern Waterthrush.

Yellow Creek State Park

From jct of US 422 and US 119 just south of Indiana, follow US 422 east about 9 mi to jct with PA 259 South. Turn right on PA 259 and proceed a short way to the park entrance on the right. Paula Ford says Yellow Creek is "one of the best birding sites in the Commonwealth" (1995, 113). The principal attraction is a high diversity of both water- and landbird migrants.

Virginia

Back Bay National Wildlife Refuge

At jct of I-64 and US 58 in Virginia Beach, take US 58 east and proceed about 12.5 mi to jct with US 60 at the beach. Turn right (south) on US 60 and continue about 7.5 mi to jct with VA 149 (Sandbridge Rd). Turn left (east) on Sandbridge Rd and follow this about 6.6 mi to the beach at jct of Sandbridge Rd and Sandpiper Rd. Turn right (south) on Sandpiper Rd and proceed 6 mi to the refuge. Back Bay, which has the usual coastal avifauna of Virginia, has been the main site for several rarities, including White Pelican, Wood Ibis, Brewer's Blackbird, and Lincoln's Sparrow. Sedge Wrens winter in Back Bay marshes.

Bear Mountain Road (CR 601)

From jct of US 220 and US 250 at Monterey, go west on US 250 about 12 mi (6.6 mi west of Hightown) to jct with CR 601 (Bear Mountain Rd). Turn left (south) onto Bear Mountain Rd into mountain meadow habitat, where breeding Savannah Sparrows can be found occasionally.

Byrd Visitors Center, Shenandoah National Park

The visitors' center is located at Milepost 51 on the Skyline Drive (51 mi from the Front Royal entrance to the drive). The site is good for highland/ northern species like the Black-capped Chickadee and White-breasted Nuthatch.

The Chesapeake Bay Bridge-Tunnel, Virginia, is as close as you can get to the open ocean without taking a boat. For this reason, it is a popular site for scanning for gannets, jaegers, storm-petrels, and sea ducks.

Carvin Cove Reservoir

In Roanoke on I-81, exit north onto VA 419 and go about 0.5 mi to jct with VA 311. Continue north on VA 311 about 2.4 mi to jct with CR 740 (Carvin Cove Rd). Turn right on CR 740 and proceed about 6.4 mi to the reservoir, a popular stopover for transient waterbirds.

Chesapeake Bay Bridge-Tunnel

From jct of I-64 and I-264 in Virginia Beach, go north on I-64 about 2 mi to the east exit for US 13. Take US 13 east and continue about 6 mi to the south entrance for the bridge-tunnel (toll, $10). Using a spotting scope from designated parking areas on the bridge provides the best opportunity for viewing pelagic species without taking a boat ride.

Chincoteague National Wildlife Refuge

At jct of US 13 with VA 175, turn east on VA 175 and follow it 10 mi to Chincoteague Island. Turn left on Main St and right on Maddox Blvd. Proceed about 2 mi to the refuge hq. Chincoteague is one of the top ten birding destinations in North America, and rightly so. About 320 species have been recorded for the refuge, many of which occur rarely or not at all elsewhere in the region.

Chincoteague National Wildlife Refuge, Virginia, lists several birds that are not found regularly anywhere else in the region, such as White Ibis, Purple Gallinule, Long-billed Curlew, and Lark Sparrow.

Clinch Valley College Campus

From jct of US 23 and US 58 Alt in western Virginia, go north on US 23 about 2 mi to the town of Wise. In Wise, turn right on CR 646, and go 0.4 mi to the second entrance to the college. Take this entrance, drive past dormitories to the picnic area, and park. Follow the cross-country trail to the football field and ponds (R. Peake in Johnston 1997, 187). Second-growth locust habitat on the campus is reputedly good for Golden- and Blue-winged warblers and their hybrids.

Colonial Parkway	From jct of I-64 and VA 199 at Williamsburg, go west on VA 199 2 mi to jct with the Colonial Parkway. Turn left (south) onto the parkway. In winter, look for Fox Sparrows on the roadside along the portion of the parkway paralleling the James River. The mix of salt, brackish, and freshwater marshes provides habitat for Clapper Rails, Boat-tailed Grackles, Common Snipe, Marsh Wrens, and many other wetland birds.
Craney Island Landfill	From jct of I-664 and VA 135 just west of Portsmouth in southeastern Virginia, take VA 135 south. Proceed 1.9 mi on VA 135 to jct with VA 164. Turn left (east) on VA 164 and go 2.2 mi to jct with Cedar Lane. Turn left on Cedar Lane and left again on River Shore Rd. Follow River Shore Rd 1 mi to jct with Hedgerow Rd. Turn right on Hedgerow Rd and go 0.3 mi to the entrance of the facility. Bear left and go to the main office of the U.S. Army Corps of Engineers. All visitors must sign in, and the facility is open only on weekdays from 7:00 A.M. to 3:30 P.M. (R. Beck, R. Ake, and R. Peake in Johnston 1997, 58). The landfill is a good locality for breeding Black-necked Stilts and transient American Avocets.
Dark Hollow Falls, Shenandoah National Park	Enter the parking lot for Dark Hollow Falls Trail just short of Milepost 51 (51 mi south of the Front Royal entrance to the Skyline Drive), park, and hike, following signs to the falls.
Dickey Ridge Visitor Center	From jct of US 340 and VA 55 in Front Royal, go south on US 340 0.2 mi and turn left (east) to enter Shenandoah National Park (fig. 30). Go 3.8 mi on the Skyline Drive to the well-marked visitor center and picnic area. Forests along this portion of the drive have breeding Cerulean and Kentucky war-

Shenandoah National Park, Virginia, stretches 105 miles from Front Royal at the northern end to Waynesboro at the southern end. In between, the Skyline Drive provides access to mountain vistas, trails, picnic areas, campgrounds, and extensive views of mature Appalachian deciduous, mixed, and coniferous forest. Dickey Ridge, located just north of Milepost 5, is one of many attractive stopping places where Cerulean Warblers, Dark-eyed Juncos, Scarlet Tanagers, Wood Thrushes, and other species typical of the park's woodlands can be readily heard and seen.

blers, Ovenbirds, Dark-eyed Juncos, and Common Ravens.

Dyke Marsh

From jct of US 1 and I-495 (Capital Beltway) south of Washington, go north on US 1 about 0.5 mi to jct with VA 236 (Duke St). Turn right (east) on Duke St and go 3 blocks to Washington St (VA 400). Turn right (south) on Washington St, which becomes Mount Vernon Memorial Hwy, for about 2 mi. Look for a left turn for Belle Haven Picnic Area, and follow this road to a chained-off trail on the right. Park and follow this trail to the marsh, watching carefully for transient songbirds. The trail is also good for wintering birds like Fox and White-throated sparrows and Northern Saw-whet Owl.

Eastern Shore of Virginia National Wildlife Refuge	At the southern tip of the Delmarva Peninsula, just before the north entrance to the Chesapeake Bay Bridge-Tunnel, exit US 13 on CR 600 to the east, and follow signs to the refuge visitor center. Not so well known as nearby Chincoteague and Assateague, this refuge is comparable in terms of bird species diversity.
False Cape State Park	At jct of I-64 and US 58 in Virginia Beach, take US 58 east and proceed about 12.5 mi to jct with US 60 at the beach. Turn right (south) on US 60 and continue about 7.5 mi to jct with VA 149 (Sandbridge Rd). Turn left (east) on Sandbridge Rd and follow this about 6.6 mi to the beach at jct of Sandbridge Rd and Sandpiper Rd. Turn right (south) on Sandpiper Rd and proceed 9 mi to the park. Fulvous Whistling-Ducks have been recorded at this site, the southernmost point on the Virginia coast.
FR 812 (Parkers Gap Rd), Blue Ridge Parkway	From the Blue Ridge Parkway's James River Visitor Center about 10 mi northwest of Lynchburg, go south on the parkway about 15.4 mi to the Sunset Field Overlook parking lot just south of Milepost 78. Turn right on FR 812 (Parkers Gap Rd) and follow the road as it winds through highland forest, old field, and second-growth habitat good for Blue-winged Warbler, Northern Parula, Black-and-white Warbler, and Song Sparrow.
George Washington Birthplace National Monument	From jct of US 301 and VA 3 about 20 mi east of Fredericksburg, go east on VA 3 about 15 mi to jct with VA 204. Turn left on VA 204 and proceed about 2 mi to the monument, a good site for herons and egrets as well as such upland birds as Blue Grosbeak and Prairie Warbler.

Grandview Beach

At jct of I-64 and US 258/VA 134 just north of Hampton, go east on US 258/VA 134 (Mercury Blvd) about 3 mi to jct with VA 169 (Fox Hill Rd). Turn left on VA 169 and continue 3 mi to jct with Beach Rd. Bear left on Beach Rd and go about 2.7 mi (through Fox Hill to Grandview), turn left on State Park Dr, park, and walk to park entrance. The park has breeding populations of Saltmarsh Sharp-tailed Sparrows and Seaside Sparrows.

Grayson Highlands State Park

From I-81 about 45 mi northeast of Bristol, VA, exit south onto VA 16 and proceed about 25 mi to jct with US 58. Turn right (west) on US 58 and go 7.5 mi to the park entrance on your right. Spruce-fir and other highland habitats at the park have summer populations of Hermit Thrush, Red-breasted Nuthatch, Cedar Waxwing, Blackburnian Warbler, and Northern Saw-whet Owl.

Great Dismal Swamp National Wildlife Refuge

From jct of US 13/32 (Main St) and CR 337 (Washington St) in downtown Suffolk, go south on US 13/32 about 3.5 mi until VA 37 branches off to the left. Turn left on VA 37 and go 3.0 mi to jct with CR 759. Turn left on CR 759 and go 0.7 mi to jct with CR 642. Continue straight ahead (east) on CR 642 0.5 mi to jct with CR 604. Turn right on CR 604 (Desert Rd) and go 1.6 mi to the refuge visitor center on the left. This northernmost extent of southern swamp forest is home to breeding Swainson's Warbler, along with other species typical of that habitat, including Yellow-throated Warbler, Prothonotary Warbler, Hooded Warbler, Carolina Wren, Louisiana Waterthrush, and Northern Parula.

Harvey's Knob

This well-known hawk-watching site is located 0.2 mi south of Mile Marker 95 on the Blue Ridge Parkway, about 16 mi northeast of Roanoke.

Hemlock Springs Overlook, Shenandoah National Park

The parking lot for the overlook is located about 0.7 mi south of Milepost 39 (39 mi from the Skyline Drive entrance at Front Royal). Excellent views of mixed highland forest habitat can be had here, home to Black-capped Chickadee, Blue-headed Vireo, Dark-eyed Junco, Blackburnian Warbler, and Eastern Towhee.

Hog Island Waterfowl Management Area

From jct of US 258 and VA 10 west of Smithfield, go north on VA 10 about 9 mi to jct with CR 650. Turn right (north) on CR 650, and proceed about 7 mi to enter the management area, which is a

The Austroriparian Biotic Province covers the entire southeastern United States from east Texas to extreme southeastern Virginia. Several species exotic elsewhere in the Mid-Atlantic are common in the bayous of the Great Dismal Swamp National Wildlife Refuge, Virginia. Along with cottonmouths and sliders, look for Swainson's Warblers, Prothonotary Warblers, and Chuck-will's-widows.

good place to search for such rarities as Yellow-headed Blackbird, Greater White-fronted Goose, Wood Stork, White Pelican, and Fulvous Whistling-Duck.

Hunting Bay

From jct of US 1 and I-495 (Capital Beltway) south of Washington, go north on US 1 about 0.5 mi to jct with VA 236 (Duke St). Turn right (east) on Duke St and go 3 blocks to Washington St (VA 400). Turn right (south) on Washington St, which becomes Mount Vernon Memorial Hwy, for about 2 mi. Look for a left turn for Belle Haven Picnic Area. Park at the picnic area, and walk north along the bike path to the bridge across Hunting Creek, which provides a good view of the bay for scanning to find and identify waterfowl.

Huntley Meadows Park

In Alexandria south of Washington, D.C., exit I-495 south onto US 1 and go 3.2 mi south to jct with Lockheed Blvd. Turn right on Lockheed Blvd and go 0.6 mi to Harrison Ln. Turn left into park (Wilds 1992, 70). The wetlands here support breeding populations of Pied-billed Grebe, Common Moorhen, King Rail, Least and American bitterns, and Common Moorhen.

James River Park, Richmond

From jct of I-195 and VA 161 in Richmond, go south on VA 161 and take the first left after crossing the river onto Riverside Dr. The park is on your left. The mixture of riparian forest, second-growth, marsh, and grassland habitats in the park attracts both upland and wetland species.

Kerr Reservoir

From jct of US 1 and US 58 in southern Virginia, head west on US 58. The highway parallels the reservoir for several miles, crossing arms of the reservoir. Occoneechee State Park on the reservoir shore is located about 20 mi from the US 1 jct with

US 58, just before the road crosses the Roanoke River. The huge lake, parts of which are in North Carolina, offers important stopover habitat for migrating waterbirds.

Kiptopeke State Park

At the southern tip of the Delmarva Peninsula, just before the north entrance to the Chesapeake Bay Bridge-Tunnel, exit US 13 on CR 704 to the west, and follow signs to the state park. The park is renowned as a major observation site for raptor and shorebird migrations—Buff-breasted Sandpipers, Peregrine Falcons, and Golden Eagles in particular.

Lake Anna

From jct of US 522 and VA 208 about 25 mi south of Culpeper, go east on VA 208 and proceed about 3.1 mi to the lake. Continue another 5 mi to jct with CR 601 North. Turn left on CR 601 and go about 3.2 mi to jct with CR 7000. Turn left on CR 7000 and go about 2.2 mi to the park and lake shore. The lake serves as an excellent inland habitat for transient and wintering waterbirds.

Long Branch Nature Center

From jct of US 50 (Arlington Blvd) and VA 120 (Glebe Rd) in Washington, D.C., proceed west on US 50 about 1.7 mi to Carlin Springs Rd. Exit right for Carlin Springs Rd, and continue on Carlin Springs Rd about 0.7 mi to a left turn just past the Northern Virginia Doctors Hospital that leads to the nature center parking lot. Trails lead to ponds, fields, and shrubby habitat that can be good for transients. Common Redpoll has been recorded here in winter.

Lucketts, Hwy 661

From jct of US 15 (Business) and VA 7 in the northern Virginia town of Leesburg, go north on US 15 about 5 mi to jct with CR 662 at Lucketts. Turn right (east) on CR 662 and go about 1 mi to jct with CR 661. Turn right on CR 661. The farmland along this road is good for open country, grassland species.

Mackay Island National Wildlife Refuge

Most of Mackay Island is in North Carolina. Only a small tip extends into Virginia. To reach the refuge, exit I-64 onto VA 165 South (Military Hwy) in Norfolk. Follow VA 165 (which becomes Princess Anne Rd) for about 13 mi to jct with VA 149 at Princess Anne. Turn left on VA 149 (Princess Anne Rd) and follow this for about 23 mi all the way south to Mackay Island (passing through the wildlife refuge) and jct with Knotts Island Rd. Turn left on Knotts Island Rd, and go 2 mi to enter the Virginia portion of Mackay Island. Common Ground-Dove has been reported from the refuge, along with other rarities reaching the northern extent of their ranges.

Mason Neck National Wildlife Refuge

At jct of US 1 and VA 242 about 12 mi south of Washington, D.C., go east on VA 242 about 3 mi to Gunston Hall, where VA 242 becomes CR 600. Continue on CR 600 another mile to High Point Rd, where there is a sign for the refuge. Bear right on High Point Rd and proceed 0.7 mi to the refuge. Established principally for the Bald Eagle, the riparian forest, marshes, and open water of the Potomac are good for many other upland and wetland species as well.

Mendota Fire Tower

This well-known hawk-watching locality can be reached as follows: "From Abington take US-19 north 7.7 miles to Route 802. Take 802 southwest for 14.7 miles to Route 612. Drive 612 north about 3.1 miles up Clinch Mountain. Park on 612 and walk 0.3 mile up a steep unmarked access road to the tower" (G. Larkins and A. Kirby in Johnston 1997, 187).

Mount Rogers, the highest point in Virginia, is the place to see and hear boreal species at the southern end of their breeding distribution. Northern Saw-whet Owl, Swainson's Thrush, Magnolia Warbler, and Purple Finch are found here during the breeding season, as are Olive-sided Flycatchers during fall migration.

Mount Rogers

You can't drive to Mount Rogers, although you can get within three miles or so of the peak on the roads for Grayson Highlands State Park. See directions to Grayson Highlands State Park above, and stop at the park's visitor center to get maps for trails leading to the top of Mount Rogers. Spruce-fir and other highland habitats on the mountain have summer populations of Hermit Thrush, Red-breasted Nuthatch, Cedar Waxwing, Blackburnian Warbler, and Northern Saw-whet Owl.

Mountain Lake

From the town of Pembroke, proceed east on US 460 a little over 2 mi to jct with CR 613 at Hoges Chapel. Turn left (north) on CR 613 and go about 5 mi to Mountain Lake. Birds typical of highland coniferous forest, like Pine Siskin and Hermit Thrush, can be found here. See David Johnston's Mountain Lake Region and Its Bird Life (2000) for a delightful account of the history and birds of the site.

Natural Chimneys Regional Park	From jct of US 250 and VA 42 about 7 mi northwest of Staunton, go north on VA 42 about 7.7 mi to jct with CR 731 at Moscow. Turn left on CR 731 about 3.2 mi to a Y where CR 731 goes left and the road to Natural Chimneys goes right. Bear right on the road to Natural Chimneys and proceed about half a mile to the site. The mixture of hardwood, coniferous, and riparian forests and fields is good habitat for Ruffed Grouse, Northern Bobwhite, American Woodcock, Merlin, and Long-eared Owl.
Northern Virginia Regional Park (Upton Hill Regional Park)	From jct of I-495 and US 50, take US 50 (Arlington Blvd) east about 3.6 mi to the Seven Corners jct with CR 613 (Sleepy Hollow Rd/Wilson Blvd). Turn left onto Wilson Blvd and go about 0.6 mi to the park, which is located on the south side of Wilson Blvd. Claudia Wilds (1992, 75) cites this park as a favorite place to watch Common Nighthawks and Chimney Swifts on late summer evenings.
Otter Creek	The creek parallels the Blue Ridge Parkway for several miles, and a hiking trail runs along the creek from the Otter Creek campground (located at jct of the Blue Ridge Parkway and VA 130 about 14 mi southeast of Lexington) about 4 mi to the James River visitor center. Belted Kingfisher, Cerulean Warbler, Louisiana Waterthrush, and Northern Parula can be found at this site.
Pocahontas State Park	At jct of I-95 and VA 288 about 5 mi south of Richmond, go west on VA 288 about 6 mi to jct with VA 10 (Iron Bridge Rd). Turn left (south) on VA 10 and proceed about 1.2 mi to jct with CR 655 (Beach Rd). Turn right on CR 655 and go about 4 mi to jct with CR 780. Turn right on CR 780 and proceed to the park entrance. The park is an excellent site for breeding birds of the Pied-

mont, including Summer Tanager, Hooded and Worm-eating warblers, Whip-poor-will, Yellow-billed Cuckoo, Northern Parula, and Yellow-breasted Chat.

Presquile National Wildlife Refuge

At jct of I-295 and CR 697 (Bermuda Hundred Rd) about 10 mi southeast of Richmond, go east on CR 697 about 3.5 mi to the James River, where the road goes north. Continue another mile or so to the channel. Access to the refuge is by ferry. Refuge visitors are asked to make reservations for the ferry with the refuge manager at (804) 733-8042. The tidal swamp, marsh, and cropland composing the refuge bring in large concentrations of waterfowl during migration, especially Canada Geese.

Ramseys Draft

From jct of US 250 and US 11 in Staunton, go west on US 250 about 25 mi to Ramseys Draft Picnic Area in the George Washington National Forest. Park at the picnic area and hike the trail, which parallels the stream. This highland site is attractive to northern transients, such as Gray-cheeked Thrush, Bicknell's Thrush, Olive-sided Flycatcher, and Bay-breasted and Blackpoll warblers.

Roanoke Sewage Treatment Plant

"From I-581 (east) [in Roanoke] take the Elm Ave Exit and go 0.8 mile via Elm Avenue (which becomes Bullit Avenue) to 13th Street and turn right onto 13th Street (watch for the 'Water Pollution Control Plant' sign). Go another 0.8 mile to Carlisle Ave (first street after the bridge over the Roanoke River) and turn left. Go one block to Spruce Street and turn left. Cross past Brownlee Avenue and watch for the open gate (chain link) on your left. Enter on your right. Park on your right, or take the first cross-dike past the 30-million-gallon concrete basin. On the left you will see a white metal box on a railing with the Roanoke Valley Bird Club logo on it. Inside is a clipboard containing notations on recent sightings at the sewage treatment plant. *On your first visit during regular working hours, go to the office and register.* (The office is located on Brownlee Avenue. Go to the left end of building, upstairs.)" (M. Purdy in Johnston 1997, 174–75). Sewage settling ponds draw transient shorebirds.

Saxis Wildlife Management Area

At jct of US 13 with VA 175, go south on US 13 about 3.6 mi to jct with CR 695 (Saxis Rd). Turn right (west) on CR 695 and proceed 8.2 mi to Saxis marshes, where Sedge Wrens have been known to breed on occasion and Henslow's Sparrows have been found in the tall grass meadows.

Seashore State Park

From jct of US 13 and US 60 in Virginia Beach, go east on US 60 (Shore Dr) about 4 mi to the turnoff on your right to the park. Lesser Black-backed Gulls have been reported fairly regularly from the park in winter.

Scott's Run Nature Preserve

From I-495 take the Great Falls exit west onto the Georgetown Pike (VA 193) and proceed about 0.7 mi. Turn right to enter the preserve (also known as Dranesville District Park). A visit to the deciduous forest, thickets, and hemlock groves of the preserve will turn up birds typical of the Piedmont.

Shenandoah River State Park

From jct of VA 55 and US 340 in the town of Front Royal, go south on US 340 about 8.1 mi to Daughter of Stars Dr. Turn right (west) on Daughter of Stars Dr and follow signs to the park. Riparian forest at the park provides habitat for Warbling Vireos, Summer Tanagers, Baltimore Orioles, Yellow-throated Vireos, Carolina Wrens, and other bottomland species.

Shenandoah Valley

The South Fork of the Shenandoah River flows north in a broad, fertile valley for more than 100 mi from south of Staunton to Front Royal, where it joins the North Fork. US 340 parallels the river along much of its length, providing several side-road access points (e.g., CR 613 in Bentonville and CR 661 at Oak Hill) where you can drive west to the river and riparian habitat. Old fields, pastures, and orchards in the valley have breeding Loggerhead Shrikes, Indigo Buntings, Chestnut-sided Warblers, Blue Grosbeaks, and Yellow-breasted Chats.

Sherando Lake Recreation Area

From jct of the Blue Ridge Parkway and CR 814 about 20 mi south of Waynesboro, go north on CR 814 about 4 mi to jct with FR 91. Turn left on FR 91 and proceed to the lake and recreation area (about 2 mi). The wetland habitat is good for waterbirds.

Silver Lake

From jct of US 15 and VA 55 just west of Haymarket, go west on VA 55 about 0.5 mi to where CR 681 (Antioch Rd) splits off to the right. Follow CR 681 about 1.5 mi to a turnoff on the right for Silver Lake.

Sky Meadows State Park

Proceeding west on I-66, take US 17 north (Exit 23, north only) at the Delaplane exit. Continue on US 17 about 7 mi to the park on the left (CR 710, Edmonds Lane). Breeding Savannah Sparrows have been recorded in grasslands along CR 710.

Southampton County

US 58 between Franklin and Emporia traverses Southampton County from east to west. Take side roads to explore mature loblolly and longleaf pine habitat for Red-cockaded Woodpeckers, Brown-headed Nuthatches, and Bachman's Sparrows.

State Arboretum	From jct of US 50/17 and US 340 about 8 mi southeast of Winchester, proceed east on US 50/17 about 1.5 mi. Look for the entrance to the arboretum (also known as the University of Virginia's Blandy Experimental Farm) on the right (south) side of the highway. Plantings of exotic, fruiting trees at the arboretum draw Cedar Waxwings and House Finches, while flowers bring Ruby-throated Hummingbirds.
Staunton River State Park	From jct of US 360 and VA 344 about 7 mi northeast of South Boston, take VA 344 east about 11 mi to the park. Riparian forest, marsh, and open water habitats of the park attract both upland species and waterbirds.
Stony Man Nature Trail	From jct of US 340 and VA 55 in Front Royal, go south on US 340 about 0.4 mi and turn left (east) to enter Shenandoah National Park. From the entrance kiosk, go about 38.6 mi to the parking lot for Stony Man Trail. For several years Stony Man has been the site to observe breeding Peregrine Falcons at the park.
Stumpy Lake	From jct of I-64 and VA 407 (Indian River Rd) south of Norfolk, go east on VA 407 about 3 mi. The shore of the lake borders VA 407 on the south side of the road for the next mile and a half or so. Stumpy Lake is one of the few sites in the region from which Anhingas have been reported.
Sussex County	From I-95 about 10 mi south of Petersburg, exit east onto VA 35. This road enters Sussex County about 4 mi east of the exit and crosses the county from north to south for the next 15 mi or so. Take side roads to explore mature loblolly and longleaf pine habitat for Red-cockaded Woodpeckers, Brown-headed Nuthatches, and Bachman's Sparrows.

Swift Creek Lake	See directions for Pocahontas State Park. Swift Creek Lake is located within the state park.
Virginia Coast Reserve Barrier Islands	The 13 islands constituting this Nature Conservancy reserve are located along the ocean side of the Delmarva Peninsula from Accomac south to Kiptopeke at the southern tip. The islands are accessible only by boat from nearby coastal launch sites, such as Quinby (VA 182 from US 13) for Parramore Island and Red Bank (CR 617 from US 13) for Hog Island. Inshore pelagic species can be viewed from the barrier islands, as well as beach-nesting shorebirds, gulls, skimmers, and terns. Many common species of herons and egrets have breeding colonies on the islands.

The bird ponds at the National Zoological Park in downtown Washington provide excellent views of a wide variety of waterfowl, most of which are not part of the formal collection. These birds fly in for the abundant food provided. During migration, it is not unusual to see several species of puddle ducks, diving ducks, and mergansers. Fish Crows, Black-crowned Night-Herons, grackles, and blackbirds also frequent the pools.

Claudia Wilds (1992, 54) notes that Glover-Archbold Park is "a good place to listen to the evening chorus in June." The park certainly offers a delightful green space of tall trees and escape from the surrounding hurly-burly of Washington.

Washington, D.C.

Bird Ponds, National Zoological Park

From jct of I-495 and VA 185 (Connecticut Ave) north of Washington, D.C., go south on VA 185 about 5.9 mi to the entrance to the National Zoo on the left (east) side of Connecticut Ave. Enter the zoo, park, and pick up a zoo map directing you to the Bird Ponds/Wetlands Exhibit. Many wild individuals of waterfowl species join the pinioned members of the zoo's collection in the ponds, allowing up-close looks at Ruddy Ducks, Bufflehead, Northern Pintail, Mallard, Hooded Merganser, and several other species.

Glover-Archbold Park

Following I-66 east into Washington, D.C., cross the Roosevelt Bridge (I-66 becomes Constitution Ave) and turn left at the second light (22nd St), then left again on C St and right on 23rd (usually no left turn allowed from Constitution Ave onto 23rd). Continue on 23rd St north about 0.6 mi to Olive St. Turn left on Olive St and go 0.6 mi to jct with Wisconsin Ave. Turn right on Wisconsin and go about 0.6 mi to jct with Reservoir Rd. Turn left on Reservoir Rd and go about 0.9 mi to jct with 44th St. A trail leads into the park from Reservoir

Although the setting is not quite on an aesthetic par with the grounds of the Imperial Palace in Vienna, the small assortment of species one can see on the Mall, Washington, D.C., are just as tame. You can get as close as you like, and probably closer than you really want, to House Sparrows, Common Grackles, Rock Doves, Fish Crows, American Crows, Ring-billed Gulls, and European Starlings.

Rd just east of 44th St (Wilds 1992, 54). Birds are those typical of eastern deciduous woodlands.

The Mall

Following I-66 east into Washington, D.C., cross the Roosevelt Bridge (I-66 becomes Constitution Ave). Proceed east on Constitution Ave about 1 mi to 14th St. The mall is on your right for the next mile (on the other side of the museums). Urban species, such as Fish Crows, Common Grackles, Rock Doves, and House Sparrows, get up close and personal here as they press tourists for handouts.

National Arboretum

From jct of I-95 and US 50 east of Washington, D.C., take US 50 west and go about 7 mi to jct with Bladensburg Rd. Turn left on Bladensburg Rd and go about 0.5 mi south to R St. Turn left on R St and proceed to the arboretum entrance. In

The strange juxtaposition of trees and shrubs brought from radically different plant communities around the world attracts avian visitors not often seen elsewhere in the city to the National Arboretum, Washington, D.C. Barred, Great Horned, and Eastern Screech-owls are resident on the site, while Northern Saw-whet, Long-eared, and Barn owls occur once or twice a winter. Cedar Waxwings and American Robins often frequent Mountain Ash and other fruiting trees, and kinglets, nuthatches, siskins, and occasional crossbills can be found in the conifers in winter.

winter, fruiting trees hold flocks of Cedar Waxwings and American Robins while ornamental conifers shelter kinglets, owls, Pine Siskins, and, on rare occasions, both species of crossbills.

National Zoological Park

See Bird Ponds, National Zoological Park, above.

Rock Creek Park

From jct of I-495 and VA 185 (Connecticut Ave) north of Washington, D.C., go south on VA 185 about 3 mi to jct with Military Rd. Go left (east) on Military Rd about 1 mi to jct with Glover Rd in the park. Turn right on Glover Rd and go 0.1 mi to the Rock Creek Nature Center on the left (east) side of the road. Made famous by the fifty years of breeding bird censuses conducted by Shirley Briggs, the forests of Rock Creek, nestled in the heart of DC, have not changed much in appearance over the last half-century. Nevertheless, several members of the nesting songbird fauna have declined for reasons unknown.

West Virginia

Babcock State Park

From jct of US 60 and US 19 about 50 mi southeast of Charleston, go south on US 60 about 8 mi to jct with WV 41. Continue south on combined US 60/WV 41 about 2.5 mi until WV 41 turns right. Go right on WV 41 about 3 mi to park entrance on the right. Look here for birds typical of deciduous woodlands, including Downy Woodpecker, White-breasted Nuthatch, Great Crested Flycatcher, and American Redstart.

Beech Fork State Park

From I-64 just east of Huntington, exit east onto US 60. Go east on US 60 1.2 mi to jct with WV 10 Alternate. Turn right (south) on 10 Alt and go 2.4 mi to jct with WV 10 (there are signs from this jct to the park). Turn right (west) on WV 10 and go 0.6 mi to jct with CR 43. Turn left (south) on CR 43 and go 3.8 mi to jct with CR 54. Turn left on CR 54 and go 1.9 mi to the park entrance on your right. The lake attracts several species of waterfowl during migration. In neighboring fields and woodlands, House Wrens, Blue-gray Gnatcatchers, Black-billed Cuckoos, Wood Thrushes, Gray Catbirds, and Brown Thrashers can be found.

Berwind Lake Wildlife Management Area

From jct of US 52 and WV 16 about 25 mi west of Bluefield, go south on WV 16 about 16 mi to the town of War. Go right on CR 12-4 in War to enter the wildlife management area, or continue another 3 mi or so on WV 16 to the village of Rift and turn

Nestled in the steep valley of Beech Fork Creek in West Virginia's western hill region, Beech Fork State Park offers excellent viewing opportunities for species typical of the forest, field, riparian, and aquatic habitats of the region, including Sharp-shinned Hawk, Spotted Sandpiper, Warbling Vireo, and Blue Grosbeak.

right on CR 9 to enter the area. Forest birds here include Ruffed Grouse; Pileated, Downy, and Hairy woodpeckers; Northern Flicker; and Acadian Flycatcher, among others.

Bluestone Wildlife Management Area

From I-64 east of Beckley, exit south onto WV 20. Go south on WV 20 about 11 mi to jct with WV 3. From jct, continue south on WV 20 another 0.5 mi to the National Park Service Visitor Center. Get checklist, brochures, and directions to attractions along the Bluestone Lake section of the New River. Check riparian habitats for Wood Duck; Belted Kingfisher; Purple Martin; Tree, Northern Rough-winged, and Barn swallows; Louisiana Waterthrush; and Eastern Phoebe.

Canaan Valley State Park

From jct of WV 32 and WV 55 at Harman, go north about 9 mi on WV 32 to the park entrance on your left. Principal habitats here are northern hardwood forest, old fields, marshes, bogs, and alder thickets. Characteristic birds include

Home to crinoids and blastoids when it formed the floor of an inland sea 315 million years ago, Canaan Valley State Park, West Virginia, is now a highland valley covered with northern hardwood and coniferous forest, meadows, and marshes and home to Ruffed Grouse, Pileated Woodpecker, Black-throated Blue Warbler, Bobolink, and Rose-breasted Grosbeak.

	Nashville Warbler, Hermit Thrush, Red-eyed Vireo, Ovenbird, American Woodcock, Swamp Sparrow, Black-throated Blue Warbler, and Rose-breasted Grosbeak.
Cheat Lake	On I-68 east of Morgantown, exit west onto CR 857 and drive to the lake. This site offers some of the best waterfowl viewing in the state.
Cranberry Back Country Wilderness, Monongahela National Forest	The Cranberry Visitor Center is located on WV 150 (Highland Scenic Highway) just west of jct of WV 55/US 219 and WV 150 on the left of the road. Stop here for directions to Cranberry Wilderness sites and Cranberry Glades Botanical Area.

Cranberry Glades Botanical Area, West Virginia, is one of the southernmost boreal bogs in eastern North America. Along with unique plant life including cranberries, sphagnum moss, cinnamon ferns, rose pogonias, and the carnivorous sundew, Cranberry Glades hosts a unique breeding bird community that includes Yellow-bellied Flycatcher, Alder Flycatcher, Willow Flycatcher, American Woodcock, Common Snipe, Sedge Wren, and Swamp Sparrow.

The bogs, thickets, and surrounding coniferous and mixed forests support a distinctive avifauna that includes Olive-sided, Yellow-bellied, Acadian, Alder, Least, and Willow flycatchers; Red-breasted Nuthatch; Winter Wren; Swainson's Thrush; and Yellow-rumped, Mourning, and Magnolia warblers.

Cranberry Glades Botanical Area, Monongahela National Forest

The Cranberry Visitor Center is located on WV 55 (Highland Scenic Highway) about 6 mi west of Mill Point and just west of jct of WV 55 and WV 150 on the left of the road. Stop here for directions to Cranberry Wilderness sites and Cranberry Glades Botanical Area. To get to the Cranberry

Glades Botanical Area, go west on WV 150 about 1.0 mi from jct of WV 150 and WV 55/US 219. Turn right and follow signs about 1 mi to the parking area for the glades.

Dolly Sods Scenic Area	From the town of Hopeville, go north on WV 56/WV 28 and bear left on CR 28-7. Turn left at jct with FR 19 and follow FR 19 about 7 mi to enter the Dolly Sods Wilderness. Take trails off from FR 19 to examine highlands habitat.
East Lynn Lake Wildlife Management Area	From the town of East Lynn in western West Virginia, proceed south on WV 37. The road goes past the dam that forms the lake and winds through the management area. Side roads and trails can be taken to scrub and old field habitat good for Black-billed Cuckoos and Blue-winged Warblers, while mature forest harbors Northern Parulas, Wood Thrushes, and many other breeding songbird migrants.
Falls of Hills Creek	The Cranberry Visitor Center is located on WV 55 (Highland Scenic Highway) about 6 mi west of Mill Point and just west of jct of WV 55 and WV 150 on the left side of the road. Stop here for directions to the Falls of Hills Creek and an information brochure. The turnoff for the falls is about 8 mi west on WV 55 from the visitor center on the left of the road. There are three sets of falls, the highest of which is 65 feet. Some birds to be found in the forest on the walk to the falls are Northern Waterthrush, Ovenbird, Black-throated Blue and Black-and-white warblers, and Hermit Thrush.
Flat Top Lake	From I-77 south of Beckley, exit east onto CR 48. Go about 0.3 mi to jct with US 19. Turn left (north) on US 19 and go about 0.5 mi. Turn right and wind back to the lake (about 0.2 mi), a good

southern West Virginia locality for transient shorebirds.

Greenbottom Wildlife Management Area

From I-64 just east of Huntington, exit west onto US 60. Go about 4 mi on US 60 to where it turns left at the river. Turn right on WV 2 and go about 15 mi on WV 2 east and north along the Ohio River. When WV 2 bends back east, the wildlife management area is on the left (north) side of WV 2 between the road and the river for the next 3 mi or so. This and other Ohio River sites are excellent for transient and wintering waterfowl.

Holly River State Park

From jct of WV 20 and CR 3 in the town of Hacker Valley, proceed east on CR 3 about 0.2 mi to the park entrance on the left (CR 10).

Lewis Wetzel Wildlife Management Area

At jct where WV 20 splits west to leave US 19/WV 20 about 8 mi north of Clarksburg, go west on WV 20 about 23 mi to jct with CR 82-1 at Jacksonburg. Turn left on CR 82-1 to enter the wildlife management area. In addition to its attractiveness for waterfowl, this refuge offers a range of marsh and upland habitats for songbirds like the Willow Flycatcher, Common Yellowthroat, Song Sparrow, Yellow-breasted Chat, Blue-winged Warbler, and Blue-gray Gnatcatcher.

McClintick Wildlife Management Area

From I-77 north of Charleston, exit west onto US 33. Go about 7 mi on US 33 to Cottageville and turn left on WV 331. Go about 4 mi on WV 331 to jct with WV 2 just west of Mount Alto. Go west on WV 2 about 10 mi to just past Flatrock at jct of WV 2 and CR 13. Turn right on CR 13 and go about 6 mi to enter the wildlife management area. McClintick is perhaps the favorite destination for waterbird enthusiasts in western West Virginia.

Ohio River, south of Parkersburg	From I-77, 30 mi south of Parkersburg, exit onto WV 2 going west. Go about 2 mi to the T where WV 2 turns south and WV 68 goes north. Turn right and follow WV 68, which follows the river for several miles. This and other Ohio River sites are excellent for transient and wintering waterfowl.
Ohio River Islands National Wildlife Refuge	From I-77, three miles north of Parkersburg, exit east onto WV 2. Go 33 mi to just south of the town of Sistersville. The refuge is on your left. This and other Ohio River sites are excellent for transient and wintering waterfowl.
Panther State Forest	From the town of Laeger on US 52 in southern West Virginia, proceed north on US 52 about 1 mi to jct with CR 1. Turn left on CR 1 and go about 7 mi to the village of Panther. Turn left (south) in Panther on CR 3. In about a half-mile, the road splits into CR 3 (right fork) and CR 3-1 (left fork). Bear left on CR 3-1. In another mile or so, the road splits again into CR 3-1 (right) and CR 3-2 (left). Follow CR 3-2 to the left. This road goes through the middle of Panther State Forest.
R. D. Bailey Lake Wildlife Management Area	From WV 80 in the town of Verner in southern West Virginia, go southeast on CR 12. Proceed about 1 mi to enter the area. Check shrub and old field habitats for Golden-cheeked Warblers.
Red Creek Recreation Area	Go north about 7.6 mi on WV 32 from to jct of WV 32 and WV 55 at Harman to jct with FR 19. Turn right (east) on FR 19 and go about 9.3 mi to jct with FR 75. Turn left (north) on FR 75 and proceed about 7.5 mi to the recreation area, a good site for bird species of the Northern Hardwoods.

Shenandoah Mountain Recreation Area

From US 220 9 mi south of Franklin, take CR 25 east 9 mi to the town of Sugar Grove, where CR 25 and CR 21 combine for about 1 mi. Turn right (east) to follow CR 25 east and continue about 4 mi to jct with FR 61. Turn right (south) on FR 61 to enter the recreation area. Look here for Acadian Flycatcher, Scarlet Tanager, Rose-breasted Grosbeak, Pine Warbler, Nothern Parula, and Great Crested Flycatcher.

Sleepy Creek Wildlife Management Area

From I-81 at Martinsburg, exit west onto CR 15. Go west on CR 15 (becomes CR 18) about 10 mi to jct with CR 7 in Shanghai. Continue west on CR 18 (now CR 7-13) about 3 mi to enter the wildlife management area. Turn right at jct with PR 826 to traverse a large portion of the park. The mixture of habitats supports a variety of species including Eastern Kingbird, Chipping Sparrow, American Redstart, Yellow-billed Cuckoo, and Barred Owl. Fish Crows have bred in the vicinity.

Spruce Knob

From jct of WV 32 and WV 55/US 33 in Harman, go west on WV 55/US 33 a little over a mile to the turnoff for CR 29 (Whitmore Rd). Turn left on CR 29 and follow signs 29 mi to Spruce Knob, the highest point in West Virginia at 4,861 feet, where the spruce, fir, and mixed Northern Hardwood habitats support breeding populations of Broad-winged Hawk, Least Flycatcher, Winter Wren, Golden-crowned Kinglet, Hermit and Swainson's thrushes, and Yellow-rumped and Blackburnian warblers.

Stonecoal Lake Wildlife Management Area

From jct of US 33 and WV 20 in Buckhannon, proceed about 3 mi south on WV 20 to jct with CR 14. Turn right (west) on CR 14 and go about 2.5 mi to jct with CR 7 in Atlas. Turn left (south) in Atlas on CR 7 and go about 2 mi to enter the

Spruce Knob (see p. 101) is the highest point in West Virginia. Readily accessible by vehicle, the spruce-fir plant community attracts an interesting assortment of birds, including Olive-sided Flycatcher, Yellow-rumped Warbler, Northern Goshawk, Yellow-bellied Sapsucker, and Blackburnian Warbler.

management area. Characteristic species of the lake and associated wetlands and uplands include Green Heron, Red-tailed Hawk, Whip-poor-will, Wild Turkey, and Mourning Dove.

Summersville Lake Wildlife Management Area

From jct of US 19 and WV 39 east of Summersville, go west on WV 39 about 6 mi to jct with CR 19-15. Turn left on CR 19-15 and go about 3.5 mi to Summersville Lake and the management area. Typical species include Red-bellied Woodpecker, Barn Swallow, Eastern Phoebe, Tufted Titmouse, and Eastern Bluebird.

Summit Lake Recreation Area

The Cranberry Visitor Center is located on WV 55 (Highland Scenic Highway) about 6 mi west of Mill Point and just west of jct of WV 55 and WV 150 on the left of the road. Stop here for directions to Summit Lake Recreation Area and an information brochure. The turnoff for the recreation area is about 16 mi west on WV 55 from the visitor center at jct of WV 55 and CR 39-5.

Tomlinson Run State Park

At jct of WV 2 and WV 8 about 1.5 mi north of New Cumberland, go east on WV 8 about 3.7 mi to a left turn for the park. Species to be found here include Indigo Bunting, Field Sparrow, Downy Woodpecker, Northern Cardinal, Scarlet Tanager, Hooded and Kentucky warblers, and Baltimore Oriole.

Woodbine Recreation Area

From the town of Richwood, go north about 1 mi to where FR 78 forks off to the left. Follow FR 78 north about 4 mi to the recreation area and Big Rick Campground. Habitat here is highland mixed forest, home to Red-breasted Nuthatch, Blackburnian Warbler, Purple Finch, Black-throated Blue Warbler, Blue-headed Vireo, and Veery.

Color Tab Index to Bird Groups

Species Accounts

Loons
(Order Gaviiformes, Family Gaviidae)

Mallard-sized or somewhat larger water birds with tapered body and chisel-shaped bill. Legs set far back for swimming and diving. Awkward on land, loons require a considerable distance of flapping and running on water to become airborne. In flight, rapid wing beats, hump-backed silhouette, and feet extending beyond tail are characteristic.

Red-throated Loon
Gavia stellata
(L-25 W-42)

Black mottled with gray on back; gray, rounded head; gray neck striped with white on nape; rufous throat; red eye; white below with barring on flanks; relatively thin, slightly upturned bill. *Winter:* Black back with indistinct white spots; head grayish with some white below eye; throat and underparts white.

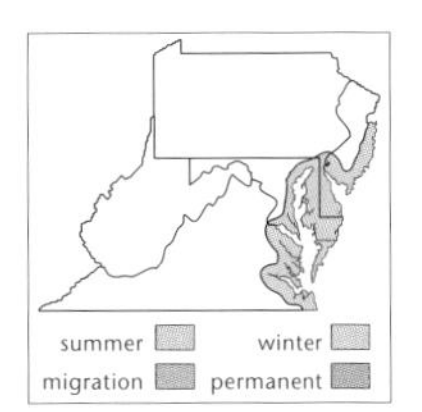

Similar Species White spotting on black back; lack of white patch on flank; thin, upturned bill and usual upward tilt of head separate this bird from other winter loons.
Habitat Marine, bays, lakes, rivers.
Abundance and Distribution Common transient and winter resident (Nov–Apr) along the coast; rare to casual transient and winter resident inland; a few birds remain through the summer.

Where to Find Brigantine National Wildlife Refuge, New Jersey; Prime Hook National Wildlife Refuge, Delaware; Mackay Island National Wildlife Refuge, Virginia.
Range Breeds in the high Arctic; winters in coastal temperate and boreal regions of the Northern Hemisphere.

Common Loon

Gavia immer

(L-31 W-54)

Checked black and white on back; black head with a white, black-streaked collar; white below with black streaking on breast and flanks; thick, heavy bill; red eye. *Winter:* Dark gray above; white below; dark gray crown and nape; white face and throat with a partial collar of white around neck.

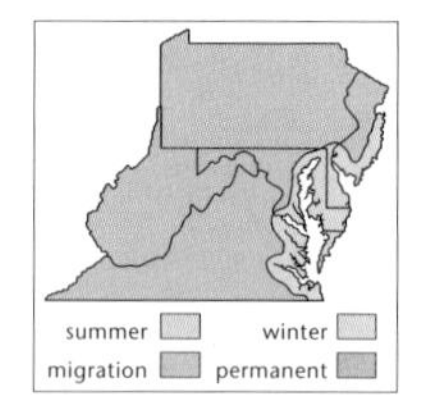

Similar Species The Common Loon in winter plumage can be distinguished from the smaller Red-throated Loon by heavy, chisel-shaped rather than thin, upturned bill and by lack of white spotting on back.
Habitat Marine, bays, lakes, rivers.
Abundance and Distribution Common transient and winter resident (Oct–May) along the coast; uncommon to rare transient and rare winter resident inland; some individuals remain through the summer; has bred.*
Where to Find Cheat Lake, West Virginia; Presque Isle State Park, Pennsylvania; Avalon, New Jersey.
Range Breeds in northern North America south to northern United States; winters coastal North America and on large lakes inland south to Baja California, Sonora, and south Texas.

Grebes
(Order Podicipediformes, Family Podicipedidae)

Loonlike birds the size of a small to medium-sized duck; dagger-shaped bill and lobed toes are distinctive. Seldom seen in flight, though most species are migratory.

Pied-billed Grebe
Podilymbus podiceps
(L-13 W-22)

Grayish brown body; short, pale bill with dark black ring; black throat; dark brown eye with white eyering. *Winter:* Whitish throat; no black ring on bill. *Juvenile:* Prominently striped in dark brown and white (Mar to as late as Oct).

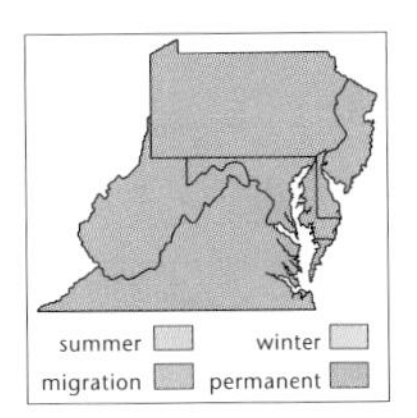

Habits This bird has the curious facility of simply sinking when disturbed. Carries newly hatched young on its back.
Voice A cuckoolike "hoodoo hoo hoo hoo hoo kow kow kow kow."
Similar Species The blunt, white bill (with dark ring in breeding plumage) is distinctive. Other grebes have sharp, dark (in Eared and Horned grebes) or horn-colored (Red-necked Grebe) bills.
Habitat Ponds, lakes, swales, marshes, estuaries, bays.
Abundance and Distribution Common to uncommon transient and winter resident (Aug–Apr) throughout; uncommon to rare and local summer resident.*
Where to Find Huntley Meadows Park, Virginia; Keystone State Park, Pennsylvania; Deal Island Wildlife Management Area, Maryland.

Range Breeds across most of Western Hemisphere from central Canada south to southern Argentina, West Indies; winters in temperate and tropical portions of breeding range.

Horned Grebe

Podiceps auritus

(L-14 W-23)

Dark back; rusty neck and sides; whitish breast; black head with buffy orange ear patch and eyebrow; red eye. *Winter:* Upper half of head dark with white spot in front of eye; cheek, throat, and breast white; nape and back dark.

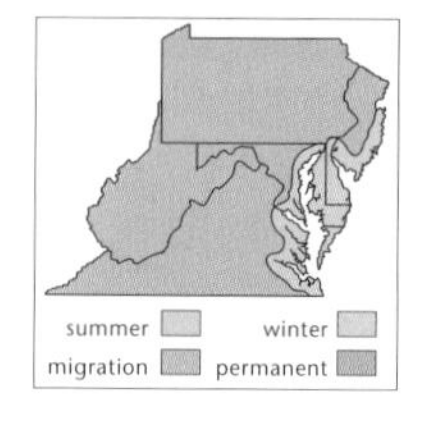

Similar Species Winter Horned Grebe has white cheek and neck, not grayish as in Eared Grebe; Horned Grebe has whitish spot in front of eye, which Eared Grebe lacks. Not all individuals of these 2 species are readily separable in winter plumage.

Habitat Bays, estuaries, larger lakes.

Abundance and Distribution Common transient and winter resident (Nov–Apr) along the coast; uncommon transient (Nov, Mar–Apr) and rare to casual winter resident (Dec–Feb) inland.

Where to Find Cheat Lake, West Virginia; Kerr Reservoir, Virginia; Barnegat National Wildlife Refuge, New Jersey.

Range Breeds in boreal and north temperate regions; winters mainly along coast in boreal and temperate regions of the Northern Hemisphere.

Red-necked Grebe

Podiceps grisegena

(L-20 W-32)

A large, long-necked grebe; dark back and upper half of head; rusty neck; lower half of head and chin white; dark brown eye; rusty breast; yellowish lower mandible. *Winter:* Dark back, posterior portion of neck, and crown; white chin and partial collar; grayish white anterior portion of neck, breast, and belly.

Voice Crowlike "aaahh"s; also high-pitched squeaks.

Similar Species Long neck, light-colored bill, and dark brown eye are distinctive.

Habitat Lakes, ponds, estuaries.

Abundance and Distribution Rare to casual transient (Nov, Mar) throughout, scarcer inland; rare winter resident along coast; more numerous in some years.

Where to Find Marsh Creek State Park, Pennsylvania; Baltimore Harbor, Maryland; Swift Creek Lake, Virginia.

Range Breeds locally in boreal and north temperate areas; winters along temperate and boreal coasts of Northern Hemisphere.

Eared Grebe

Podiceps nigricollis

(L-12 W-22)

Dark back, black neck, and rusty sides; whitish breast; black head with buffy orange ear tufts; red eye. *Winter:* Dark gray throughout except grayish chin and ear patch, whitish on breast and throat.

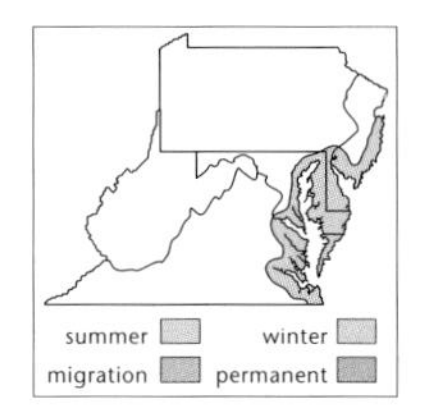

Voice "Kip kip kip kuweep kuweep."
Similar Species Winter Eared Grebe has grayish cheek and neck not as white as in Horned Grebe; Eared Grebe lacks whitish spot in front of eye, which is present in Horned Grebe. Not all individuals of these 2 species are readily separable in winter plumage.
Habitat Lakes and ponds (breeding); bays, estuaries, larger lakes (winter).
Abundance and Distribution Rare to casual winter resident (Dec–Mar) along the coast.
Where to Find Brigantine National Wildlife Refuge, New Jersey; Back Bay National Wildlife Refuge, Virginia; Delaware Seashore State Park, Delaware.
Range Breeds locally in southern boreal, temperate, and tropical regions of the world; in North America mainly in the West.

Fulmars and Shearwaters
(Order Procellariiformes, Family Procellariidae)

Dove- or crow-sized seabirds with long, narrow wings, short legs and tail, tubed nose, and hooked bill; rapid fluttering flight with frequent glides.

Northern Fulmar
Fulmarus glacialis
(L-19 W-42)

Light Phase (predominates in Atlantic): Plump, large-bodied shearwater, uniformly white head and underparts with grayish wings; thick, yellow bill with prominent nostril tubes typical of procellariiforms. *Dark Phase:* Uniformly dark brownish gray throughout.

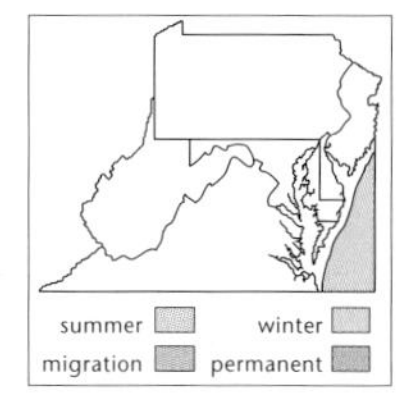

Similar Species Gull-like in size, but with short neck and stubby, tube-nosed bill. Characteristic stiff-winged flap-and-glide behavior.
Habitat Pelagic.
Abundance and Distribution Rare winter visitor (Oct–Apr) offshore.
Where to Find Pelagic birding trips (e.g., Patteson in Johnston 1997, 203).
Range Breeds in coastal Arctic and sub-Arctic regions; winters in Arctic and north temperate oceans.

Cory's Shearwater

Calonectris diomedea

(L-18 W-44)

Plump, large-bodied shearwater; gray-brown above; brownish head; pale throat; white below; white wing linings with dark tips; pale bill.

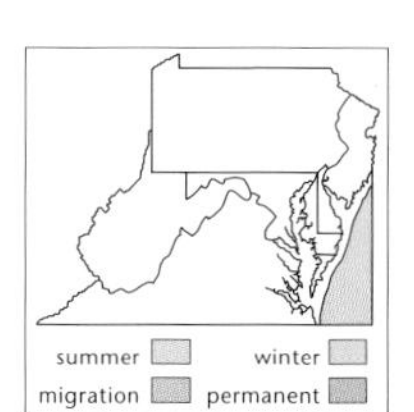

Similar Species Greater Shearwater has contrasting sharp black cap and white throat; also has black (not pale) bill, white rump, and black smudge on belly.
Habitat Pelagic.
Abundance and Distribution Uncommon transient and summer visitor (May–Oct) offshore.
Where to Find Pelagic birding trips.
Range Mainly temperate and tropical areas of Atlantic Ocean, Caribbean and Mediterranean seas, and Gulf of Mexico.

Greater Shearwater

Puffinus gravis

(L-19 W-44)

Plump; dark gray above; white below with dark belly smudge; contrasting dark cap and white throat; white collar, base of tail, and wing linings; black bill; pinkish legs.

summer winter
migration permanent

Similar Species Cory's Shearwater lacks the contrasting sharp black cap and white throat of the Greater Shearwater; also has pale (not black) bill, dark rump, and white belly.

Habitat Pelagic.

Abundance and Distribution Common to uncommon transient and summer visitor (May–Aug) offshore.

Where to Find Pelagic birding trips.

Range Atlantic Ocean.

Sooty Shearwater

Puffinus griseus

(L-19 W-42)

Entirely dark brownish gray except for pale wing linings; dark bill and legs.

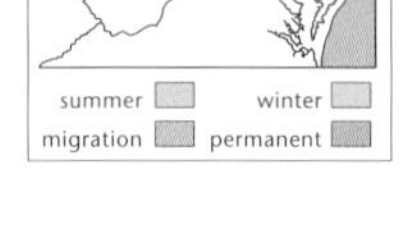

Habits Flaps and glides like other shearwaters; when foraging, drops from a yard or so above the water, wings out; often in large, loose flocks.

Similar Species Entirely dark body with pale wing linings contrasting with dark flight feathers and long bill separate the Sooty Shearwater from other shearwaters in the region, which are dark above and white below with shorter bills.

Habitat Pelagic; coastal waters.

Abundance and Distribution Common to uncommon spring transient (May–Jun) offshore.
Where to Find Pelagic birding trips.
Range Breeds on islands near Cape Horn and New Zealand during austral summer (Oct–Mar); disperses widely across the oceans of the world during migration; spends boreal summer mainly in oceans of Northern Hemisphere.

Manx Shearwater

Puffinus puffinus
(L-14 W-33)

A small shearwater; dark above, white below with white wing linings (primaries and secondaries) and undertail coverts; black bill; pink legs.

Similar Species Audubon's Shearwater has dark primary linings and undertail coverts.
Habitat Pelagic.
Abundance and Distribution Rare offshore visitor, mainly in spring (May–Jun).
Where to Find Pelagic birding trips.
Range Atlantic Ocean, Caribbean and Mediterranean seas.

Audubon's Shearwater

Puffinus lherminieri
(L-12 W-27)

Black above, white below; dark primary linings and undertail coverts; pale legs contrast with dark undertail coverts.

Similar Species Small size and contrasting plumage (dark above, white below) separate this species

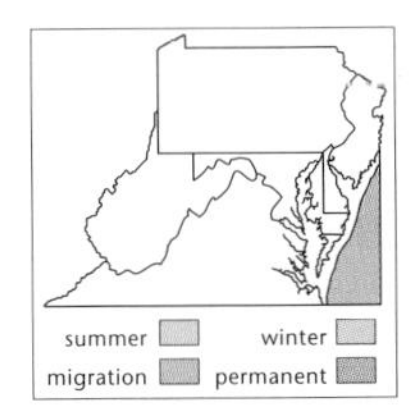

from other shearwaters found regularly in the region. Manx, the other small shearwater, has white (not dark) primary linings and white (not dark) undertail coverts that extend to the end of the short tail.

Habitat Pelagic.

Abundance and Distribution Uncommon to rare offshore visitor, mainly in fall (Aug–Sep).

Where to Find Pelagic birding trips.

Range Warm temperate and tropical seas of the world.

Storm-Petrels
(Family Hydrobatidae)

Dark, swallowlike seabirds that feed by hopping and skipping over the waves.

Wilson's Storm-Petrel
Oceanites oceanicus
(L-7 W-16)

Dark grayish brown with white rump and undertail coverts; pale stripe runs diagonally on secondaries of upper wing; square tail; feet extend beyond tail.

Habits Skims and flutters over water surface, often dabbling feet in water.

Similar Species Dabbling behavior is distinctive. Leach's Storm-Petrel has slightly forked tail; feet do not extend beyond tail in either Leach's or Band-rumped storm-petrels.

Habitat Pelagic.

Abundance and Distribution Common summer visitor (May–Sep) offshore.
Where to Find Pelagic birding trips.
Range Atlantic, Indian, and southern Pacific oceans.

Leach's Storm-Petrel
Oceanodroma leucorhoa
(L-8 W-19)

Dark gray; white rump with dark central stripe; slightly forked tail.

Habits Flight is erratic, with sudden alterations of speed and direction, low over water.

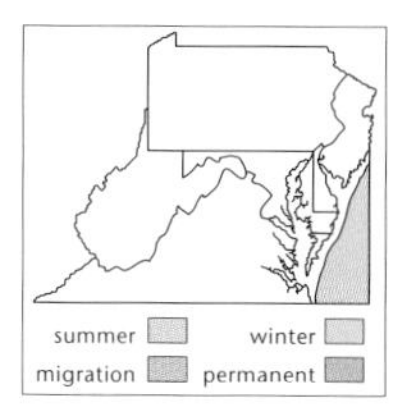

Similar Species Leach's Storm-Petrel has a dark central stripe across the white rump and slightly forked tail, which the Wilson's Storm-Petrel lacks. These 2 species are quite different in behavior. The Leach's Storm-Petrel glides and darts erratically over the water surface, whereas the Wilson's Storm-Petrel skims and flutters, often dabbling feet in water.
Habitat Pelagic.
Abundance and Distribution Uncommon to rare summer visitor (May–Oct) offshore.
Where to Find Pelagic birding trips.
Range Breeds on islands in the northern Atlantic and Pacific oceans; winters in temperate and tropical seas of the world.

Boobies and Gannets
(Order Pelecaniformes, Family Sulidae)

Boobies and gannets are large, heavy-bodied seabirds with long, cone-shaped bills; narrow, tapered wings; and webbed feet. They dive for prey from considerable heights.

Northern Gannet

Morus bassanus

(L-38 W-75)

Largest of the boobies; entirely white except for black primaries and rusty wash on head and neck; bill bluish; feet dark. *Immature:* Brown except for whitish belly and white spots on upper wing and back; bluish bill.

Voice A series of crowlike "caw"s.
Similar Species Immature gannets lack white wing lining patterns of other boobies and have a whitish belly.
Habitat Pelagic and occasionally immediately offshore.
Abundance and Distribution Common winter visitor (Dec–Apr) offshore. Peak numbers occur in Feb–Mar.
Where to Find Chesapeake Bay Bridge-Tunnel, Virginia; when gannet is diving for fish, can occasionally be seen by spotting scope from such barrier island beaches as Assateague Island.
Range North Atlantic Ocean, Gulf of Mexico, Mediterranean Sea.

Pelicans

(Family Pelecanidae)

Extremely large, heavy-bodied water birds with long bill and gular pouch.

American White Pelican

Pelecanus erythrorhynchos

(L-62 W-105)

White body; black primaries and secondaries; enormous orange bill and gular pouch; often with a hornlike growth on upper mandible (breeding); orange-yellow feet.

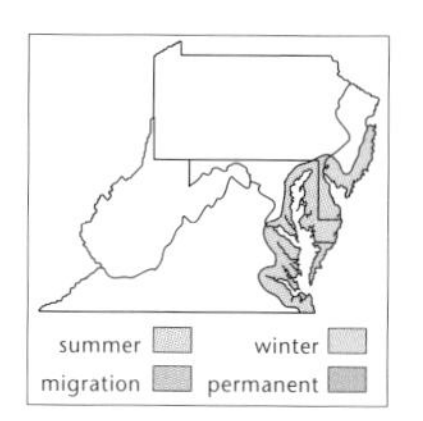

Habits Does not dive from the air for fish like Brown Pelican. Forages by dipping for prey from the surface of the water, often in groups. Migrates in large flocks.

Voice Various coughs and croaks.

Similar Species Immature Brown Pelicans can appear a dusty brownish white, but the American White Pelican is bright white in all plumages.

Habitat Large lakes, impoundments, coastal waters.

Abundance and Distribution Rare (Virginia) to casual (elsewhere) visitor along the coast in fall, winter, and spring (Sep–May).

Where to Find Back Bay National Wildlife Refuge, Virginia; Chincoteague National Wildlife Refuge, Virginia.

Range Breeds locally at large lakes and marshes in central and western Canada south through western half of the United States; winters southwestern United States to Nicaragua and from Florida around the Gulf of Mexico to Yucatan.

Brown Pelican

Pelecanus occidentalis

(L-48 W-78)

Grayish above; brownish below with dark chestnut nape and neck; whitish head tinged with yellow; enormous bill and gular pouch. *Immature:* Entirely brownish gray.

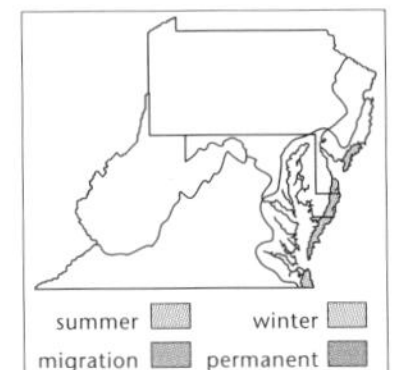

Habits Dives from a considerable height to catch fish.
Voice Occasional croaks.
Similar Species The immature Brown Pelican can appear a dusty brownish white. The American White Pelican is bright white in all plumages.
Habitat Marine.
Abundance and Distribution Uncommon to rare summer resident* (Apr–Oct) along the immediate coast north to Cape May. A few individuals remain through the winter.
Where to Find Hereford Inlet, New Jersey; Bombay Hook National Wildlife Refuge, Delaware; Chesapeake Bay Bridge-Tunnel, Virginia.
Range Coastal Western Hemisphere from southern New Jersey to eastern Brazil, including West Indies in east and California to southern Chile in west.

Cormorants
(Family Phalacrocoracidae)

Dark water birds the size of small geese, with tapered body; long, hooked bill; small gular pouch; and webbed feet. They generally fly in small flocks with necks extended.

Double-crested Cormorant
Phalacrocorax auritus
(L-32 W-51)

Entirely iridescent black; whitish or dark ear tufts during breeding season; gular pouch of bare orange skin. *Immature:* Brown with buffy head and neck.

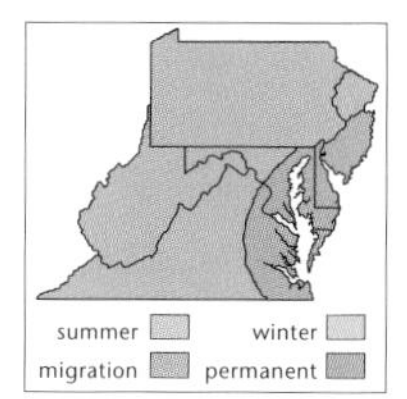

Habits Forages by swimming low in water and diving for long periods. Sits on snags and posts with wings spread to dry.
Voice Various croaks.
Similar Species The smaller immature Double-crested Cormorant has lighter throat and upper breast contrasting with darker belly; the larger, heavy-bodied immature Great Cormorant shows the reverse pattern, with darker throat and upper breast and lighter belly.
Habitat Estuaries, marine, lakes, ponds.
Abundance and Distribution Common transient (Apr, Sep–Oct) along the coast, rare inland; uncommon to rare and local summer* and winter resident along the coast.
Where to Find Barnegat National Wildlife Refuge, New Jersey; Fairmount Park, Philadelphia; Chesapeake Bay Bridge-Tunnel, Virginia.
Range Breeds locally across central and southern Canada and northern and central United States, along both coasts from Alaska to southern Mexico on the west and from Newfoundland to east Texas on the east, also Cuba; winters in coastal breeding range to south Texas, Greater Antilles, Yucatan Peninsula, and Belize.

Great Cormorant
Phalacrocorax carbo
(L-36 W-63)

Breeding: Black body with stubby, black tail; yellowish gular pouch; white throat; white flank patches. *Winter:* Lacks white flank patches. *Immature:* Dark brown above; brown neck; lighter belly.

Similar Species The smaller immature Double-crested Cormorant has lighter throat and upper breast contrasting with darker belly; the larger, heavy-bodied immature Great Cormorant shows the reverse pattern, with darker throat and upper breast and lighter belly.

Habitat Inlets, bays, and coastal waters.

Abundance and Distribution Rare winter resident (Oct–May) along the immediate coast.

Where to Find Chesapeake Bay Bridge-Tunnel, Virginia; Assateague Island National Seashore, Maryland; Brigantine National Wildlife Refuge, New Jersey.

Range Resident in temperate, tropical, and sub-Arctic coastal regions throughout much of the Old World; breeds in southwestern Greenland and winters along eastern coast of North America (Newfoundland to Florida).

Anhingas
(Family Anhingidae)

Dark water birds similar to cormorants but with serpentine neck, long, pointed bill, and long, wedge-shaped tail.

Anhinga
Anhinga anhinga
(L-35 W-45)

Snakelike neck and long triangular tail with terminal buffy band; long, sharp, yellow-orange bill; black body with iridescent green sheen; mottled white on shoulders and upper wing coverts. *Female and Immature:* Light buff head, neck, and breast; lack white wing spotting.

Habits Often forages with only the sinuous neck and head protruding above water. Sits with wings spread to dry.
Voice A series of metallic "kaakk"s, some reminiscent of a cicada.
Similar Species Cormorants have hooked rather than stiletto-shaped bill; tail is shorter than that of Anhinga and lacks terminal buffy band.
Habitat Rivers, lakes, ponds.
Abundance and Distribution Rare summer visitor in southeastern Virginia.
Where to Find Great Dismal Swamp National Wildlife Refuge, Virginia; Stumpy Lake, Virginia.
Range Breeds from the southeastern United States south through the lowlands of the subtropics and topics to southern Brazil, Cuba; winters throughout breeding range except inland in Gulf states.

Herons, Egrets, and Bitterns
(Order Ciconiiformes, Family Ardeidae)

Long-billed, long-necked, long-legged wading birds.

American Bittern
Botaurus lentiginosus
(L-26 W-39)

A chunky, relatively short-legged heron; buffy brown above and below; streaked with white and brown on throat, neck, and breast; dark brown streak on side of neck; white chin; greenish yellow bill and legs; yellow eyes.

Habits A secretive bird; often, rather than fly when approached, it will "freeze" with its neck extended

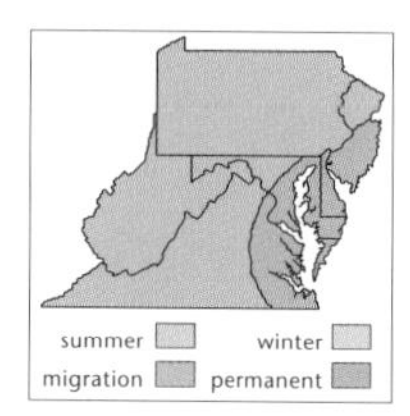

in an attempt to blend in with the reeds and rushes of the marsh.
Voice A deep "goonk glunk-a-lunk," like blowing on an empty coke bottle.
Similar Species Could be mistaken for an immature night-heron in flight, but dark brown primaries and secondaries contrast with light brown back and upper wing coverts; night-herons lack prominent dark streak on side of neck.
Habitat Marshes.
Abundance and Distribution Common to uncommon transient (Apr, Oct–Nov) throughout; rare and local summer* (May–Sep) and winter (Nov–Mar) resident, mainly along the coast.
Where to Find Canaan Valley State Park, West Virginia; Presque Isle State Park, Pennsylvania; Blackwater National Wildlife Refuge, Maryland.
Range Breeds central and southern Canada south to southern United States and central Mexico; winters southern United States to southern Mexico and Cuba.

Least Bittern

Ixobrychus exilis

(L-14 W-17)

The smallest of our herons; dark brown with white streaking on back; tan wings, head, and neck; dark brown crown; white chin and throat streaked with tan; white belly; yellowish legs and bill; yellow eyes; extended wings are half tan (basally) and half dark brown.

Habits Like the American Bittern, this bird will often "freeze" with neck extended when approached.
Voice A rapid, whistled "coo-co-co-co-coo."

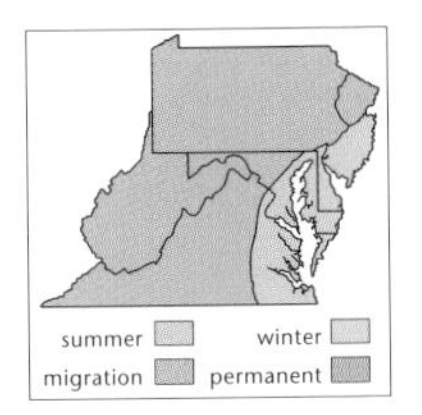

Similar Species Immature Green Heron is heavily streaked below and lacks buff shoulders.
Habitat Marshes.
Abundance and Distribution Uncommon transient and summer resident* (May–Sep), mainly along the Coastal Plain; rare and local inland. A few can be found in winter in coastal marshes.
Where to Find Back Bay National Wildlife Refuge, Virginia; Great Swamp National Wildlife Refuge, New Jersey; McClintick Wildlife Management Area, West Virginia.
Range Breeds in the eastern half of the United States and southeastern Canada; locally in the western United States, south through lowlands to southern Brazil, West Indies; winters through breeding range from the southern United States southward.

Great Blue Heron

Ardea herodias

(L-48 W-72)

A very large heron; slate gray above and on neck; white crown bordered with black stripes that extend as plumes (breeding); white chin and throat streaked with black; gray breast and back plumes; white below streaked with chestnut; chestnut thighs; orange-yellow bill; dark legs. *Immature:* Dark cap; brownish gray back; buffy neck. *White Phase* ("Great White Heron" of some authors): Entirely white or mixed white and blue with yellow bill and legs; this phase is extremely rare except in southern Florida and the Caribbean.

Voice A low "krarrrk."
Habitat Lakes, rivers, marshes, bays, estuaries.
Abundance and Distribution Common to uncom-

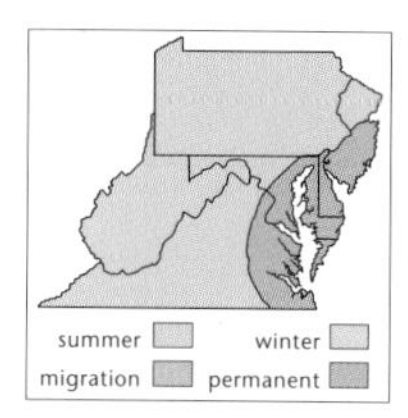

mon summer resident* (Mar–Sep), scarcer inland, especially in the mountains; uncommon (Coastal Plain) to rare or casual (inland) winter resident (Oct–Feb).

Where to Find Bombay Hook National Wildlife Refuge, Delaware; Deal Island Wildlife Management Area, Maryland; Brucker Great Blue Heron Sanctuary, Mercer County, Pennsylvania.

Range Breeds from central and southern Canada south to coastal Colombia and Venezuela, West Indies; winters southern United States southward through breeding range.

Great Egret

Ardea alba

(L-39 W-57)

Entirely white, with shaggy plumes on breast and back in breeding season; long, yellow bill; long dark legs that extend well beyond tail in flight.

Voice Various "krrank"s and "krronk"s.

Similar Species The Great Egret has yellow bill and black legs. The smaller Snowy Egret has black bill, black legs, and yellow feet. Immature Little Blue Heron has 2-tone bill (dark tip, pale base), pale legs, and usually some gray smudging on white plumage. Cattle Egret is half the size; has short, thick yellow bill and yellowish legs that barely extend beyond tail in flight.

summer winter migration permanent

Habitat Lakes, ponds, rivers, marshes, estuaries, bays.

Abundance and Distribution Common summer resident* (Apr–Sep) along the Coastal Plain; rare to casual and local inland, mainly as a transient or postbreeding wanderer (Apr–May, Aug–Oct) inland. A few winter on the coast.

Where to Find George Washington Birthplace National Monument, Virginia; Prime Hook National Wildlife Refuge, Delaware; McClintick Wildlife Management Area, West Virginia.
Range Breeds in temperate and tropical regions of the world; winters mainly in subtropical and tropical portions of breeding range.

Snowy Egret

Egretta thula

(L-23 W-45)

Small, entirely white heron; black legs with yellow feet; black bill; yellow lores; white plumes off neck and breast (breeding). *Immature:* Yellow line up back of leg.

Habits Occasionally arches wings to form a canopy while foraging; also puts foot forward and shakes it on bottom substrate.
Voice A crowlike "caaah."

Similar Species The Snowy Egret has black bill, black legs, and yellow feet; immature Little Blue Heron has 2-tone bill (dark tip, pale base), pale legs, and usually some gray smudging on white plumage; Cattle Egret has short, thick yellow bill and yellowish legs that barely extend beyond tail in flight; Great Egret has yellow bill and black legs.
Habitat Marshes, ponds, lakes, estuaries, bays.
Abundance and Distribution Common summer resident* (Apr–Sep) along the Coastal Plain; rare transient (Apr–May, Aug–Oct) and late summer wanderer in Piedmont, casual farther inland. Rare in winter on the coast.
Where to Find Island Beach State Park, New Jersey; Assateague Island National Seashore, Maryland; Chincoteague National Wildlife Refuge, Virginia.

Range Breeds locally across United States and extreme southern Canada south in lowlands to southern South America, West Indies; winters from coastal southern United States southward through breeding range.

Little Blue Heron

Egretta caerulea

(L-23 W-39)

A smallish heron; dark blue body; maroon neck and head; 2-tone bill, black at tip, pale gray or greenish at base; plumes on neck, breast, and head; dark legs. *Winter:* Mainly navy blue on neck with maroon tinge; no plumes. *Immature:* Almost entirely white in first year with some gray smudging and blue-gray wing tips; greenish legs; 2-tone bill; more gray smudging in second year.

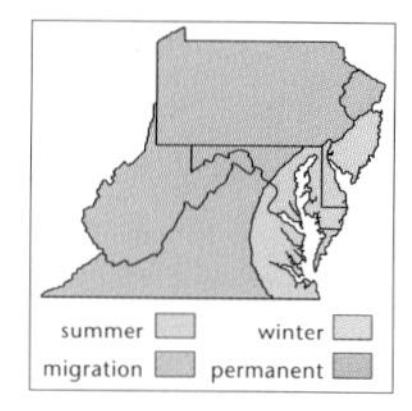

Voice Piglike squawks.
Similar Species Immature Little Blue Heron has 2-tone bill (dark tip, pale base), pale legs, and usually some gray smudging on white plumage; Snowy Egret has black bill, black legs, and yellow feet; Cattle Egret has short, thick yellow bill and yellowish legs that barely extend beyond tail in flight; Great Egret has yellow bill and black legs; rare white phase of Great Blue is much larger and has yellow bill and yellow legs (not dark).
Habitat Mainly freshwater marshes, lakes, ponds.
Abundance and Distribution Common to uncommon summer resident* (Apr–Sep) along the Coastal Plain; rare to casual transient (Apr–May, Aug–Oct) and late summer wanderer inland.
Where to Find Bombay Hook National Wildlife Refuge, Delaware; Brigantine National Wildlife

Refuge, New Jersey; Virginia Coast Reserve Barrier Islands, Virginia.
Range Breeds along Coastal Plain of eastern United States from Maine to Texas south through lowlands to Peru and southern Brazil, West Indies; winters from southern United States south through breeding range.

Tricolored Heron
Egretta tricolor
(L-26 W-36)

Slate blue body; red eye; greenish lores; rusty chin; white central stripe down neck streaked with dark blue; maroon on breast; white belly and thighs; buffy plumes on back (breeding). *Immature:* Rusty and slate above, whitish below.

Voice Crowlike "krraww," repeated.
Similar Species Similar pattern to the much larger Great Blue Heron but note white (not chestnut) thighs and lack of black on head.
Habitat Bays, estuaries, lakes, ponds.

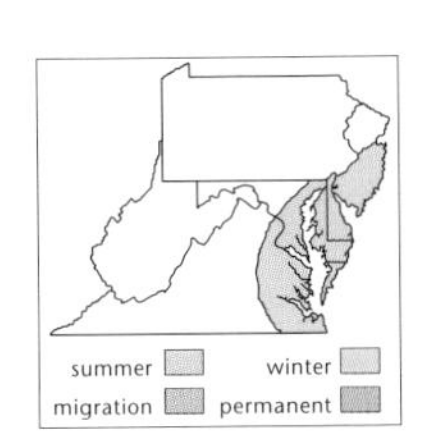

Abundance and Distribution Common to uncommon summer resident* (Apr–Sep) along the Coastal Plain, scarcer northward (New Jersey); rare to casual transient (Apr–May, Aug–Oct) and late summer wanderer inland. Rare in winter on the coast.
Where to Find Deal Island Wildlife Management Area, Maryland; Stone Harbor, New Jersey; Chincoteague National Wildlife Refuge, Virginia.
Range Breeds in the Coastal Plain of the eastern United States from Maine to Texas, south through coastal lowlands to Peru and northern Brazil, West Indies; winters from Gulf states south through breeding range.

Cattle Egret
Bubulcus ibis
(L-20 W-36)

A small, entirely white heron; yellow-orange bill and legs; buff coloration and plumes on crest, breast, and back (breeding). *Immature:* Like adult but lacks buff coloration, and legs are dark.

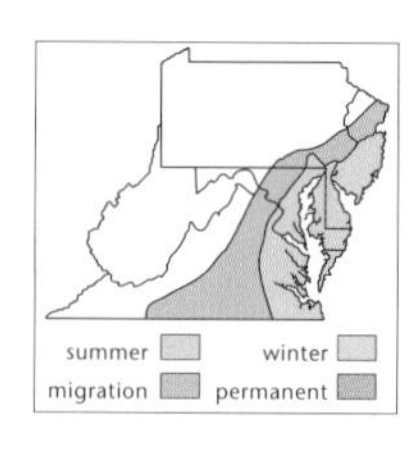

Similar Species The stubby Cattle Egret has short, thick, yellow bill and yellowish legs that barely extend beyond tail in flight; immature Little Blue Heron has 2-tone bill (dark tip, pale base), pale legs, and usually some gray smudging on white plumage; Snowy Egret has black bill, black legs, and yellow feet.

Habitat The Cattle Egret is not an aquatic species, although it occasionally nests on bay islands. It prefers grasslands, where it feeds on insects, mainly grasshoppers.

Abundance and Distribution Common to uncommon summer resident* (Apr–Sep) along the Coastal Plain, scarcer northward (New Jersey); rare transient (Apr–May, Aug–Oct) and rare late summer wanderer inland in the Piedmont, casual farther inland. Rare to casual in winter along coast.

Where to Find George Washington Birthplace National Monument, Virginia; Barnegat National Wildlife Refuge, Delaware; Assateague Island National Seashore, Maryland.

Range Formerly strictly an Old World species, the Cattle Egret appeared in South America in the late 1800s and has expanded steadily northward, reaching Virginia in 1953. Current distribution includes most temperate and tropical regions of the world. Northern populations are migratory.

Green Heron
Butorides virescens
(L-19 W-28)

A small, dark heron; olive back; black cap; chestnut neck; white throat with chestnut striping; yellow eye and yellow lores; white stripe from base of lower mandible along cheek (malar region); grayish belly; greenish legs; bill dark above, yellowish below. *Immature:* Heavily streaked below.

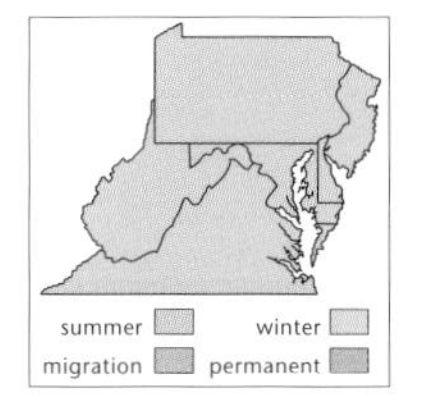

Habits Forages in a very slow, deliberate manner.
Voice A loud "kyoook."
Similar Species Least Bittern is buffy, not dark, and has 2-tone wings; wings of Green Heron are uniformly dark in flight.
Habitat Streams, rivers, lakes, marshes.
Abundance and Distribution Common to uncommon summer resident* (May–Sep) throughout.
Where to Find Summersville Lake Wildlife Management, West Virginia; Erie National Wildlife Refuge, Pennsylvania; Sherando Lake Recreation Area, Virginia.
Range Breeds from southern Canada south to northern Argentina (including Green-backed Heron form, *B. striatus*), West Indies; winters from southern United States southward through breeding range.

Black-crowned Night-Heron
Nycticorax nycticorax
(L-26 W-45)

A rather squat heron; black on back and crown with long, trailing white plumes (breeding); gray wings; pale gray breast and belly; white forehead, cheek, and chin; red eye; dark beak; pale legs. *Immature:* Dark brown above, heavily streaked with white; whitish below streaked with brown; red eye; bluish lores; pale legs.

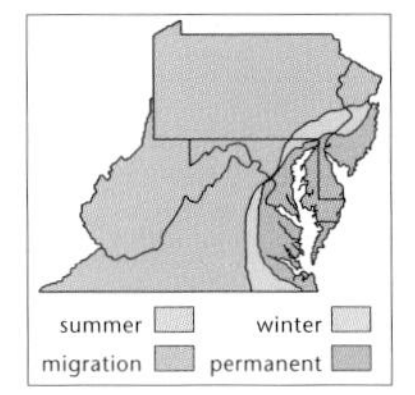

Habits Mostly nocturnal.
Voice "Kwark."
Similar Species Immature Yellow-crowned Night-Heron has longer legs that extend well beyond end of tail in flight; Black-crowned Night-Heron's legs barely reach beyond end of tail; also the black-crown is streaked on back, not spotted as in yellowcrown.
Habitat Bays, lakes, marshes.
Abundance and Distribution Common to uncommon summer resident* (Apr–Sep) along the immediate coast, rare to casual inland; uncommon to rare transient (Apr–May, Aug–Oct) and late summer wanderer inland; uncommon to rare in winter along the coast.
Where to Find Bird Ponds, National Zoological Park, Washington, D.C.; Island Beach State Park, New Jersey; Chincoteague National Wildlife Refuge, Virginia.
Range Breeds locally in temperate and tropical regions of the world; withdraws from seasonally cold portions of breeding range in winter.

Yellow-crowned Night-Heron
Nyctanassa violacea
(L-24 W-42)

Gray body, streaked with black above and on wings; black head with white cheek patch and creamy crown and plumes (breeding); red eye; dark bill; pale legs. *Immature:* Dark brown above, finely spotted with white; whitish below streaked with brown; dark bill; yellow legs; red eye.

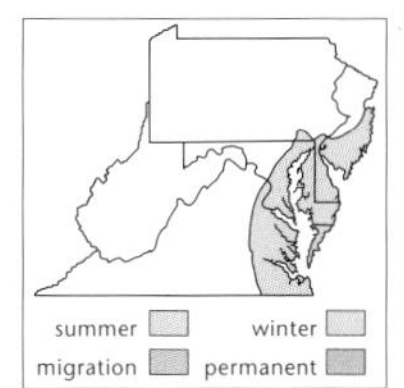

Voice "Aaak."
Similar Species Immature Yellow-crowned Night-Heron has longer legs that extend well beyond end of tail in flight; blackcrown's legs barely reach beyond end of tail; also the blackcrown is streaked on back, not spotted as in yellowcrown.
Habitat Wetlands.
Abundance and Distribution Uncommon summer resident* (Apr–Sep) along the immediate coast, scarcer northward (New Jersey); rare to casual transient (Apr–May, Aug–Oct) and late summer wanderer inland. Rare in winter on the coast.
Where to Find Prime Hook National Wildlife Refuge, Delaware; Martin National Wildlife Refuge, Maryland; Stumpy Lake, Virginia.
Range Breeds from the southeastern United States south in coastal regions to Peru and southern Brazil, West Indies; winters from coastal Gulf states south through breeding range.

Ibises and Spoonbills
(Family Threskiornithidae)

Ibises are small to medium-sized, heronlike birds with long, decurved bills.

White Ibis
Eudocimus albus
(L-25 W-39)

Entirely white with long, pink, decurved bill; pink facial skin; yellow eye; pink legs. *Immature:* Brown above; white below; neck and head mottled brown and whitish; pinkish bill and legs.

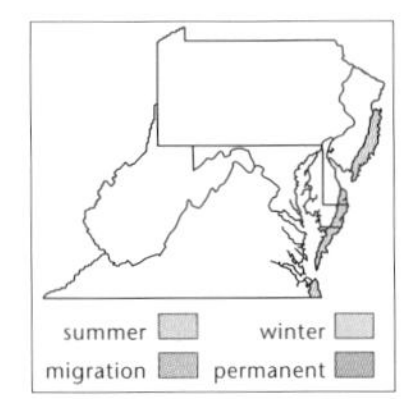

Voice Various "aaaah"s and "aaaww"s.
Similar Species Immature White Ibis has pinkish bill and legs; immature Glossy Ibis has dark bill and legs; immature White Ibis has white neck and underparts; immature Glossy Ibis is entirely dark with some streaking on the head and neck.
Habitat Bays, rivers, estuaries.
Abundance and Distribution Uncommon to rare summer visitor* (Apr–Sep) to coastal Virginia, especially the Eastern Shore; rare to casual elsewhere.
Where to Find Chincoteague National Wildlife Refuge, Virginia.
Range Breeds from coastal southeastern United States south to French Guiana and Peru, West Indies; withdraws from northern portions in winter.

Glossy Ibis
Plegadis falcinellus
(L-22 W-36)

Body entirely dark purplish brown with green sheen on wings and back; bare dark skin on face, edged with bluish skin (breeding); dark bill; dark legs; brown eyes. *Immature:* Brownish throughout with white streaking on head and neck; dark bill and legs.

Similar Species Immature Glossy Ibis is entirely dark with some streaking on the head and neck; immature White Ibis has pinkish bill and legs; immature Glossy Ibis has dark bill and legs; immature White Ibis has white neck and underparts.

Habitat Bays, marshes, lakes, ponds.
Abundance and Distribution Common to uncommon summer resident* (Mar–Sep) along the Coastal Plain, scarcer northward (New Jersey); rare to casual transient (Mar–May, Aug–Oct) and late summer wanderer inland—not recorded from West Virginia. Rare in winter on the Virginia coast.
Where to Find Stone Harbor, New Jersey; Assateague Island National Seashore, Maryland; Chincoteague National Wildlife Refuge, Virginia.
Range Breeds along coast of eastern United States from Maine to Louisiana south through West Indies to northern Venezuela and in Old World temperate and tropical regions; withdraws from colder portions of breeding range in winter.

Storks
(Family Ciconiidae)

Tall, long-legged, heavy-billed, long-necked wading birds.

Wood Stork
Mycteria americana
(L-41 W-66)

ENDANGERED A large, white-bodied bird with long, heavy bill, downturned toward the tip; black primaries and secondaries; naked, black-skinned head and neck; pale legs with pinkish feet. *Immature:* Patterned like adult but with grayish feathering on neck and head; yellowish bill.

Habitat Lakes, coastal marshes, bays.
Abundance and Distribution Rare summer and fall

visitor (Jun–Sep) along the southern Virginia coast; casual elsewhere.

Where to Find Back Bay National Wildlife Refuge, Virginia.

Range From Georgia, Florida, and the Gulf states south in coastal regions to central Argentina, West Indies.

New World Vultures

(Family Cathartidae)

Large, black, diurnal raptors with long, hooked bills and featherless heads; they forage mostly from the air for carrion.

Black Vulture

Coragyps atratus

(L-26 W-57)

Entirely black with naked, black-skinned head; primaries silvery from below.

Habits Black Vultures do not normally soar; they alternate flapping and gliding, usually at low levels.

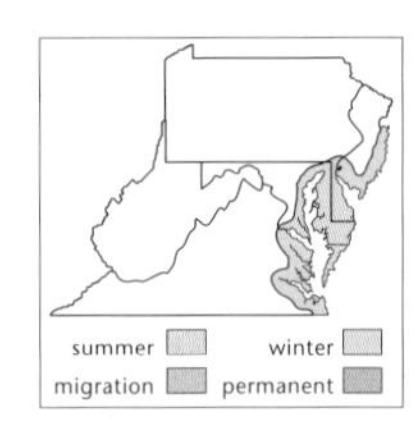

Similar Species Turkey Vultures soar, seldom flapping, and normally hold their wings at an angle, while Black Vultures hold their wings horizontally during glides; Black Vultures have silvery primaries, while Turkey Vultures have silvery primaries and secondaries.

Habitat Mainly open areas.

Abundance and Distribution Uncommon to rare resident,* scarcer in winter throughout except northern and western West Virginia, northern and western Pennsylvania, northern New Jersey, and eastern shore of Virginia, where rare to casual or absent.

Where to Find French Creek State Park, Pennsylvania; Dickey Ridge Visitor Center, Shenandoah National Park, Virginia; Patuxent Research Refuge, Maryland.

Range Breeds from eastern (New Jersey) and southwestern (Arizona) United States south to central Argentina; winters from southern United States south through breeding range.

Turkey Vulture

Cathartes aura

(L-26 W-69)

Black with naked, red-skinned head; relatively long tail; silvery flight feathers (outlining black wing lining). *Immature:* Black head.

Habits Soars for long periods with wings held at an angle above horizontal.

Similar Species Turkey Vultures soar, seldom flapping, and normally hold their wings at an angle, while Black Vultures hold their wings horizontally during glides; Black Vultures have silvery primaries, while Turkey Vultures have silvery primaries and secondaries.

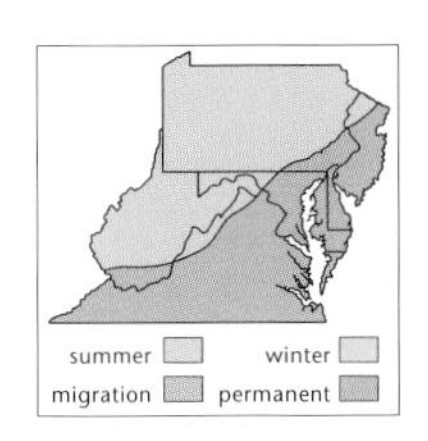

Habitat Nearly ubiquitous except in extensive agricultural areas.

Abundance and Distribution Common summer resident* (Mar–Oct) nearly throughout, though scarcer in northwestern West Virginia; uncommon in winter except northern and western West Virginia, northern and western Pennsylvania, northern New Jersey, and mountains throughout, where rare or absent.

Where to Find Norman G. Wilder Wildlife Area, Delaware; Mendota Fire Tower, Virginia; Cape May Point, New Jersey.
Range Breeds from southern Canada south to southern South America, Bahamas, and Cuba; winters from central and southern United States south through breeding range.

Ducks, Geese, and Swans
(Order Anseriformes, Family Anatidae)

Heavy-bodied water birds with short legs, webbed feet, and broad, flat bills.

Fulvous Whistling-Duck
Dendrocygna bicolor
(L-20 W-36)

A long-necked, long-legged duck; body a rich, tawny buff; mottled black and tan on back; dark streak running from crown down nape to back; whitish streaking on throat; white on flanks; gray bill and legs; dark wings; dark tail with white base.

Voice A shrill, whistled "ker-chee."
Habitat Lakes, marshes.
Abundance and Distribution Rare summer visitor (Mar–Oct), mainly in extreme southeastern Virginia.
Where to Find False Cape State Park, Virginia.
Range Breeds locally from southern California, Arizona, Texas, Louisiana, and Florida south in coastal regions to Peru and central Argentina, West Indies; also in the Old World from East Africa to Madagascar and India.

Greater White-fronted Goose

Anser albifrons

(L-28 W-57)

Grayish brown body; barred with buff on back; speckled with black on breast and belly; white lower belly and undertail coverts; pinkish bill with white feathering at base, edged in black; orange legs; in flight, white rump and gray wings are key. *Immature:* Lacks white feathering at base of bill and speckling on breast and belly; pale legs and bill.

Voice A tremulous, high-pitched "ho-ho-honk."
Similar Species Other dark geese of the region have white heads or chin straps clearly visible in flight.
Habitat Lakes, marshes, grasslands, croplands.
Abundance and Distribution Rare winter resident (Oct–Mar), mainly along the coast of Delaware, Maryland, and Virginia.
Where to Find Bombay Hook National Wildlife Refuge, Delaware; Hog Island Waterfowl Management Area, Virginia; Blackwater National Wildlife Refuge, Maryland.
Range Breeds in Arctic regions; winters in temperate regions of Northern Hemisphere.

Snow Goose

Chen caerulescens

(L-28 W-57)

A medium-sized white goose with black primaries; red bill with dark line bordering mandibles; red legs. *Immature:* Patterned like adult but with gray bill and legs and grayish wash on back. *Blue Phase:* Dark gray body with white head, neck, and belly; red legs and bill. *Immature:* Dark brownish gray throughout with white chin and belly; dark legs and bill.

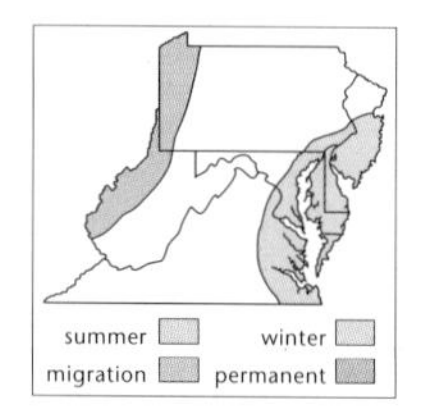

Voice A shrill, high-pitched "honk."
Similar Species White phase Snow Goose has dark line on mandible edges ("grin patch"), which Ross' Goose lacks; Snow Goose bill is longer than head whereas Ross' is shorter. The immature Greater White-fronted Goose has no white on chin and has pale legs and bill (not dark as in blue phase immature).
Habitat Lakes, marshes, grasslands, croplands.
Abundance and Distribution Common winter resident (Nov–Mar) along the coast; rare to casual transient and winter resident inland; most West Virginia and Pennsylvania records are from the west (Ohio River Valley and Lake Erie, respectively), presumably stragglers from the Mississippi Valley flyway.
Where to Find Delaware Bay, New Jersey; Blackwater National Wildlife Refuge, Maryland; Back Bay National Wildlife Refuge, Virginia.
Range Breeds in Arctic Canada; winters in the west along Pacific Coast from southwestern Canada to central Mexico and in the east from Chesapeake Bay south through the southeastern United States to northeastern Mexico; also east Asia.

Canada Goose

Branta canadensis
(L-26 to 48 W-54 to 84)

Subspecies of this goose vary considerably in size; grayish brown above; grayish below; black neck and head with white chin strap; white rump, belly, and undertail coverts; black tail.

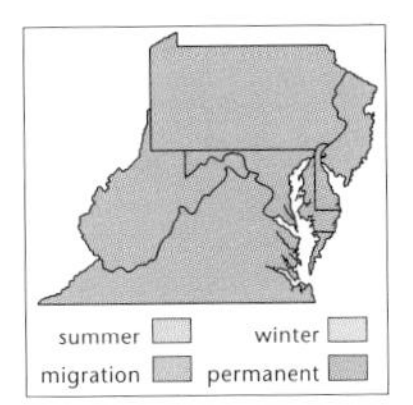

Habits These geese often fly in large Vs.
Voice A medium- or high-pitched, squeaky "honk," given at different pitches by different flock members.
Similar Species Brant has black across breast and lacks white chin strap of Canada Goose.
Habitat Lakes, estuaries, grasslands, croplands.
Abundance and Distribution Common winter resident (Oct–Mar) and increasingly common summer resident* (Apr–Sep) nearly throughout; mainly a transient (Feb–Mar, Oct–Nov) in West Virginia, but widely introduced (22 counties) and increasing as a resident.
Where to Find McClintick Wildlife Management Area, West Virginia (breeding population); Pymatuning Wildlife Management Area, Pennsylvania; Prime Hook National Wildlife Refuge, Delaware.
Range Breeds across northern half of North America; winters from northern United States south to northern Mexico, farther north along coasts; also introduced in various Old World localities.

Brant
Branta bernicla
(L-24 W-39)

A small goose; dark above, barred with brown; white upper tail coverts nearly obscure dark tail; white below with gray-barred flanks; black breast, neck, and head with white bars on side of throat; dark bill and legs. *Immature:* Lacks barring on throat.

Habits Feeds by dabbling for sea grasses in shallows of bays and estuaries.

Voice A rolling, guttural "krrrronk."
Similar Species Black breast contrasting with white belly and white upper tail coverts are distinctive even in flight.
Habitat Bay shores, estuaries.
Abundance and Distribution Common winter resident (Oct–Apr) along the coast; rare in summer; uncommon to rare transient (Oct–Nov, Apr–May) in inland New Jersey and eastern Pennsylvania; rare to casual transient and winter visitor elsewhere in the region.
Where to Find Brigantine National Wildlife Refuge, New Jersey; Assateague Island State Park, Maryland; Chincoteague National Wildlife Refuge, Virginia.
Range Breeds in high Arctic; winters along northern coasts of Northern Hemisphere.

Mute Swan
Cygnus olor
(L-60 W-88)

White body; orange bill with large, black knob at base. *Immature:* Brownish body with grayish bill. Young birds begin molt into white plumage by midwinter; bill becomes pinkish.

Habits Holds neck in a graceful curve; often holds wings arched above the back; can be very aggressive, advancing on intruders with hisses and croaks.
Similar Species The Tundra Swan keeps its neck straight, while Mute Swans hold their necks in a graceful curve. Adult Mute Swans have a large, black knob at the base of a bright orange bill, which the Tundra Swan lacks.
Habitat Lakes, bays.

Abundance and Distribution Uncommon to rare and local resident* along the coast and locally inland (domesticated).
Where to Find Chincoteague National Wildlife Refuge, Virginia; Eastern Neck National Wildlife Refuge, Maryland.
Range Breeds in northern and temperate Eurasia; winters in southern portions of breeding range south to the Mediterranean and Caspian seas, northern India, and eastern China. Introduced widely throughout North America, but scarce in most parts except along the coast of the northeastern United States.

Tundra Swan

Cygnus columbianus
(L-52 W-81)

Very large, entirely white bird; rounded head; black bill, often with yellow preorbital spot. *Immature:* Brownish gray with orangish bill.

summer winter
migration permanent

Voice A mellow "hoonk," repeated.
Similar Species The Tundra Swan keeps its neck straight, while Mute Swans hold their necks in a graceful curve. Adult Mute Swans have a large, black knob at the base of a bright orange bill, which the Tundra Swan lacks.
Habitat Lakes, bays.
Abundance and Distribution Uncommon to rare transient (Feb–Mar, Nov–Dec) nearly throughout, scarcer in western Virginia and West Virginia; common winter resident in Chesapeake Bay.
Where to Find Back Bay National Wildlife Refuge, Virginia; Point Lookout State Park, Maryland.
Range Breeds in Arctic regions of Northern Hemisphere; winters in coastal boreal and north temperate areas.

Wood Duck

Aix sponsa

(L-19 W-29)

Green head and crest streaked with white; red eye, face plate, and bill; white throat; purplish breast; iridescent dark bluish back; beige belly with white flank stripes. *Female:* Deep iridescent blue on back; brownish flanks; grayish belly; brownish gray head and crest; white eyering and postorbital stripe; white chin.

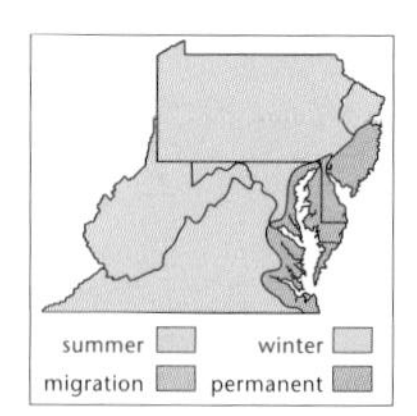

Voice A high, whistling "aweek aweek aweek."
Similar Species The shrill flight call, relatively short neck, large head, and long square tail are distinctive for the birds in flight.
Habitat Mainly rivers and swamps.
Abundance and Distribution Common to uncommon summer resident* (Mar–Oct) throughout; uncommon (Coastal Plain) to rare or casual (inland) in winter (Nov–Feb).
Where to Find Great Swamp National Wildlife Refuge, New Jersey; Beaver Creek Wetlands/Wildlife Project, Pennsylvania; bird ponds, National Zoological Park, Washington, D.C.
Range Breeds across southeastern Canada and the eastern half of the United States and in the west from southwestern Canada to central California, also Cuba and Bahamas; winters in southeastern United States south through northeastern Mexico, Cuba, and Bahamas and in the west from Oregon, California, and New Mexico south through northwestern Mexico.

Gadwall

Anas strepera

(L-20 W-34)

A dapper, medium-sized duck; gray above; scalloped gray, black, and white on breast and flanks; brownish head; white belly; black hindquarters; chestnut, black, and white patches on wing. *Female:* Brown body mottled with buff and dark brown; orange bill marked with black; white wing patch, which, when visible, is distinctive.

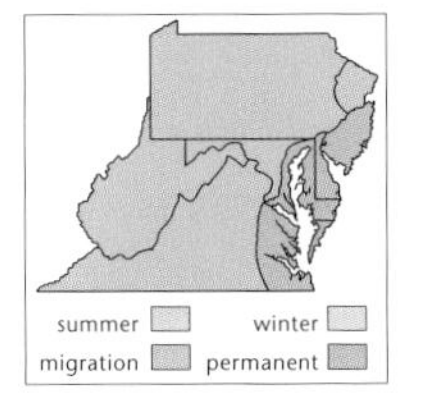

Voice A nasal "ack."

Similar Species Female Mallard has mottled (not white) belly, white tail, and lacks white patch on wing.

Habitat Lakes, estuaries, ponds, bays.

Abundance and Distribution Common to uncommon transient (Oct–Nov, Mar–Apr) throughout; common to rare winter resident (Nov–Feb) and locally uncommon summer resident,* mainly along the coast.

Where to Find Chincoteague National Wildlife Refuge, Virginia; Brigantine National Wildlife Refuge, New Jersey; Bombay Hook National Wildlife Refuge, Delaware.

Range Breeds in boreal and north temperate prairie regions; winters in temperate and northern tropical areas of Northern Hemisphere.

Eurasian Wigeon

Anas penelope

(L-19 W-32)

Medium-sized duck; gray above and on flanks; pinkish breast; chestnut head with creamy forehead and crown; white belly; black hindquarters; white upper wing and black and green secondaries (speculum) visible as patches on side when wings are folded. *Female (gray phase):* Brown body mottled with buff and dark brown; grayish bill; black and green secondaries often visible as patches on side (speculum) when wings are folded. *Female (rufous phase):* Similar to gray phase, but with a rufous rather than a grayish head.

Similar Species Eurasian Wigeon has gray axillaries visible in flight (white in American Wigeon) and the female has a brownish or reddish head (paler in female American Wigeon, contrasting with brown back).
Habitat Estuaries, lakes, marshes, crops, fields.
Abundance and Distribution Rare winter resident (Oct–Mar) along the coast.
Where to Find Deal Island Wildlife Management Area, Maryland; Brigantine National Wildlife Refuge, New Jersey; Chincoteague National Wildlife Refuge, Virginia.
Range Breeds in northern Eurasia; winters in temperate and subtropical Eurasia, but regular along coasts of North America.

American Wigeon

Anas americana
(L-19 W-33)

Medium-sized duck; gray above; purplish brown on breast and flanks; white crown and forehead; broad, iridescent green stripe through and past eye; densely mottled black and white on cheek, chin, and throat; pale blue bill with black tip; white belly; black hindquarters; white upper wing and black and green secondaries, visible as patches on the side (speculum) when wings are folded. *Female:* Brown body mottled with buff and dark brown; grayish head; grayish bill; black and green secondaries visible as patches on the side (speculum) when wings are folded.

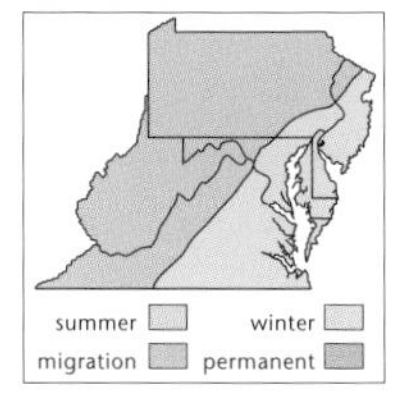

Voice A wheezy "whip" or "wheep."
Similar Species American Wigeon (both sexes) has white axillaries visible in flight (gray in Eurasian Wigeon), and female American Wigeon has pale head contrasting with brown back (brown or reddish head in female Eurasian showing little contrast with brown back).
Habitat Lakes, estuaries, bays, ponds, crops, fields.
Abundance and Distribution Common to uncommon transient (Oct–Nov, Mar–Apr) throughout; common (coast) to uncommon or rare (inland) winter resident (Oct–Apr); rare to casual summer resident* along the coast.
Where to Find Staunton River State Park, Virginia; McClintick Wildlife Management Area, West Virginia; Prime Hook National Wildlife Refuge, Delaware.
Range Breeds from northern North America south to the northern United States; winters along Atlantic and Pacific coasts and inland from southern United States to northwestern Colombia and in West Indies.

American Black Duck

Anas rubripes

(L-23 W-36)

A large duck; dark brown body, mottled with light brown; violet secondaries, often visible as a patch on the side (speculum) when wing is folded; reddish legs; yellowish or greenish bill; light brown head and neck finely streaked with dark brown.

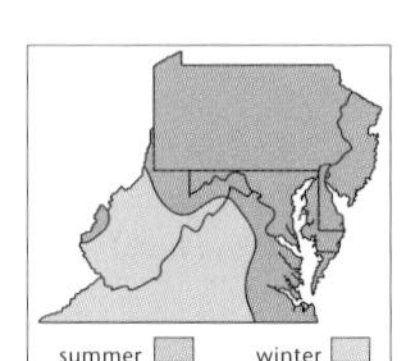

Voice A relatively high-pitched "quack," repeated.
Similar Species The female Mallard has a white tail (dark brown in Black Duck).
Habitat Rivers, lakes, ponds, estuaries.
Abundance and Distribution Common to uncommon winter resident (Oct–Mar) throughout; uncommon to rare summer resident* (Apr–Sep).
Where to Find Brigantine National Wildlife Refuge, New Jersey; Erie National Wildlife Refuge, Pennsylvania; Blackwater National Wildlife Refuge, Maryland.
Range Breeds across northeastern North America south to North Carolina; winters in the eastern United States south to northern portions of the Gulf states.

Mallard

Anas platyrhynchos
(L-23 W-36)

A large duck; iridescent green head; yellow bill; white collar; rusty breast; grayish brown body; purple secondaries, often visible as a patch on the side (speculum) when wing is folded; black rump and undertail coverts; curling black feathers at tail; white tail. *Female:* Brown body mottled with buff; orange bill marked with black; white outer tail feathers; blue speculum.

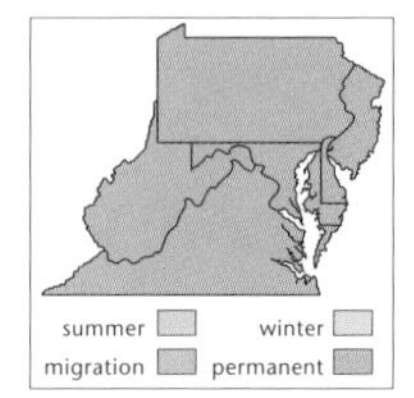

Voice A nasal "quack" (male) or series of "quack"s (female).
Similar Species The female Mallard differs from the Black Duck in showing less contrast between pale throat and brown breast than Black Duck

(dark brown breast) and differs from the female Gadwall in having a mottled (not white) belly, white tail, and no white patch on wing.
Habitat Ponds, lakes, marshes, estuaries, bays.
Abundance and Distribution Common resident* throughout; more common during migration and in winter (Oct–Mar).
Where to Find Stumpy Lake, Virginia; McClintick Wildlife Management Area, West Virginia; Prime Hook National Wildlife Refuge, Delaware.
Range Breeds across boreal and temperate regions of the Northern Hemisphere; winters in temperate and subtropical regions.

Blue-winged Teal

Anas discors

(L-16 W-25)

A small duck; brown mottled with dark brown above; tan marked with spots below; light blue patch on wing; green secondaries visible as patch on side (speculum) when wings are folded; dark gray head with white crescent at base of bill. *Female:* Mottled brown and dark brown above and below; tan undertail coverts spotted with brown; yellowish legs.

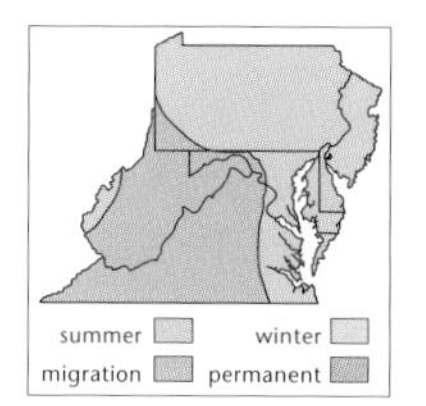

Voice A high-pitched "eeeee" (male); a series of soft "quack"s (female).
Similar Species Female Blue-winged Teal often shows some light blue on wing, has yellowish legs (not grayish); female Green-winged Teal has smaller bill than other teals and white (not spotted) undertail coverts.
Habitat Lakes, estuaries, marshes, ponds.
Abundance and Distribution Common transient (Apr–May, Aug–Sep) throughout; uncommon to

rare and local in summer* (May–Aug) and winter (Oct–Mar), mainly along the coast.
Where to Find Mason Neck National Wildlife Refuge, Virginia; McClintick Wildlife Management Area, West Virginia; Prime Hook National Wildlife Refuge, Delaware.
Range Breeds across boreal and temperate North America south to central United States; winters in southern United States to northern South America, West Indies.

Northern Shoveler

Anas clypeata

(L-19 W-31)

Medium-sized duck with large, spatulate bill; green head; golden eye; black back; rusty sides; white breast and belly; black rump and undertail coverts; blue patch on wing; green secondaries visible as patch on side (speculum) when wings are folded. *Female:* Mottled brown and dark brown above and below; brown eye; orange "lips" on dark bill.

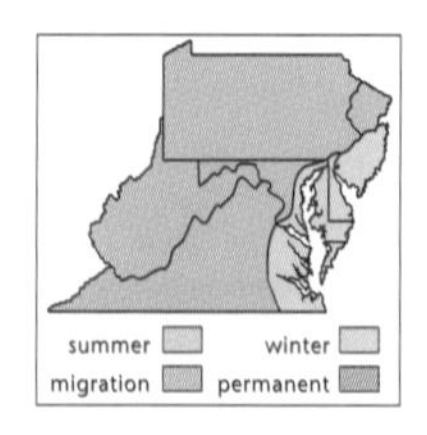

Voice A low, hoarse "kuk kuk."
Similar Species The heavy, flattened bill is distinctive.
Habitat Lakes, estuaries, bays, ponds.
Abundance and Distribution Common (Coastal Plain) to uncommon or rare (inland) transient (Sep–Oct, Mar–Apr); common to uncommon winter resident (Nov–Feb) and rare summer resident* (May–Aug) along the coast.
Where to Find Gifford Pinchot State Park, Pennsylvania; West Ocean City Pond, Maryland; Brigantine National Wildlife Refuge, New Jersey.
Range Breeds across boreal and north temperate

regions (mainly in west in North America); winters along temperate coasts south to subtropics and tropics of Northern Hemisphere.

Northern Pintail

Anas acuta

(L-26 W-36)

A long-necked, long-tailed duck; gray back and sides; brown head; white neck, breast, and belly; black rump, undertail coverts, and tail with extremely long central feathers; secondaries iridescent brown, visible as patch on side (speculum) when wings are folded. *Female:* Brownish mottled with dark brown throughout; grayish bill; pointed tail.

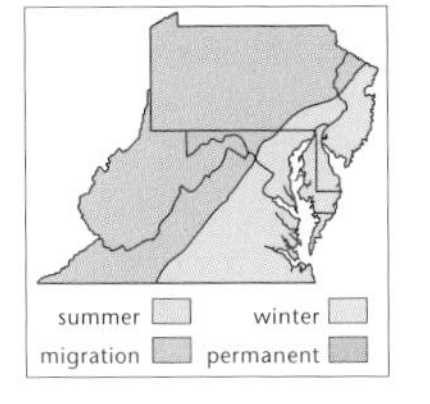

Voice A high-pitched "quip" (male); a series of "quack"s (female).

Similar Species Female resembles other female dabblers but shape is distinctive (long neck and relatively long, pointed tail).

Habitat Flooded fields, swales, shallow ponds, bays.

Abundance and Distribution Common to uncommon transient (Nov, Mar) throughout; common to uncommon winter resident (Dec–Feb) along coastal regions and in the Piedmont. Rare in summer* (May–Aug), mainly along the coast.

Where to Find MacNamara Wildlife Management Area, New Jersey; Yellow Creek State Park, Pennsylvania; Blackwater National Wildlife Refuge, Maryland.

Range Breeds in Arctic, boreal, and temperate grasslands and tundra; winters in temperate, subtropical, and tropical areas of Northern Hemisphere.

Green-winged Teal

Anas crecca

(L-15 W-24)

A small, fast-flying duck; chestnut head with broad, iridescent green stripe above and behind eye; gray body; beige breast with black spots; white bar on side of breast; white tail; black rump and undertail coverts. *Female:* Mottled brown and white above and below; whitish undertail coverts; green secondaries appear as patch on side (speculum) when wings are folded.

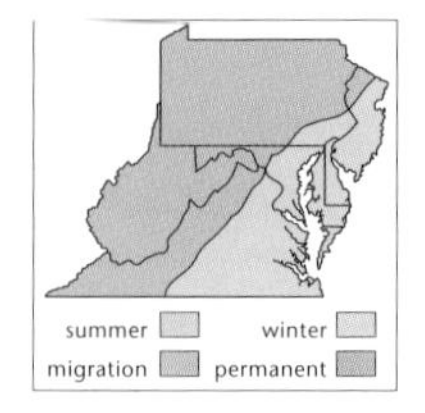

Voice A high-pitched "teet" or a nasal "kik quiik kik kik."

Similar Species Female Green-winged Teal has smaller bill than other teals and white (not spotted) undertail coverts; female Blue-winged Teal often shows some light blue on wing, has yellowish (not grayish) legs; female Cinnamon Teal is a dark rusty brown, rather than the grayish brown of the bluewing and the buffy brown of the greenwing.

Habitat Lakes, estuaries, marshes, ponds.

Abundance and Distribution Common to uncommon transient (Mar–Apr, Aug–Oct) throughout; common (along coast) to rare (inland) winter resident (Nov–Feb); rare summer resident* (May–Aug), mainly on the Atlantic side of the southern Delmarva Peninsula.

Where to Find Chincoteague National Wildlife Refuge, Virginia; McClintick Wildlife Management Area, West Virginia; Bombay Hook National Wildlife Refuge, Delaware.

Range Breeds in boreal and Arctic areas; winters in temperate and subtropical regions of Northern Hemisphere.

Canvasback

Aythya valisineria

(L-21 W-33)

A medium-sized, heavy-bodied duck with steeply sloping forehead; rusty head; red eye; black breast; gray back and belly; black hindquarters. *Female:* Grayish body, brownish neck and head.

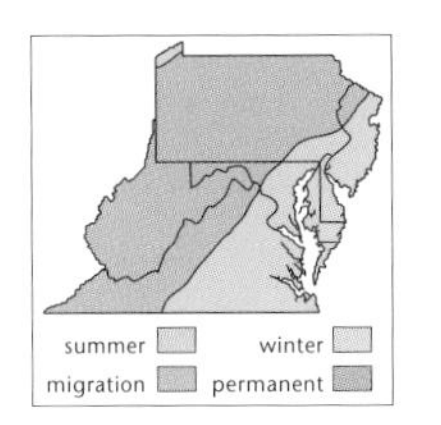

Voice A gabbling "kup kup kup."
Similar Species The sloping forehead is distinctive for both sexes; female Redhead has bluish bill with black tip (Canvasback is all black); Canvasback female shows contrast between brown head and gray body that is lacking in the all-brown female Redhead.
Habitat Lakes, bays.
Abundance and Distribution Common transient (Nov, Feb–Mar) throughout; common to uncommon winter resident (Nov–Mar) in the Piedmont, Coastal Plain, and Lake Erie.
Where to Find Dyke Marsh, Virginia; West Ocean City Pond, Maryland; Barnegat National Wildlife Refuge, New Jersey.
Range Breeds across northwestern North America south to California and Iowa; winters locally from southern Canada south through the United States to southern Mexico.

Redhead

Aythya americana

(L-20 W-33)

Rusty head; gold eye; bluish bill with white ring and black tip; black breast; gray back and belly; dark brown hindquarters. *Female:* Brownish throughout; bluish bill with black tip.

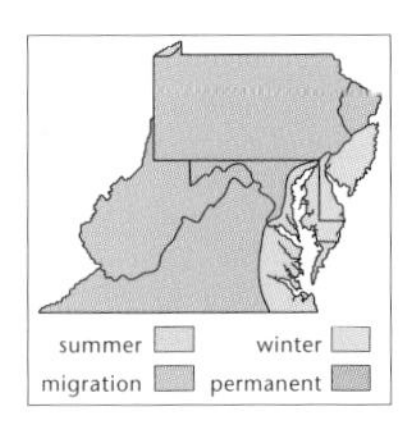

Voice A soft, catlike "yow," repeated.
Similar Species Redheads lack the sloping forehead of the Canvasback; female Redhead has bluish bill with black tip (Canvasback bill is all black); Canvasback female shows contrast between brown head and gray body that is lacking in the all-brown female Redhead.
Habitat Bays, lakes, ponds.
Abundance and Distribution Common to uncommon transient (Oct–Nov, Mar–Apr) throughout; common to uncommon winter resident (Nov–Feb) along the coast and on Lake Erie; rare to casual summer resident* (May–Sep) along the coast.
Where to Find McClintick Wildlife Management Area, West Virginia; Presque Isle State Park, Pennsylvania; Blackwater National Wildlife Refuge, Maryland.
Range Breeds in western Canada, the northwestern United States, and locally in the Great Lakes region; winters central and southern United States south to Guatemala, also Greater Antilles.

Ring-necked Duck

Aythya collaris
(L-17 W-28)

A smallish duck with characteristically pointed (not rounded) head; black back, breast, and hindquarters; dark head with iridescent purple sheen; gray flanks with white bar edging breast; golden eye; white feather edging at base of bill; white band across dark bill. *Female:* Brown body and head; white eyering; bill with whitish band.

Voice "Caah," repeated.
Similar Species The pointed head shape of Ring-necked Ducks is distinctive. Scaup have rounded

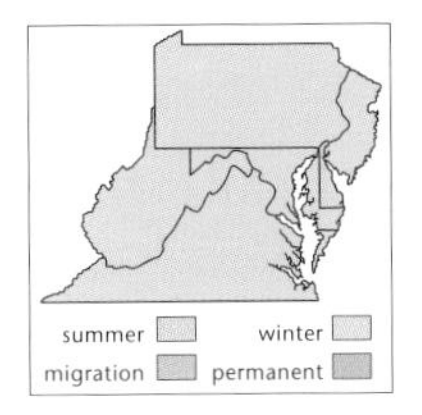

heads; male scaup have light (not dark) backs; female scaup have distinct white face patch at base of bill (dark in female Ring-necked Duck).

Habitat Lakes, ponds.

Abundance and Distribution Common transient and common to uncommon winter resident (Oct–Apr) throughout; rare in summer along the coast.

Where to Find Assunpink Wildlife Management Area, New Jersey; Bird Ponds, National Zoological Park, Washington, D.C.; Mason Neck National Wildlife Refuge, Virginia.

Range Breeds across central and southern Canada, northern United States; winters along both U.S. coasts, southern United States south to Panama, West Indies.

Greater Scaup

Aythya marila

(L-19 W-31)

Rounded head; back gray mottled with black; black breast and hindquarters; dark head with iridescent green sheen; gray flanks; golden eye; bluish bill with dark tip. *Female:* Brown body and head; white patch at base of bill; bill bluish gray.

Voice A soft cooing or rapid, whistled "week week week" (male); a guttural "caah" (female).

Similar Species Male Greater Scaup has green sheen on a more rounded head (male Lesser has purple sheen on a more pointed head); Lesser Scaup has white band on secondaries only, Greater Scaup has white band on primaries and secondaries (both sexes, visible only in flight).

Habitat Large lakes, bays.

Abundance and Distribution Common transient and winter resident (Oct–Apr) along the Coastal Plain and Lake Erie; rare transient (Nov, Mar) inland.
Where to Find Brigantine National Wildlife Refuge, New Jersey; Presque Isle State Park, Pennsylvania; Chesapeake Bay Bridge-Tunnel, Virginia.
Range Breeds in Old and New World Arctic; winters along temperate and northern coasts and large lakes in Northern Hemisphere.

Lesser Scaup

Aythya affinis

(L-17 W-28)

A smallish duck; gray back; black breast and hindquarters; dark head with iridescent purple sheen (sometimes greenish); gray flanks; golden eye; bluish bill with dark tip. *Female:* Brown body and head; white patch at base of bill; bill grayish.

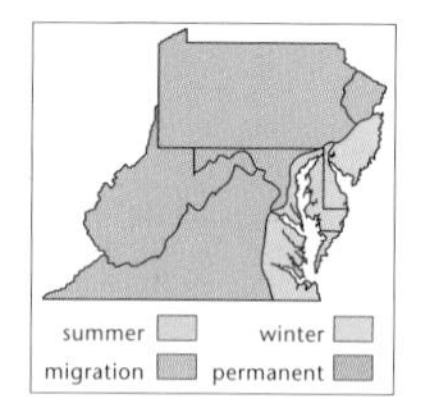

Voice A soft "wheeooo" or single whistled "weew" (male); a weak "caah" (female).
Similar Species Male Lesser Scaup has purple sheen on a more pointed head (male Greater has green sheen on a more rounded head); Lesser Scaup has white band on secondaries only, Greater Scaup has white band on primaries and secondaries (both sexes, visible only in flight). The pointed head shape of Ring-necked Ducks is distinctive. Scaup have rounded heads; male scaup have light backs (dark in male Ring-necked Duck); female scaup have a distinct white face patch at base of bill (dark in female Ring-necked Duck).
Habitat Bays, lakes.
Abundance and Distribution Common to uncommon transient (Nov, Mar–Apr) throughout; common to uncommon (Coastal Plain) to rare (inland)

winter resident (Oct–Apr); rare in summer along coast.

Where to Find Presque Isle State Park, Pennsylvania; Lake Anna, Virginia; MacNamara Wildlife Management Area, New Jersey.

Range Breeds in Alaska, western and central Canada, and northern United States; winters coastal and central inland United States south to northern South America, West Indies.

King Eider

Somateria spectabilis

(L-22 W-36)

A large duck with distinct orange protuberance (frontal shield) from forehead (male); pale blue crown and nape; blood-orange bill; beige cheek and chin; white neck and breast; black back and sides; white crissum; black hindquarters. *Female:* Entirely brown mottled with dark brown; dark brown markings are wedge shaped on sides; feathering on side of bill extends nearly to nares.

Similar Species Female Common Eider has barring rather than wedging on sides and is more reddish than the King Eider female.

Voice Various "cooo"s and "crok"s.

Habitat Coastal waters, often at rocky shores and jetties.

Abundance and Distribution Rare winter resident (Nov–Mar) along the immediate coast.

Where to Find Chesapeake Bay Bridge-Tunnel, Virginia; Island Beach State Park, New Jersey; Cape Henlopen State Park, Delaware.

Range Breeds in Old and New World high Arctic; winters along northern coasts, south to Alaska and New York in the United States.

Common Eider

Somateria mollissima

(L-24 W-38)

A large duck with white back and creamy breast; black sides and tail; peculiar, wedge-shaped head; greenish stripe along bill and forehead; black patch through and behind eye; greenish back of neck and white throat. *Male, First Winter:* Brownish head and body with white breast. *Female:* Reddish brown throughout with barring on sides and scalloping on breast.

Similar Species Female Common Eider has barring rather than wedging on sides, as in female King Eider, and is more reddish.

Habitat Coastal waters, often near rocky jetties.

Abundance and Distribution Rare winter resident (Oct–Mar) along the immediate coast.

Where to Find Chesapeake Bay Bridge-Tunnel, Virginia; Island Beach State Park, New Jersey; Cape Henlopen State Park, Delaware.

Range Breeds in the high Arctic of the Old and New World. Winters in Arctic, North Atlantic, and North Pacific oceans and coasts.

Harlequin Duck

Histrionicus histrionicus

(L-17 W-21)

Dark gray head with white, black, and chestnut markings; dark gray back, neck, and breast with white collar and breast bar; chestnut flanks and belly. *Female:* Brown with whitish belly; white patches on ear, forehead, and base of bill.

Voice Various squeeks, whistles, and trills.

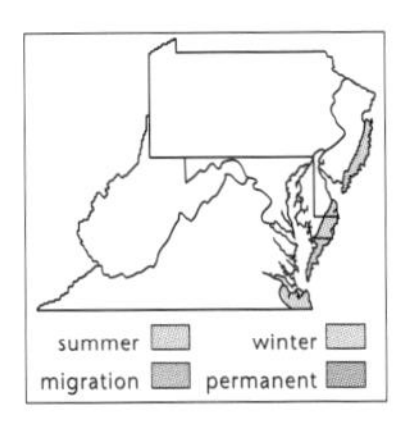

Habitat Rocky coasts.
Abundance and Distribution Rare winter resident (Nov–Mar) along the immediate coast.
Where to Find Chesapeake Bay Bridge-Tunnel, Virginia; Island Beach State Park, New Jersey; Cape Henlopen State Park, Delaware.
Range Breeds in northwestern North America from Alaska to Idaho; in the northeast south to Quebec; also Iceland, Greenland, and Siberia; winters mainly along northern coasts, south to California and New York in the United States.

Surf Scoter

Melanitta perspicillata
(L-19 W-34)

Black with white patches on nape and forehead; orange bill with bull's-eye on side (white circle, black center); white eye. *Female:* Entirely brown with white patches in front of and behind eye; dark bill.

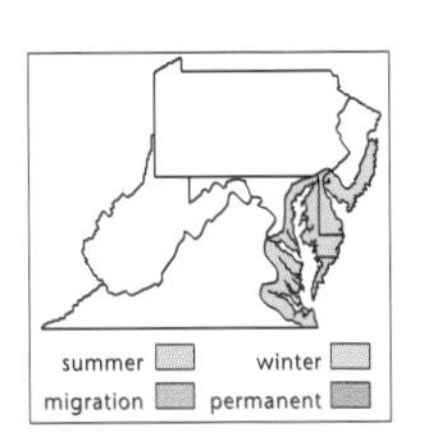

Voice A guttural croak.
Similar Species Surf Scoter female has distinct white pre- and postorbital patches on otherwise brown head and brown throat; female White-winged Scoter has feathering almost to nostrils on bill (lacking on Black and Surf scoters); female Black Scoter has white cheek and throat contrasting with dark brown head and neck.
Habitat Marine, bays, lakes.
Abundance and Distribution Common winter resident (Oct–Apr) along coast.
Where to Find Chesapeake Bay Bridge-Tunnel, Virginia; Barnegat National Wildlife Refuge, New Jersey; Prime Hook National Wildlife Refuge, Delaware.

Range Breeds northern North America south to central Canada; winters mainly along coasts from Alaska to northwestern Mexico and from Nova Scotia to Florida, also Great Lakes.

White-winged Scoter

Melanitta fusca

(L-22 W-39)

Black with white eye and wing patches; dark bill with black knob at base and orange tip. *Female:* Entirely dark brown (sometimes with whitish pre- and postorbital patches); bill dark orange with black markings; feathering on bill extends nearly to nostrils; white secondaries sometimes visible on swimming bird.

Voice A plaintive whistle or low growl.
Similar Species Female White-winged Scoter has feathering almost to nostrils on bill (lacking on Black and Surf scoters).
Habitat Bays, lakes.
Abundance and Distribution Common winter resident (Oct–Apr) along coast.
Where to Find Chesapeake Bay Bridge-Tunnel, Virginia; Assateague National Wildlife Refuge, Maryland; Avalon, New Jersey.
Range Breeds in boreal and Arctic regions of the Old and New World; winters mainly along northern coasts, south to northwestern Mexico and South Carolina in North America.

Black Scoter

Melanitta nigra

(L-19 W-33)

Entirely black with orange knob at base of bill. *Female:* Dark brown with whitish cheek and throat contrasting with dark crown and nape. *Immature:* Patterned similarly to female but whitish belly.

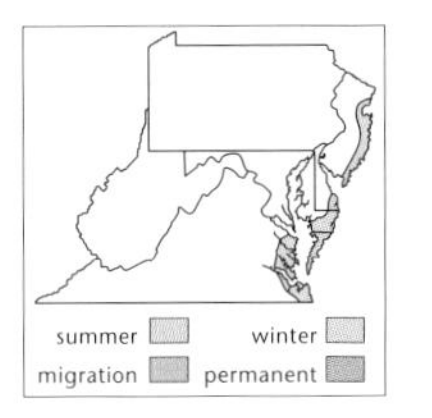

Voice A rattlelike "quack."
Similar Species Female White-winged Scoter has feathering almost to nostrils on bill (lacking on Black and Surf scoters); female Black Scoter has white cheek and throat contrasting with dark brown head and neck; Surf Scoter female has distinct white pre- and postorbital patches on otherwise brown head and brown throat.
Habitat Marine, bays, lakes.
Abundance and Distribution Common winter resident (Oct–Apr) along coast.
Where to Find Chesapeake Bay Bridge-Tunnel, Virginia; Barnegat National Wildlife Refuge, New Jersey; Prime Hook National Wildlife Refuge, Delaware.
Range Breeds locally in tundra regions of Eurasia and North America; winters in northern and temperate coastal waters of Northern Hemisphere, south to California and South Carolina in the United States.

Long-tailed Duck

Clangula hyemalis

(L-19 W-29)

Dark head with white face; dark neck and breast; gray and brown on back; dark wings; white belly and undertail coverts; black, extremely long, pointed central tail feathers. *Female and Winter Male:* White head with dark brown cheek patch and crown; white neck; grayish brown breast; white belly and undertail coverts; short, sharply pointed tail.

Voice "Ah ah ah-ah-ee-ah."
Similar Species Combination of white head and all dark wings is distinctive, as is the call.
Habitat Marine, bays, large lakes.
Abundance and Distribution Common (along coast and Lake Erie) to rare (inland) transient and winter resident (Nov–Mar).
Where to Find Presque Isle State Park, Pennsylvania; Chesapeake Bay Bridge-Tunnel, Virginia; Ocean City inlet, Maryland.
Range Breeds in high Arctic of Old and New World; winters mainly along northern coasts of Northern Hemisphere, in United States south to California and South Carolina.

Bufflehead

Bucephala albeola

(L-14 W-23)

A small, plump, short-billed duck; head white from top of crown to nape, the rest iridescent purple; black back; white breast, belly, and sides; gray bill; pink legs. *Female and Immature:* Dark back; grayish white below; dark head with large white patch extending below and behind eye.

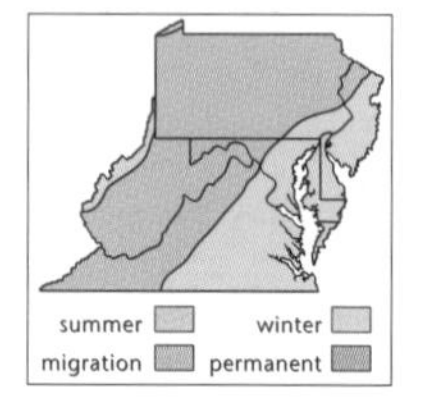

Voice A weak, nasal "eeh."
Similar Species The larger male Hooded Merganser also has white back of crown and nape, but the white is edged in black; Hooded Merganser has golden eye (dark in Bufflehead); thin, pointed bill; rusty sides.

Habitat Bays, lakes, estuaries.
Abundance and Distribution Common to uncommon transient (Nov–Dec, Feb–Apr) throughout; common to uncommon winter resident (Nov–Apr) along the Coastal Plain and Piedmont, scarce inland except along the Ohio River and Lake Erie, where it can be locally common; rare or casual in summer along the coast.
Where to Find Kettle Creek State Park, Pennsylvania; Silver Lake, Virginia; Bird Ponds, National Zoological Park, Washington, D.C.
Range Breeds across Canada and extreme northern United States; winters from sub-Arctic along both coasts of North America and inland from central United States south to central Mexico.

Common Goldeneye

Bucephala clangula
(L-18 W-30)

Iridescent green head (sometimes purplish); golden eye; white patch at base of bill; black back and hindquarters; white breast, sides, and belly; black and white scapulars; white wing patch visible in flight. *Female:* Gray body; brown head; white neck; golden eye; gray bill yellowish at tip (mostly yellow in some birds).

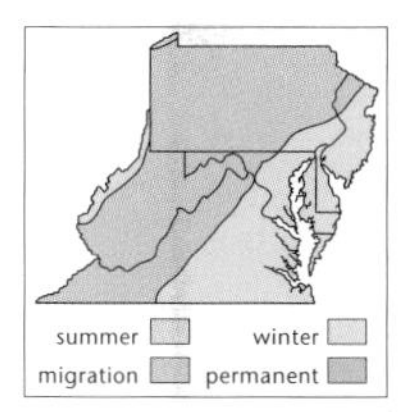

Voice A high-pitched, nasal "eeh."
Habitat Bays, lakes.
Abundance and Distribution Common to uncommon transient (Nov–Dec, Feb–Mar) throughout; common to uncommon winter resident (Nov–Mar) along the Coastal Plain and Piedmont, scarce inland except along the Ohio River and Lake Erie, where it can be locally common; rare or casual in summer along the coast.

Where to Find Greenbottom Wildlife Management Area, West Virginia; Presque Isle State Park, Pennsylvania; Indian River Inlet, Delaware.
Range Breeds across boreal and north temperate regions of Old and New World; winters along northern coasts south to temperate and subtropical regions of the Northern Hemisphere.

Hooded Merganser

Lophodytes cucullatus
(L-18 W-26)

White crest from top of crown to nape broadly edged in black, the rest of head black; golden eye; black back and tail; white breast with prominent black bar; rusty sides; sharp, black bill. *Female and Immature Male:* Body brownish; head pale orange with dusky crown; pale orange crest off back of crown and nape; upper mandible dark; lower mandible orangish.

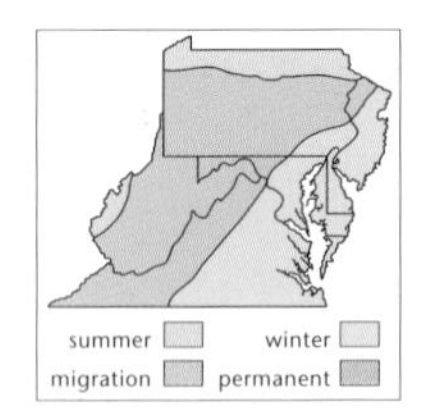

Voice A trilled "crrroooo" (male); low grunt (female).
Similar Species Other female mergansers (Common and Red-breasted) are much larger and have dark russet heads; long, orange bills; grayish bodies.
Habitat Ponds, lakes, estuaries, bays; for breeding sites, the bird prefers cavities located in heavily wooded bottomlands with swift, clear-running streams nearby.
Abundance and Distribution Common transient (Oct–Dec, Feb–Mar) throughout; common winter resident (Oct–Apr) along the Coastal Plain and Piedmont; uncommon to rare and local in summer,* mainly in northern Pennsylvania and the Ohio River Valley in western West Virginia.

Where to Find Bird Ponds, National Zoological Park, Washington, D.C.; Deep Creek Lake State Park, Maryland; Hopatcong State Park, New Jersey.
Range Breeds across central and southern Canada and northern United States, farther south in Rockies and Appalachians; winters mainly along coasts from southern Canada to northern Mexico, West Indies.

Common Merganser

Mergus merganser
(L-25 W-36)

Iridescent green head; sharp, red-orange bill; black back; gray rump and tail; white breast, sides, and belly. *Female and Immature Male:* Rufous, crested head; white chin; rufous throat and neck ending abruptly at white breast; gray back and sides; orange bill.

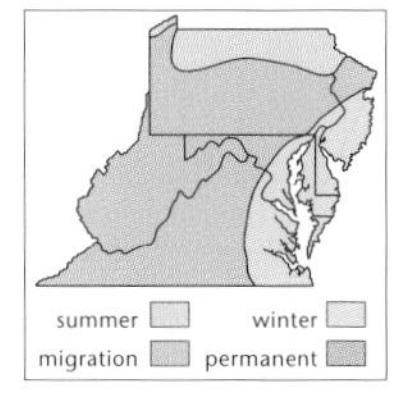

Voice A low "uu-eek-wa" (male); a harsh "karr" (female).
Similar Species The female Common Merganser lacks white wing patch evident in female Red-breasted Merganser; female Common has distinct boundary between white throat and rufous neck; female Red-breasted has a whitish neck and throat with no abrupt line between rufous neck and throat.
Habitat Lakes, rivers, bays.
Abundance and Distribution Common winter resident (Nov–Mar) along the coast and on Lake Erie; uncommon to rare and local transient (Nov–Dec, Feb–Mar) and rare or casual winter resident inland; uncommon and local summer resident* (Apr–Oct) in northern Pennsylvania.

Where to Find Bird Ponds, National Zoological Park, Washington, D.C.; Greenbottom Wildlife Management Area, West Virginia; Presque Isle State Park, Pennsylvania; Indian River Inlet, Delaware.
Range Breeds in Old and New World sub-Arctic and boreal regions south in mountains into temperate areas; winters from northern coasts south inland through temperate and subtropical zones of the Northern Hemisphere.

Red-breasted Merganser

Mergus serrator

(L-22 W-32)

Iridescent green, crested head; sharp, red-orange bill; white collar; buffy breast streaked with brown; gray back, rump, and tail; black shoulder with white chevrons; white scapulars; grayish sides. *Female and Immature Male:* Rufous, crested head; white chin and throat; gray back and sides; white wing patch.

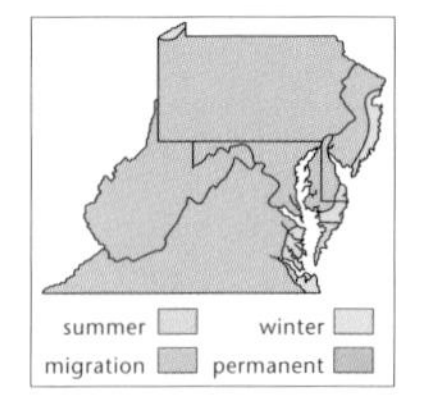

Voice "Eeoww" (male); a harsh "karr" (female).
Similar Species Female Red-breasted Merganser has a whitish foreneck and throat gradually merging into rufous hindneck; female Common has distinct boundary between white throat and rufous neck; female Red-breasted has a white wing patch, which the female Common Merganser lacks.
Habitat Bays, lakes, marine.
Abundance and Distribution Common winter resident (Nov–Mar) and rare summer resident* along the coast and on Lake Erie; uncommon to rare and local transient (Nov, Feb–Mar) and rare or casual winter resident inland.

Where to Find Kerr Reservoir, Virginia; Ocean City Inlet, Maryland; Presque Isle State Park, Pennsylvania.
Range Breeds in Arctic and boreal regions of Old and New World; winters mainly along coasts in southern boreal and temperate areas.

Ruddy Duck

Oxyura jamaicensis
(L-15 W-23)

A small duck; chestnut body; stiff black tail held at a 45-degree angle; black cap; white cheek; blue bill. *Winter:* Grayish brown body; dark cap; white cheek. *Female:* Mottled grayish and white body; stiff black tail; dark cap and dark line below eye.

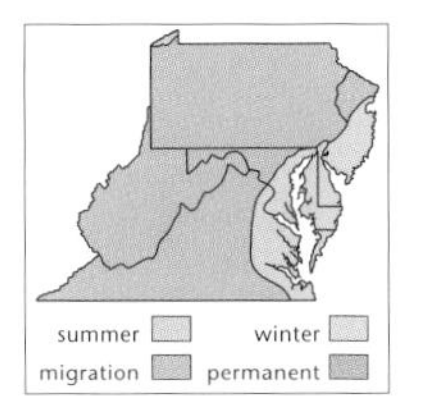

Voice A staccato, cicada-like "tsk-tsk-tsk-tsk quark."
Habitat Lakes, ponds, bays.
Abundance and Distribution Common winter resident (Oct–Apr) and rare summer resident* along the coast and on Lake Erie; uncommon to rare and local transient (Oct–Nov, Mar–Apr) and rare or casual winter resident inland.
Where to Find Bird Ponds, National Zoological Park, Washington, D.C.; Deep Creek Lake State Park, Maryland; Hopatcong State Park, New Jersey.
Range Breeds locally from northern Canada through the United States to central Mexico, West Indies, and South America; winters coastal and southern United States south to Nicaragua and elsewhere in tropical breeding range.

Kites, Hawks, Ospreys, and Eagles
(Order Falconiformes, Family Accipitridae)

A diverse assemblage of diurnal raptors, all of which have strongly hooked bills and powerful talons. Clark and Wheeler (1987) and Dunne and Sutton (1989) provide identification guides for advanced students of this difficult group.

Osprey
Pandion haliaetus
(L-23 W-63)

Dark brown above; white below, often with dark streaking (females); white crown with ragged crest from back of head; broad, dark line extending behind yellow eye; white chin and cheek; extended wings are white, finely barred with brown, with prominent dark patches at wrist.

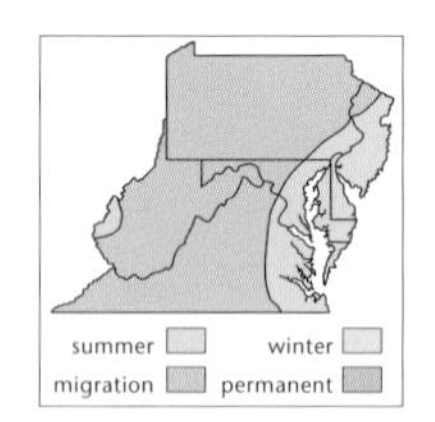

Habits Feeds mainly on fish snatched from water surface.
Voice A shrill "kew," repeated.
Habitat Estuaries, lakes, rivers, bays.
Abundance and Distribution Common to uncommon transient (Mar–May, Sep–Oct) throughout, more common in spring; common to uncommon and local summer resident* (May–Sep) along the coast, in eastern Pennsylvania, and along the southern Ohio River in West Virginia; rare to casual winter resident along the coast.
Where to Find Eastern Neck National Wildlife Refuge, Maryland; Mason Neck National Wildlife Refuge, Virginia; Delaware Bay, New Jersey.
Range Breeds in boreal, temperate, and some tropical localities of Old and New World, particularly

along coasts; winters mainly in tropical and subtropical zones.

Bald Eagle

Haliaeetus leucocephalus

(L-35 W-84)

ENDANGERED Huge; dark brown body; white head and tail; yellow beak and legs. *Immature:* Entirely brown with whitish wing linings and base of tail.

Habits Feeds primarily on fish.
Voice A descending "keee chip-chip-chip-chip."
Similar Species Immature Golden Eagle has well-defined white (not whitish) base of tail and white wing patches at base of primaries.
Habitat Lakes, rivers, estuaries.
Abundance and Distribution Uncommon and local resident* along the coast and in northwestern Pennsylvania; rare to casual elsewhere.
Where to Find Bombay Hook National Wildlife Refuge, Delaware; Mason Neck National Wildlife Refuge, Virginia; Blackwater National Wildlife Refuge, Maryland.
Range Breeds across Canada and northern United States, south along coasts to Florida, California, and Texas; winters throughout breeding range from southern Canada southward, particularly along the coasts and at larger inland lakes.

Northern Harrier

Circus cyaneus

(L-19 W-42)

A slim, long-tailed, long-winged hawk; gray above; pale below with dark spots; white rump; yellow

eye and cere (skin at base of bill); long, yellow legs. *Female:* Streaked dark brown and tan above; whitish below, heavily streaked with brown; yellow eyes; pale yellowish cere; yellow legs; barred tail; white rump.

Habits Flies low over open areas, usually within a few feet of the ground, alternately flapping and gliding; wings at an angle during glides; often hovers just above the ground.

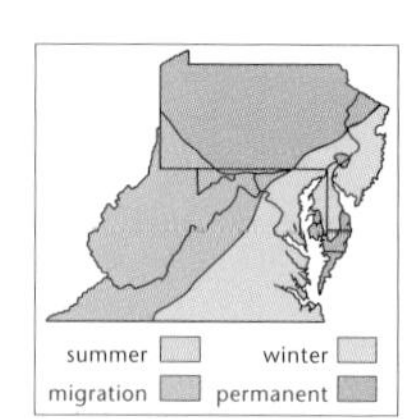

Voice A rapid, descending "cheek-cheek-cheek-cheek-cheek-cheek."

Similar Species Similar haunts and habits as Short-eared Owl, but white rump is distinctive.

Habitat Marshes, grasslands, estuaries, agricultural areas.

Abundance and Distribution Uncommon to rare and local resident* in Pennsylvania, the Delmarva Peninsula, and scattered localities elsewhere throughout the region; uncommon transient (Oct, Apr) throughout; common to uncommon winter resident (Nov–Mar), mainly along the coast and in the Piedmont.

Where to Find McClintick Wildlife Management, West Virginia; Curllsville strip mines, Clarion Co., Pennsylvania; Assateague Island National Seashore, Maryland.

Range Breeds across boreal and temperate regions of Northern Hemisphere; winters in temperate and tropical zones.

Sharp-shinned Hawk

Accipiter striatus

(L-12 W-24)

Slate gray above; barred rusty and white below; gray crown with rusty face; red eye; yellow cere

(skin at base of bill); barred tail; long, yellow legs; as in other accipiters, the female is much larger than the male. *Immature:* Brown above; streaked brown and white on head, breast, and belly.

Habits All 3 accipiters (sharpshin, Cooper's, and goshawk) are distinguished from other hawks by their relatively short, broad wings and long tails and by their behavior in flight, which is characterized by a series of rapid wing beats followed by a short, flat-winged glide. They seldom soar except during migration. All 3 species are forest bird hunters.

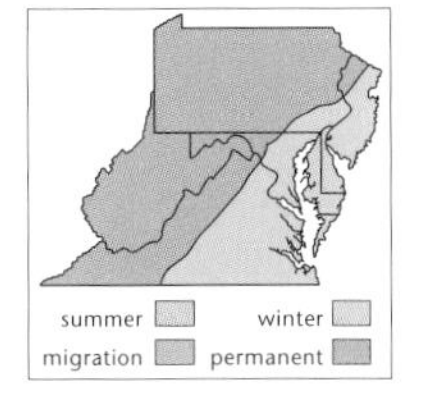

Voice A rapid, high-pitched "kew-ki-ki-ki-ki-ki."
Similar Species The calls of the 3 similar species of accipiters are different; sharpshin has a squared tail (slightly rounded in the larger Cooper's Hawk); immature sharpshin tail has narrow and indistinct terminal white band (immature Cooper's terminal white tail band is broader and more distinct); head does not extend forward much beyond bend of wing in gliding sharpshin (head extends well beyond bend of wing in gliding Cooper's); immature sharpshin has heavy streaking on breast (immature Cooper's has finer streaking). The immature Merlin is also similar to the immature sharpshin but has dark eyes (not red) and dark tail with thin, light bands (not broad, light bands as in sharpshin).
Habitat Forests.
Abundance and Distribution Uncommon resident* in highlands except central West Virginia, where scarce; uncommon winter resident (Sep–Apr) along coast and in Piedmont, rare in summer.
Where to Find Beech Fork State Park, West Virginia; Sherando Lake Recreation Area, Virginia; Savage River State Forest, Maryland.

Range Breeds from sub-Arctic Alaska and Canada to northern Argentina (except prairie regions and most of southern United States), also in Greater Antilles; winters from northern coastal regions and southern Canada south through breeding range.

Cooper's Hawk

Accipiter cooperii

(L-18 W-32)

Slate gray above; barred rusty and white below; gray crown with rusty face; red eye; yellow cere (skin at base of bill); barred tail; long, yellow legs. *Immature:* Brown above; streaked brown and white on head, breast, and belly.

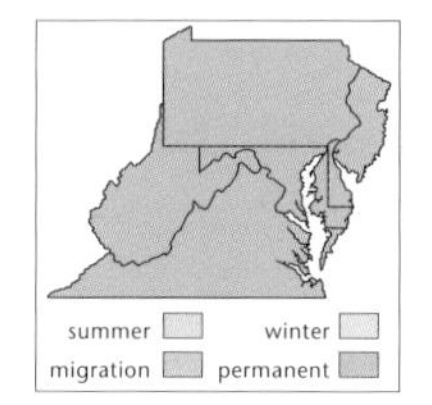

Habits All 3 accipiters (sharpshin, Cooper's, and goshawk) are distinguished from other hawks by their relatively short, broad wings and long tails and by their behavior in flight, which is characterized by a series of rapid wing beats followed by a short, flat-winged glide. They seldom soar except during migration. All 3 species are forest bird hunters.

Voice A wheezy "peeew," repeated.

Similar Species The calls of the 3 similar species of accipiters are different; Cooper's has a rounded tail (square in the smaller sharpshin); immature Cooper's terminal white tail band is broad and distinct; immature sharpshin tail has narrow and indistinct terminal white band; head does not extend forward much beyond bend of wing in gliding sharpshin (head extends well beyond bend of wing in gliding Cooper's); immature Cooper's has relatively fine streaking on breast (immature sharpshin has heavier streaking). Adult Northern Goshawk is barred gray and white below and has a prominent white eyebrow; immature goshawk closely resembles immature Cooper's but has white eyebrow.

Habitat Forests.

Abundance and Distribution Uncommon to rare resident* nearly throughout; less numerous along Coastal Plain in summer.

Where to Find Hawk Mountain Sanctuary, Pennsylvania; Cape May Point, New Jersey; Harvey's Knob, Virginia.

Range Breeds from southern Canada to northern Mexico; winters from northern United States to Honduras.

Northern Goshawk

Accipiter gentilis

(L-23 W-39)

Slate gray above; barred gray and white below; gray crown with prominent white eyebrow; dark patch behind eye; red eye; yellow cere (skin at base of bill); unevenly barred tail; long, yellow legs. *Immature:* Brown above; streaked brown and white on head, breast, and belly; white eyebrow; uneven tail barring.

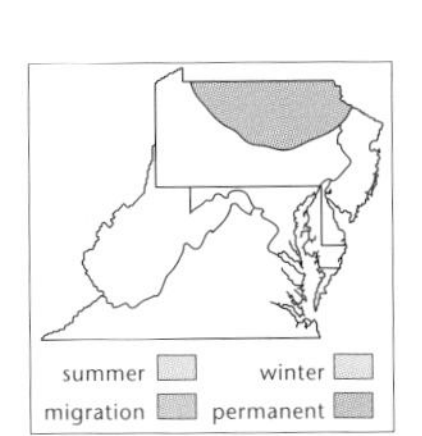

Habits All 3 accipiters (sharpshin, Cooper's, and goshawk) are distinguished from other hawks by their relatively short, broad wings and long tails and by their behavior in flight, which is characterized by a series of rapid wing beats followed by a short, flat-winged glide. They seldom soar except during migration. All 3 species are forest bird hunters.

Voice "Tew tew tew tew tew tew."

Similar Species The calls of the 3 similar species of

accipiters are different; adult Northern Goshawk is barred gray and white below and has a prominent white eyebrow; immature goshawk closely resembles immature Cooper's but has white eyebrow.
Habitat Northern mixed hardwood and coniferous forests.
Abundance and Distribution Uncommon to rare resident* in northern Pennsylvania and scattered localities in northwestern New Jersey, western Maryland, and eastern West Virginia; rare transient and winter visitor (Nov–Apr) elsewhere.
Where to Find Hawk Mountain Sanctuary, Pennsylvania; Cape May Point, New Jersey; Harvey's Knob, Virginia.
Range Breeds in boreal regions of Northern Hemisphere, south in mountains to temperate and tropical zones; winters in breeding range and irregularly southward and in lowlands.

Red-shouldered Hawk
Buteo lineatus
(L-19 W-39)

Mottled dark brown and white above with rusty shoulders; barred rusty and white below; tail dark with 3 or 4 narrow, whitish bars; brown eye; pale yellowish cere (skin at base of bill); yellow legs; in flight, note brown-and-white barring on flight feathers, pale white patch at base of primaries ("window"), and rusty wing linings. *Immature:* Brown mottled with white above; buff streaked with brown on breast, barred on belly; rusty shoulders.

Voice A strident "ki-cheek ki-cheek ki-cheek keeew."
Similar Species Voice is distinctive. Immature is

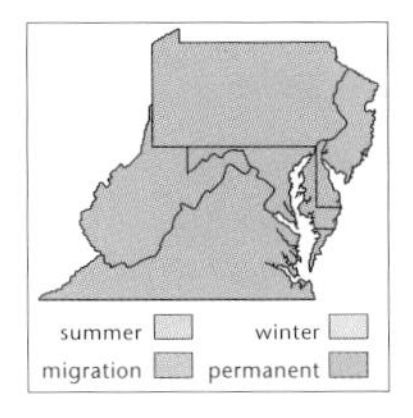

similar to other immature buteos but usually shows rusty shoulders; it is the only species in the region with a crescent-shaped wing "window" visible in flight; tail appears dark with light bars. The Red-tailed Hawk immature has unstreaked upper breast, and tail appears light with narrow, dark bars. Immature Rough-legged Hawk shows dark belly and light tail with dark, subterminal band. The immature Red-shouldered Hawk is also similar to the immature Broad-winged Hawk. Immature redshoulder often has reddish tinge on shoulders. Also, the redshoulder tail appears to be dark with light bands, while the broadwing tail appears to be white with dark bands.

Habitat Riparian forest, woodlands, swamps.

Abundance and Distribution Uncommon to rare resident* throughout, more common along the Coastal Plain.

Where to Find Kiptopeke State Park, Virginia; Cedarville State Forest, Maryland; Beech Fork State Park, West Virginia.

Range Breeds from southeastern Canada and the eastern United States south to central Mexico, also California; winters from central United States south through breeding range.

Broad-winged Hawk

Buteo platypterus

(L-16 W-35)

Dark brown above; barred rusty and white below; dark tail with 2 white bands equal in width to dark bands; brown eye; yellowish cere (skin at base of bill); yellow legs; in flight, note whitish flight feathers with black tips; buffy wing linings. *Immature:* Brown above; streaked brown and

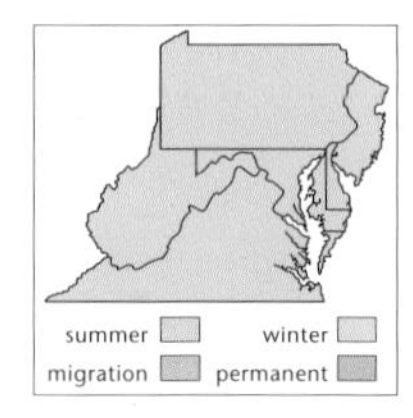

white below; tail narrowly barred with brown and white.

Voice A high-pitched "ki-cheeeee," occasionally imitated by Blue Jays.
Similar Species The immature Red-shouldered Hawk is very similar to the immature Broad-winged Hawk. Immature redshoulder often has reddish tinge on shoulders. Also, the redshoulder tail appears to be dark with light bands, while the broadwing appears to be white with dark bands. Additionally, the soaring broadwing lacks the pale, crescent-shaped "wing window" near the outer tips of the primaries characteristic of the redshoulder.
Habitat Deciduous forests but seen in loose flocks (kettles) during migration over any habitat.
Abundance and Distribution Common to uncommon transient and summer resident* (Apr–Sep) nearly throughout; uncommon to rare as a summer resident along the coast. This species is the most abundant transient at hawk watch sites in the region.
Where to Find Hawk Mountain Sanctuary, Pennsylvania; Cape May Point, New Jersey; Harvey's Knob, Virginia.
Range Breeds across southern Canada from Alberta eastward and the eastern half of the United States south to Texas and Florida; winters from southern Mexico south to Brazil; resident in West Indies.

Red-tailed Hawk

Buteo jamaicensis

(L-22 W-53)

A large hawk, extremely variable in plumage; most common adult plumage is mottled brown and

white above; white below with dark streaks and speckling across belly; rusty tail (appears whitish from below); in flight, dark forewing lining contrasts with generally light underwing. *Immature:* Mottled brown and white above; streaked brown and white below; brown tail finely barred with grayish white. *Light Phase (Krider's):* Much paler; pale orange tale. *Dark Phase (Harlan's):* Dark throughout with some white speckling; dark tail, whitish at base, darker at tip with rusty wash.

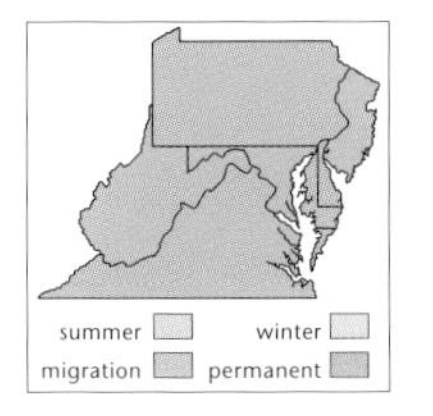

Voice A hoarse, drawn-out screech, "ke-aaaaaaaah" (as in Buick commercials on TV).
Similar Species Rusty tail is distinctive for adults. Immature is similar to other immature buteos but usually shows unstreaked upper breast and light-colored tail with several dark bars; immature red-shoulder has streaked breast, rusty shoulders, crescent-shaped wing "window" in flight, and tail that appears dark with light bars. Immature Rough-legged Hawk shows dark belly and light tail with dark, subterminal band.
Habitat Open areas, woodlands.
Abundance and Distribution Common resident* nearly throughout; uncommon to rare in summer along the coast.
Where to Find Summit Lake Recreation Area, West Virginia; Trap Pond State Park, Delaware; Washington Monument Knob, Maryland.
Range Breeds from sub-Arctic of Alaska and Canada to Panama; winters northern United States south through breeding range; resident in West Indies.

Rough-legged Hawk

Buteo lagopus

(L-22 W-51)

A large hawk with legs feathered all the way to the toes; mottled brown and white above; buffy with brown streaks on breast; dark brown belly; white tail with dark subterminal band. *Dark Phase:* Entirely dark except whitish flight feathers and tail with broad, dark, subterminal band. *Immature:* Similar to adult—whitish or finely barred tail with broad, dark, subterminal band; base of primaries shows white from above in flight.

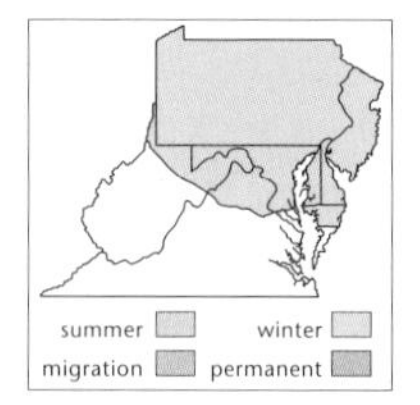

Habits Flies low over open areas, often hovering.

Voice A catlike "keeeeew," dropping in pitch.

Similar Species Northern Harrier forages in similar fashion but has a white rump (not tail), is much slimmer, and lacks dark band contrasting with buffy breast.

Habitat Open areas—farm fields, salt marshes, pastures, strip mine sites, grasslands.

Abundance and Distribution Uncommon to rare winter resident (Nov–Mar) in Pennsylvania and New Jersey, scarcer in West Virginia and the more southern portions of the region.

Where to Find Hawk Mountain Sanctuary, Pennsylvania; Raccoon Ridge, New Jersey; Washington Monument Knob, Maryland.

Range Breeds in Arctic and sub-Arctic regions of Old and New World; winters mainly in temperate zone.

Golden Eagle

Aquila chrysaetos

(L-34 W-71)

Huge; dark brown with golden wash on head and shaggy neck; legs feathered to feet; appears entirely dark in flight. *Immature:* Dark brown with white patch at base of primaries and basal half of tail white.

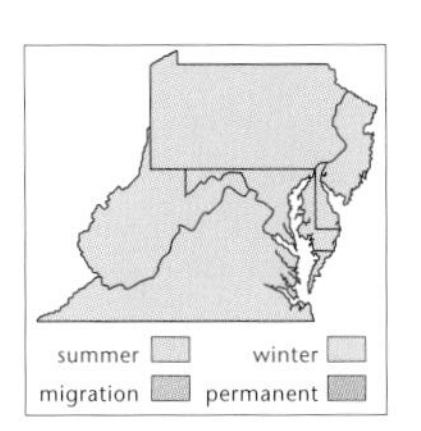

Voice A rapid series of nasal chips.

Similar Species Immature Golden Eagle has well-defined white (not whitish as in Bald Eagle) base of tail and white wing patches at base of primaries; also has golden head.

Habitat Open areas.

Abundance and Distribution Rare transient (Nov, Mar), more common in fall, and rare winter visitor (Dec–Feb) throughout.

Where to Find Hawk Mountain Sanctuary, Pennsylvania; Raccoon Ridge, New Jersey; Washington Monument Knob, Maryland.

Range Breeds primarily in open and mountainous regions of boreal and temperate zones in Northern Hemisphere; winters in central and southern portions of breeding range.

Falcons

(Family Falconidae)

With the exception of the caracara (not in this book), the falcons are sleek birds with pointed wings and long, square tails.

American Kestrel

Falco sparverius

(L-10 W-22)

A small falcon; gray crown; black-and-white facial pattern; orange-buff back and underparts spotted with brown; blue-gray wings; orange tail with black subterminal bar edged in white. *Female:* Rusty back and wings barred with brown.

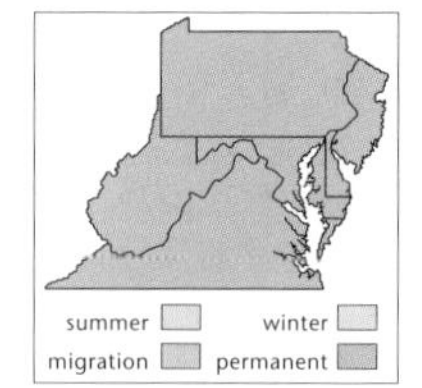

Habits Perches or hovers a few feet off the ground while foraging; often seen on telephone wires along the road.
Voice A rapid, high-pitched "kle-kle-kle-kle-kle."
Similar Species Merlin shows a distinctly dark tail barred with white; lacks facial pattern of kestrel.
Habitat Open areas.
Abundance and Distribution Common resident* throughout, somewhat more common in winter.
Where to Find Beech Fork State Park, West Virginia; Mendota Fire Tower, Virginia; Bombay Hook National Wildlife Refuge, Delaware.
Range Breeds nearly throughout Western Hemisphere from sub-Arctic Alaska and Canada to southern Argentina, West Indies; winters from north temperate regions south through breeding range.

Merlin

Falco columbarius

(L-11 W-25)

A small falcon; slate above, buffy below with dark brown spots and streaks; white throat; brown eyes, yellow cere (skin at base of bill) and legs; black tail, white at base, with 2 white bars and white terminal edging. *Female:* Dark brown above.

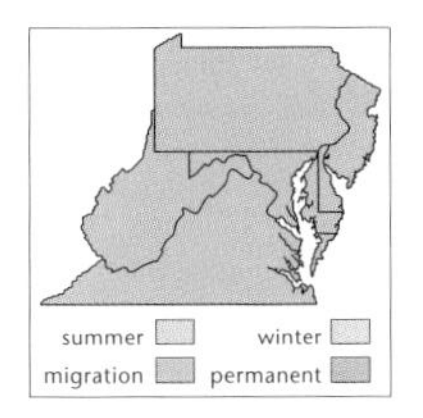

Habits A bird hunter; sallies from low perches, using surprise and speed to capture prey.
Voice "Kwe kwe kwe kwe kwe."
Similar Species Merlin shows a distinctly dark tail barred with white; lacks facial pattern of kestrel.
Habitat Open woodlands, hedgerows, second growth; often hunts small birds in trees or shrubs bordering water in winter.
Abundance and Distribution Uncommon to rare transient (Sep–Oct, Mar–Apr) throughout; rare winter visitor (Nov–Feb) along the coast and in Piedmont.
Where to Find Back Bay National Wildlife Refuge, Virginia; Prime Hook National Wildlife Refuge, Delaware; Cape May Point, New Jersey.
Range Breeds across boreal and north temperate regions of Old and New World; winters in south temperate and tropical zones.

Peregrine Falcon

Falco peregrinus

(L-18 W-40)

THREATENED A large falcon; dark gray above; white below with spotting and barring on belly and thighs; black crown and cheek with white neck patch; brown eye; yellow eyering, cere (skin at base of bill), and legs; black and white facial pattern; in flight, note large size; long, pointed wings; long tail; whitish underparts finely barred and spotted. *Immature:* Dark brown above; buffy below spotted and streaked with brown; has facial pattern of adult.

Habits Forages by flying with swift, powerful wing beats, well up in the air, then stooping on prey.
Voice A rapid "kee kee kee kee."

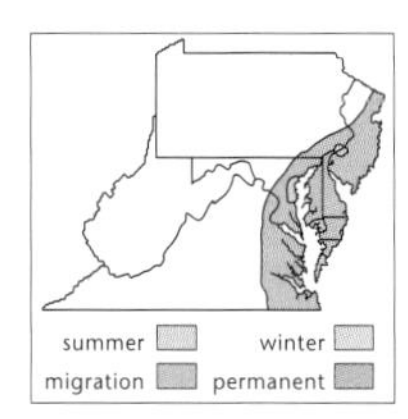

Habitat Open areas, usually near water.
Abundance and Distribution Uncommon to rare transient (Sep–Oct, Mar–Apr) and rare winter visitor (Nov–Feb) along the coast, scarcer inland. Rare and local summer resident* at reintroduction sites in New Jersey, Virginia, Maryland, and Pennsylvania.
Where to Find Breeding birds have been reintroduced at several areas in the region, including Barnegat National Wildlife Refuge, New Jersey; Stony Man Nature Trail, Shenandoah National Park, Virginia; and Francis Scott Key Bridge in Baltimore, Maryland. During migration, Peregrine Falcon can be seen at coastal sites, such as Cape Henlopen State Park, Delaware, and Chincoteague National Wildlife Refuge, Virginia.
Range Breeds (at least formerly) in boreal and temperate regions of Northern Hemisphere; winters mainly in the tropics.

Pheasants, Grouse, and Turkeys
(Order Galliformes, Family Phasianidae)

Terrestrial, heavy-bodied, mostly seed- and fruit-eating birds with short, curved wings for quick takeoff.

Ring-necked Pheasant
Phasianus colchicus
(Male L-33 W-31)

A chicken-sized bird with long legs and extremely long, pointed tail; green head with naked red skin on face; white collar; body rich chestnuts, grays, golds, and bronzes, spotted with white and brown; grayish brown, long, pointed tail barred with brown; has spurs on tarsi. Female: Pale grayish brown spotted with dark brown; long, pointed tail.

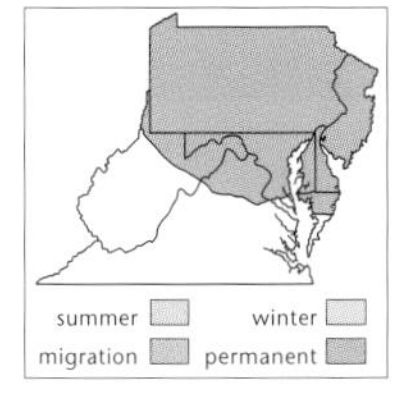

Voice A loud, hoarse "keow-kuk."
Habitat Pastures, agricultural fields.
Abundance and Distribution Introduced repeatedly throughout the region. Uncommon resident* in most areas, though rare in the mountains and absent from West Virginia (except northern and eastern panhandles, where still present).
Where to Find Great Swamp National Wildlife Refuge, New Jersey; Bombay Hook National Wildlife Refuge, Delaware; Erie National Wildlife Refuge, Pennsylvania.
Range Native of central Asia. Introduced throughout North America. Self-sustaining populations exist in cool, temperate regions with extensive grain crops and moderate hunting.

Ruffed Grouse
Bonassa umbellus
(L-17 W-28)

Body mottled brown and white; head with a crest; white line through eye; black shoulder patches; tail chestnut, barred with black and white with broad black subterminal band and white terminal band. *Female:* Similar to male, but black-and-white terminal bands are incomplete (central tail feathers grayish at tip).

Habits "Explodes" in a burst of sound when flushed.
Voice Male makes a distinct, loud booming sound with his wings; the booms gradually increase in

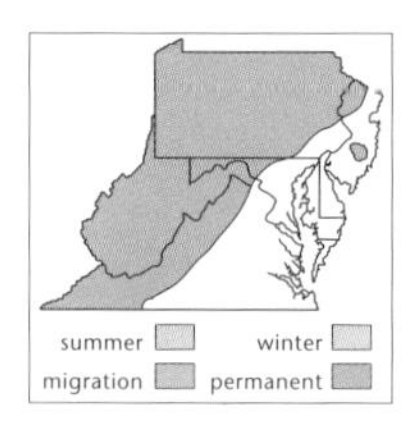

frequency until they almost run together. The female gives various chickenlike clucks to the brood.
Similar Species Pheasant female has a relatively long, pointed tail, not square as in the grouse.
Habitat Deciduous and mixed woodlands, especially near aspen groves (feeding) and dense conifer stands (roosting).
Abundance and Distribution Uncommon resident* inland; scarce or absent from the Coastal Plain though found in the southern New Jersey Pine Barrens.
Where to Find Grayson Highlands State Park, Virginia; Summit Lake Recreation Area, West Virginia; Rocky Gap State Park, Maryland.
Range Boreal and north temperate regions of North America, south in mountains to northern California and Utah in the west and north Georgia in the east.

Wild Turkey

Meleagris gallopavo

(Male L-47 W-63)

Dark brown body with iridescent bronze highlights; naked red head and neck; blue skin on face with dangling red wattles; hairy beard hanging from breast; tail dark barred with buff; tarsi with spurs. Female: Naked facial skin is grayish; lacks beard and wattles.

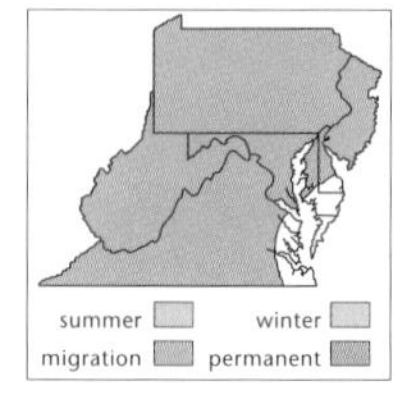

Habits Males associate in groups of 2–3, females in larger flocks.
Voice A rapid, high pitched gobble.
Habitat Deciduous forest, oak woodlands.
Abundance and Distribution Uncommon resident,* mainly west of the Coastal Plain; apparently absent from some southwestern West Virginia counties.

Where to Find Dickey Ridge Visitor Center, Shenandoah National Park, Virginia; Barnegat National Wildlife Refuge, New Jersey; Bombay Hook National Wildlife Refuge, Delaware.
Range Formerly from southern Canada to central Mexico, now extirpated from many areas but being reintroduced.

Quail
(Family Odontophoridae)

Northern Bobwhite
Colinus virginianus
(L-10 W-15)

Brown and gray above mottled with dark brown and white; chestnut sides spotted with white; white breast and belly scalloped with black; chestnut crown with short, ragged crest; white eyebrow and throat. *Female:* Tawny eyebrow and throat.

Habits Coveys fly and scatter in a burst when approached.
Voice Familiar, whistled "hoo WHIT" ("bob white"); several other calls including a whistled "perdeek."

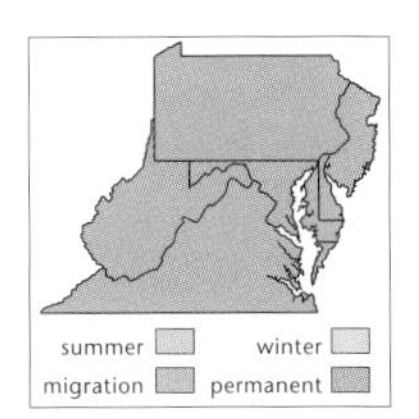

Habitat Brushy fields, pastures, grasslands, agricultural areas.
Abundance and Distribution Uncommon and declining resident* of grasslands throughout the region; scarce in central West Virginia, extreme northern Pennsylvania, and highlands throughout.
Where to Find Friendship Hill National Historic Site, Pennsylvania; Blackwater National Wildlife

Refuge, Maryland; Beech Fork State Park, West Virginia.
Range Eastern and central United States south to Guatemala; isolated populations in Arizona and Sonora.

Rails, Coots, and Gallinules
(Order Gruiformes, Family Rallidae)

Except for the ubiquitous, ducklike coot, these are secretive marsh birds, heard more often than seen. They have cone-shaped or long, narrow bills; short, rounded wings; short tails; and long toes for support in walking on floating vegetation.

Yellow Rail
Coturnicops noveboracensis
(L-7 W-13)

A small (cowbird-sized) rail; dark brown with tawny stripes above; dark brown crown; broad tawny eyebrow; dark brown mask; whitish chin; tawny underparts barred with black on the flanks; short, yellowish bill and pale legs; in flight (a rarely observed event), shows white secondaries.

Habits Extremely secretive.
Voice "Tik tik tik-tik-tik tik tik tik-tik-tik," like hitting 2 stones together.
Habitat Wet prairies, marshes.
Abundance and Distribution Rare to casual transient (Oct, Apr–May), mainly in fall, along the coast. Some summer and winter records as well.
Where to Find Great Swamp National Wildlife Refuge, New Jersey; Chincoteague National Wildlife Refuge, Virginia.

Range Breeds in east and central Canada and northeastern and north-central United States; winters along the southern Atlantic and Gulf coast of United States; resident in central Mexico.

Black Rail

Laterallus jamaicensis

(L-6 W-10)

A tiny (sparrow-sized) rail; dark grayish black with chestnut nape and black-and-white barring on flanks; short, black bill.

Habits Secretive.
Voice "Tic-ee-toonk."
Habitat Wet prairies, marshes.
Abundance and Distribution Uncommon to rare transient and local summer resident* (Apr–Oct) along the immediate coast; some winter records as well.
Where to Find Bombay Hook National Wildlife Refuge, Delaware; Blackwater National Wildlife Refuge, Maryland; Chincoteague National Wildlife Refuge, Virginia.
Range Breeds locally in California, Kansas, and along east coast from New York to Texas, also West Indies, Central and South America; winters along Gulf coast and in tropical breeding range.

Clapper Rail

Rallus longirostris

(L-13 W-20)

A large rail; streaked brown and tan above; buffy below with gray-and-white barring on flanks; head and neck buffy with dark crown; long, pinkish or yellowish bill; pale greenish legs.

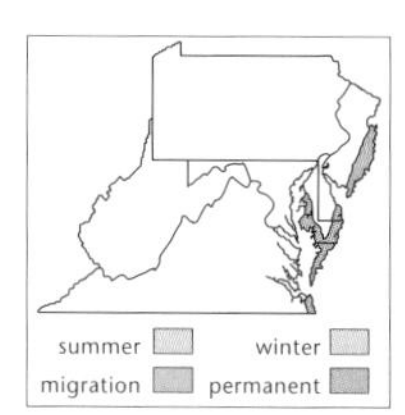

Voice "Chik chik chik chik chik," like hitting a rock with a metal rod.
Similar Species King Rail inhabits freshwater (occasionally brackish) marshes, not salt marshes, and is rustier overall, darker barred on flanks, with black lores. However, plumage in both species is variable, and they are known to interbreed.
Habitat Salt marshes.
Abundance and Distribution Common summer resident* along the immediate coast; uncommon to rare in winter.
Where to Find Brigantine National Wildlife Refuge, New Jersey; Prime Hook National Wildlife Refuge, Delaware; Chincoteague National Wildlife Refuge, Virginia.
Range Resident along coast from Connecticut to Belize, California to southern Mexico, West Indies and much of coastal South America to southeastern Brazil and Peru.

King Rail

Rallus elegans
(L-15 W-22)

A large rail; streaked brown and rusty above; tawny below with black-and-white barring on flanks; head and neck tawny with dark crown and darkish stripe through eye; long bill with dark upper and pinkish or yellowish lower mandible; pale, reddish legs.

Voice A series of low grunts, "ih ih ih ih ih."
Similar Species King Rail inhabits freshwater (occasionally brackish) marshes, not salt marshes, and

is rustier overall, darker barred on flanks, with black lores. However, plumage in both species is variable, and they are known to interbreed.

Habitat Freshwater marshes.

Abundance and Distribution Common (south) to rare (north) and local summer resident* (Apr–Oct) and rare winter visitor (Nov–Mar) along the immediate coast; rare to casual transient and summer resident* inland.

Where to Find Back Bay National Wildlife Refuge, Virginia; Eastern Neck National Wildlife Refuge, Maryland; Bombay Hook National Wildlife Refuge, Delaware; Huntley Meadows Park, Virginia.

Range Breeds across the eastern half of the United States south to central Mexico; winters along coast from Georgia to Texas and south to southern Mexico; resident in Cuba.

Virginia Rail

Rallus limicola

(L-10 W-14)

Similar to the King Rail but about half the size; streaked brown and rusty above; tawny below with black-and-white barring on flanks; gray head with rusty crown; white throat; long reddish bill; pale legs.

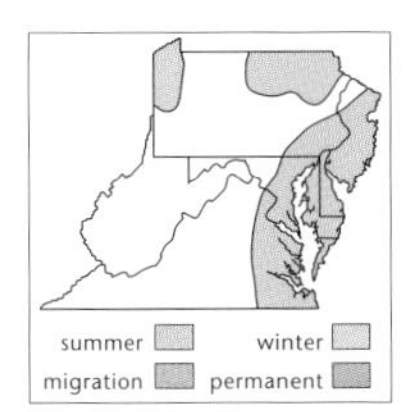

Voice "Kik kik kik ki-deek ki-deek ki-deek."

Similar Species The smaller Virginia Rail has gray face (tawny in larger King Rail) and tawny throat (white in King Rail).

Habitat Freshwater and brackish marshes.

Abundance and Distribution Common (along coast) to uncommon or rare (inland) or casual

(West Virginia) summer resident* (Mar–Oct); uncommon to rare in winter along the coast.
Where to Find Prince Gallitzin State Park, Pennsylvania; Blackwater National Wildlife Refuge, Maryland; Great Swamp National Wildlife Refuge, New Jersey.
Range Breeds locally from southern Canada to southern South America; winters along coast and in subtropical and tropical portions of breeding range.

Sora
Porzana carolina
(L-9 W-14)

A medium-sized rail; streaked brown and rusty above; grayish below with gray-and-white barring on flanks; gray head with rusty crown; black at base of bill; black throat and upper breast; short, yellow bill; greenish legs. *Female:* Amount of black on throat and breast reduced. *Immature:* Browner overall, lacks black on throat and breast.

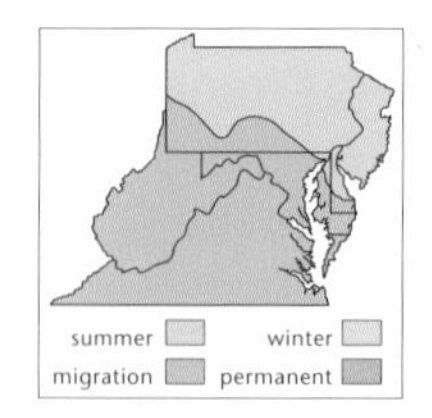

Voice "Ku-week," also a long descending series of whistles.
Similar Species Same size and habitat preferences as Virginia Rail but note short, yellowish, cone-shaped bill in Sora (reddish and long in Virginia Rail).
Habitat Fresh and brackish marshes.
Abundance and Distribution Uncommon to rare and local summer resident* (Apr–Sep) in northern and southeastern Pennsylvania, New Jersey, and Delaware, rare to casual elsewhere; uncommon to rare transient (Apr–May, Sep–Oct) throughout.
Where to Find Great Swamp National Wildlife Refuge, New Jersey; Deal Island Wildlife Manage-

ment Area, Maryland; Bombay Hook National Wildlife Refuge, Delaware.
Range Breeds from central Canada to southern United States; winters along coast and from southern United States to northern South America and West Indies.

Purple Gallinule
Porphyrula martinica
(L-13 W-23)

A large rail with short, cone-shaped bill and extremely long toes; iridescent green back; purple head and underparts; blue frontal shield of naked skin on forehead; red bill with yellow tip; yellow legs. *Immature:* Buffy overall, darker on back.

Habits The extremely long toes enable this rail to walk on lily pads and other floating vegetation without sinking.
Voice A chickenlike "cuk cuk cuk cuk cuk-kik cuk-kik."
Similar Species Clapper and King rails have long, pointed (not short, cone-shaped) bills; immature Common Moorhen is grayish rather than brownish and shows white edging along folded wing.
Habitat Freshwater marshes.
Abundance and Distribution Rare summer resident* (Apr–Sep) in coastal Virginia; rare to casual elsewhere.
Where to Find Back Bay National Wildlife Refuge; Chincoteague National Wildlife Refuge.
Range Breeds from the eastern United States south to northern Argentina, West Indies; winters from southern Gulf states south through the breeding range.

Common Moorhen

Gallinula chloropus

(L-14 W-23)

A large rail with short, cone-shaped bill and extremely long toes; entirely sooty gray with white edging along wing and white undertail coverts; red frontal shield of naked skin on forehead; red bill with yellow tip; greenish legs. *Winter:* Similar to summer but with olive bill, frontal shield, and legs.

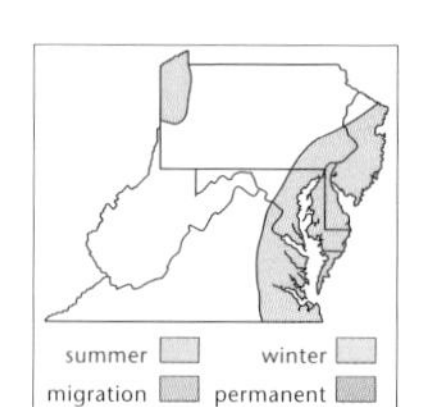

Habits The extremely long toes enable this rail to walk on lily pads and other floating vegetation without sinking; swims more than most other rails.
Voice Low croaks and whiney, high-pitched squeaks.
Similar Species Winter and juvenile Common Moorhen show white edging along folded wing, which American Coot lacks.
Habitat Freshwater marshes.
Abundance and Distribution Uncommon to rare summer resident* (Apr–Oct) along the Coastal Plain and locally inland (e.g., northwestern Pennsylvania); scattered summer records elsewhere; uncommon to rare in extreme southeastern Virginia in winter.
Where to Find McClintick Wildlife Management Area, West Virginia; Pymatuning Reservoir, Pennsylvania; Huntley Meadows Park, Virginia.
Range Breeds locally in temperate and tropical regions of the world; winters mainly in subtropical and tropical zones.

American Coot

Fulica americana

(L-16 W-26)

A large, black rail, more ducklike than raillike in appearance and behavior; white bill with dark tip; red eye; greenish legs and lobed toes. *Immature:* Paler; pale gray bill.

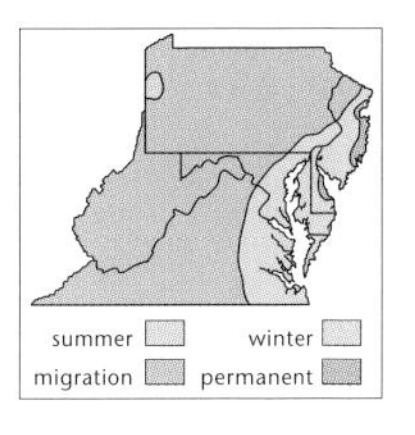

Habits Swims in open water, tipping and diving for food instead of skulking through reeds like most rails; pumps head forward while swimming.
Voice Various croaks and catlike mews.
Similar Species Juvenile coot lacks white wing edging of Common Moorhen.
Habitat Ponds, lakes, marshes, bays.
Abundance and Distribution Common to uncommon transient (Apr–May, Oct–Nov) throughout; local summer resident* (May–Aug) along the coast, in northwestern Pennsylvania, and in scattered localities elsewhere; common to uncommon winter resident, mainly along the Coastal Plain but at open water elsewhere in the region.
Where to Find Prime Hook National Wildlife Refuge, Delaware; Barnegat National Wildlife Refuge, New Jersey; Lake Anna, Virginia.
Range Breeds from central Canada to Nicaragua and West Indies; winters along coast and from central United States to northern Colombia, West Indies.

Plovers
(Order Charadriiformes, Family Charadriidae)

Plovers are rather compact shorebirds with relatively long legs, short necks, and short, thick bills.

Black-bellied Plover
Pluvialis squatarola
(L-12 W-25)

A Killdeer-sized plover; checked black and white on back; white crown, nape, and shoulder; black face, throat, and breast; white belly and undertail coverts. *Winter:* Heavy black bill contrasts with whitish at base of bill; pale eyebrow; crown and back brownish mottled with white; breast mottled with brown; belly white; black legs.

Voice A gull-like "keee keee kee-a-wee keee."
Similar Species The Black-bellied Plover has black axillaries (visible only in flight), which golden-plover species lack.
Habitat Beaches, bays, mud flats, estuaries.
Abundance and Distribution Common to uncommon transient (Aug–Oct, Apr–May) and winter resident (Aug–Apr) along the coast; uncommon to rare in summer; rare or casual transient inland—most records are from the fall.
Where to Find Ocean City, Maryland; John Heinz National Wildlife Refuge, Pennsylvania; Cape Henlopen State Park, Delaware.
Range Breeds in the high Arctic of Old and New World; winters along temperate and tropical coasts of the world.

American Golden-Plover

Pluvialis dominica

(L-11 W-23)

A Killdeer-sized plover; checked black, white, and gold on back and crown; white "headband" running from forehead, over eye, and down side of neck; black face and underparts. *Winter:* Heavy black bill contrasts with white at base of bill; pale eyebrow; crown and back brown and white, often flecked with gold; underparts white, speckled with brown on breast; gray legs.

Voice A whistled "kew-wee."

Similar Species The American Golden-Plover lacks the black axillaries that show clearly on the Black-bellied Plover in flight.

Habitat Intertidal mud flats; beaches.

Abundance and Distribution Uncommon fall transient (Aug–Oct) and rare spring transient (Apr–May) along the coast; rare to casual transient elsewhere.

Where to Find Bombay Hook National Wildlife Refuge, Delaware; Brigantine National Wildlife Refuge, New Jersey.

Range Breeds in high Arctic of New World; winters in South America.

Wilson's Plover

Charadrius wilsonia

(L-8 W-16)

A smallish plover with a heavy, black bill; brown back and head; white forehead; black lores; white chin and collar; black band across throat; white underparts; pinkish legs. *Female and Immature:* Brown breast band.

Voice "Peet peet peet."
Similar Species The smaller, short-billed Semipalmated Plover has orange (not pinkish) legs.
Habitat Bays, mud flats.
Abundance and Distribution Uncommon to rare and local summer resident (Apr–Sep) along the Virginia barrier islands; rare to casual elsewhere.
Where to Find Chincoteague National Wildlife Refuge, Virginia; Virginia Coast Reserve, Virginia.
Range Breeds along both coasts from northwestern Mexico in the west and Virginia in the east south to northern South America, West Indies; winters mainly in tropical portions of breeding range.

Semipalmated Plover

Charadrius semipalmatus

(L-7 W-15)

A small plover; brown back and head; white forehead; white stripe over and behind eye; orange eyering; orange bill, black at tip; black lores; white chin and collar; black band across throat; white underparts; orange or yellow legs. *Winter:* Brown breast band; dull, dark bill (may show some orange at base). *Immature:* Similar to adult but eyering yellow; bill black at tip, brown at base; legs brown anteriorly, yellow posteriorly.

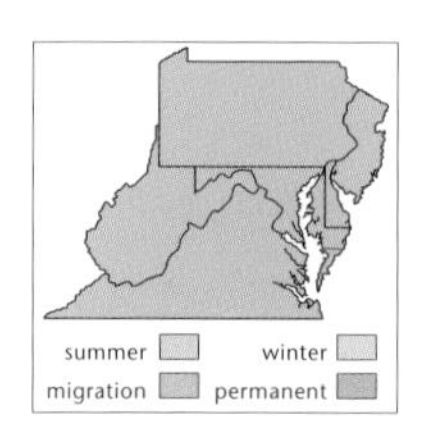

Voice A whistled "tew-wee."
Similar Species The other small plovers occurring here are paler and have incomplete breast bands. Wilson's Plover is larger, is heavy-billed, and has flesh-colored legs (orange legs in adult Semipalmated Plover, grayish in juvenile).
Habitat Estuarine mud flats, salt flats, beaches, bayshores.

Abundance and Distribution Common to uncommon transient (Apr–May, Jul–Oct) throughout; rare in summer and winter along coast.
Where to Find Erie National Wildlife Refuge, Pennsylvania; Roanoke Sewage Treatment Plant, Virginia; Eastern Neck National Wildlife Refuge, Maryland.
Range Breeds in high Arctic of North America; winters mainly along temperate and tropical coasts from Georgia and California to southern South America, West Indies.

Piping Plover

Charadrius melodus
(L-7 W-15)

THREATENED A small plover; grayish brown crown, back, and wings; white forehead with distinct black band across forecrown; white chin, throat collar, and underparts; partial or complete black band across breast; orange bill tipped with black; orange legs; white rump. *Winter:* Lacks black breast and crown bands; dark bill.

Voice "Peep," repeated.
Similar Species Semipalmated is dark brown on back (not pale) and has a complete breast band, even in winter.
Habitat Beaches.
Abundance and Distribution Uncommon transient and summer resident* (Mar–Sep) along the immediate coast; rare to casual inland except along the Lake Erie shore, where uncommon to rare as a transient and formerly a summer resident.* Reintroduction efforts are under way.
Where to Find Cape Henlopen State Park, Delaware; Presque Isle State Park, Pennsylvania;

Hereford Inlet, New Jersey; Assateague Island National Seashore, Maryland.
Range Breeds locally in southeastern and south-central Canada, northeastern and north-central United States; winters along coast from South Carolina to Veracruz, also in West Indies.

Killdeer

Charadrius vociferus

(L-10 W-20)

A medium-sized plover; brown back and head; orange rump; white forehead; white postorbital stripe; orange eyering; dark bill; black lores; white chin and collar; 2 black bands across throat and upper breast; white underparts; pale legs.

Habits Feigns broken wing when young or nest is approached.
Voice "Kill de-er"; also various peeps.
Similar Species Killdeer is the only plover in this region with 2 black bars across the breast; others have one or none.
Habitat Open areas.

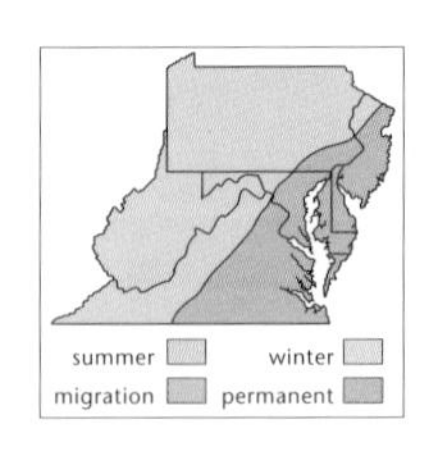

Abundance and Distribution Common summer resident* (Mar–Oct) throughout; common to uncommon in winter (Nov–Feb) in the Coastal Plain and Piedmont, rare elsewhere.
Where to Find Bluestone Wildlife Management Area, West Virginia; Dyke Marsh, Virginia; Killens Pond State Park, Delaware.
Range Breeds from sub-Arctic Canada and Alaska south to central Mexico, Greater Antilles, western South America; winters along coast from Washington and Massachusetts, and inland from southern United States, south to northern and western South America, West Indies.

Oystercatchers
(Family Haematopodidae)

Stout, gull-like birds; black or black and white with a long, brilliant orange bill.

American Oystercatcher
Haematopus palliatus
(L-19 W-35)

Gull-sized bird; black hood; long, bright orange bill; red eyering; brown back; white breast and belly; pinkish legs; broad white bars on secondaries and white base of tail show in flight. *Immature:* Brown head; dull orange bill.

Habits Feeds on oysters and other mollusks by prying open the shells.
Voice A strident "weeep."
Habitat Beaches, bays.
Abundance and Distribution Common summer resident* (Mar–Nov), mainly on the barrier islands; common to uncommon winter resident (Dec–Feb) from southern Maryland south.
Where to Find Brigantine National Wildlife Refuge, New Jersey; Assateague Island National Seashore, Maryland; Chincoteague National Wildlife Refuge, Virginia.
Range Breeds locally along coast from northwestern Mexico and Massachusetts south to southern South America, West Indies.

Stilts and Avocets
(Family Recurvirostridae)

A small family of medium-sized, long-legged, long-billed, long-necked shorebirds.

Black-necked Stilt
Himantopus mexicanus
(L-14 W-27)

Spindly shorebird with long neck, needlelike bill, and long legs; black head, nape, and back; white at base of bill, throat, and underparts; white patch behind eye; pink legs. *Female:* Similar to male but paler. *Immature:* Brown rather than black.

Voice A rapid, buzzy "keer-keer-keer."
Habitat Mud flats, marshes, estuaries, ponds.
Abundance and Distribution Uncommon to rare summer resident* (Apr–Sep) in coastal Delaware, Maryland, and Virginia.
Where to Find Chincoteague National Wildlife Refuge, Virginia; Craney Island Landfill, Virginia.
Range Breeds locally in western United States and along coast in east from Massachusetts south locally along coasts of Middle and South America to northern Argentina, West Indies; winters from North Carolina south through breeding range.

American Avocet
Recurvirostra americana
(L-18 W-32)

Long-legged, long-necked shorebird with long, upturned bill; black-and-white back and wings; white belly; rusty orange head; whitish at base of bill;

gray legs. *Winter:* Head and neck mostly whitish with little rusty tinge.

Habits Forages by swinging submerged bill from side to side as it walks along.
Voice "Kleet kleet kleet."
Habitat Estuaries, ponds, lakes, mud flats, flooded pastures, bays.

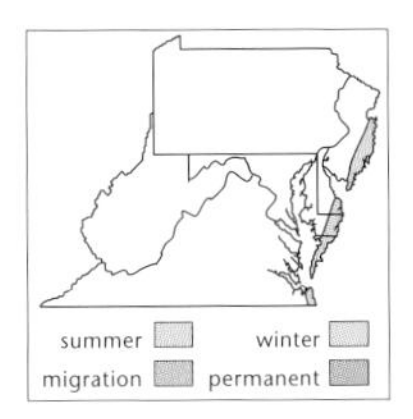

Abundance and Distribution Uncommon to rare and local summer resident (Apr–Oct) along the immediate coast; most numerous during fall migration (Sep); bred in historical times in New Jersey; rare to casual in winter.
Where to Find Bombay Hook National Wildlife Refuge, Delaware; Chincoteague National Wildlife Refuge, Virginia; Craney Island Landfill, Virginia.
Range Breeds western and central Canada south through western United States to northern Mexico; winters southern United States to southern Mexico.

Sandpipers
(Family Scolopacidae)

Scolopacids constitute a diverse assemblage of shorebirds. Breeding plumage for sandpipers is worn for only a short time, most of which is spent on Arctic breeding grounds. It is the winter plumage that is most often critical for identification of shorebirds in the Mid-Atlantic. Many second-year birds and other nonbreeders remain on the wintering ground or migration route through the summer.

Greater Yellowlegs
Tringa melanoleuca
(L-14 W-25)

Mottled grayish brown above; head, breast, and flanks white and heavily streaked with grayish brown; white belly and rump; long, yellow legs; bill long, often slightly upturned and darker at tip than at base. *Winter:* Paler overall; streaking on breast and head is reduced.

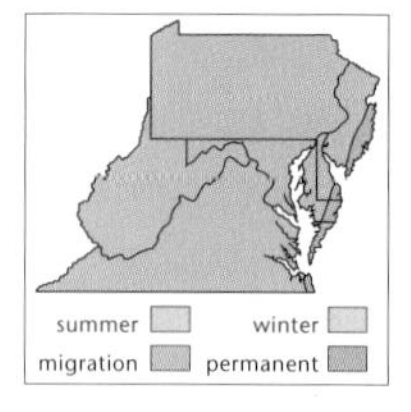

Voice Call—a rapid sequence of 3, descending notes, "tew tew tew."
Similar Species The Greater Yellowlegs is twice the size of the Lesser Yellowlegs; the descending 3-note call of the Greater is quite different from the Lesser's "pew" call, which is generally given as a single note or series of notes on the same pitch; Greater's bill is longer, thicker, slightly upturned, and dusky (rather than dark) for basal third of lower mandible.
Habitat Mud flats, estuaries, marshes, prairies, flooded agricultural fields.
Abundance and Distribution Common to uncommon transient (Apr–May, Aug–Sep) throughout; uncommon to rare resident (nonbreeding) along the coast.
Where to Find McClintick Wildlife Management Area, West Virginia; Prince Gallitzin State Park, Pennsylvania; Eastern Neck National Wildlife Refuge, Maryland.
Range Breeds in northern Canada and Alaska; winters coastal United States south to southern South America, West Indies.

Lesser Yellowlegs

Tringa flavipes

(L-11 W-20)

Mottled grayish brown above; head, breast, and flanks white and heavily streaked with grayish brown; white belly and rump; yellow legs. *Winter:* Paler overall; streaking on breast and head is reduced.

Voice Single "pew" alarm note, repeated.

Similar Species The Lesser Yellowlegs is half the size of the Greater Yellowlegs; the "pew" call of the Lesser is quite different from the Greater's 3-note, descending sequence; Lesser's bill is shorter, thinner, straight, and dark to the base (rather than yellowish or dusky for the basal third of lower mandible).

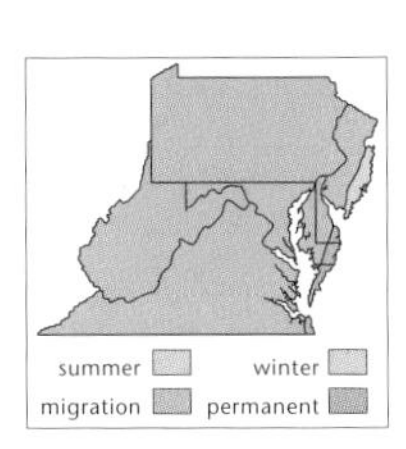

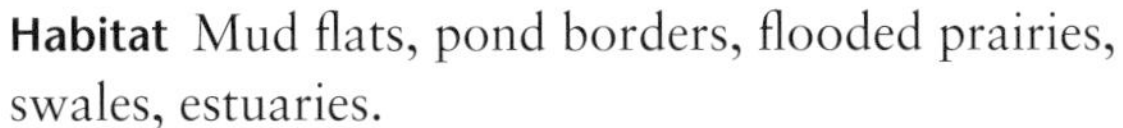

Habitat Mud flats, pond borders, flooded prairies, swales, estuaries.

Abundance and Distribution Common to uncommon transient (Apr–May, Aug–Sep) throughout, more numerous in fall; uncommon to rare summer resident (nonbreeding) in coastal areas, and winter resident (Oct–Mar) north to Delaware.

Where to Find Buzzard Swamp Wildlife Management Area, Pennsylvania; Flat Top Lake, West Virginia; South Cape May Meadows, New Jersey.

Range Breeds northern Canada and Alaska; winters from Atlantic (South Carolina) and Pacific (southern California) coasts of United States south to southern South America, West Indies.

Solitary Sandpiper

Tringa solitaria

(L-9 W-17)

Dark gray back and wings with white spotting; heavily streaked with grayish brown on head and breast; white eyering; long, greenish legs; tail dark in center, barred white on outer portions. *Winter:* Streaking on head and breast reduced or faint.

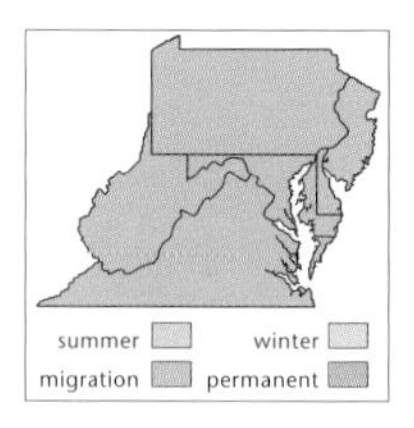

Habits Often bobs while foraging.
Voice Call—a high-pitched "tseet-eet."
Similar Species Spotted Sandpiper is found in similar habitats; however, it bobs almost continuously, shows white on wings in flight (no white on wings in Solitary), has a pale eyebrow (Solitary has eyering) and no white spotting on folded wings; tail barring is much more prominent in the Solitary. Stilt Sandpiper and yellowlegs show white rump in flight (dark in Solitary).
Habitat Ponds, rivers, lakes.
Abundance and Distribution Common to uncommon transient (Aug–Sep, Apr–May) throughout.
Where to Find Meadowside Nature Center, Maryland; Lake Anna, Virginia; Ridley Creek State Park, Pennsylvania.
Range Breeds across Arctic and boreal North America; winters southern Atlantic (Georgia, Florida) and Gulf coasts of United States south to southern South America, West Indies.

Willet

Catoptrophorus semipalmatus

(L-15 W-27)

Large, heavy-bodied, long-billed shorebird; mottled grayish brown above and below; whitish

belly; broad, white stripe on wing is conspicuous in flight. *Winter:* Gray above, whitish below.

Voice Shrill "will will-it," repeated.
Similar Species Large size, white wing stripes, and flight call separate this from other shorebirds.
Habitat Estuaries, beaches, coastal ponds, marshes, mud flats.

Abundance and Distribution Common summer resident* (Apr–Oct) along the immediate coast; rare in winter.
Where to Find Brigantine National Wildlife Refuge, New Jersey; Assateague Island National Seashore, Maryland; Chincoteague National Wildlife Refuge, Virginia.
Range Breeds southwestern and south central Canada and northwestern and north central United States, along Atlantic and Gulf coasts (Nova Scotia to Texas), West Indies; winters coastal North America (California and Virginia) south to northern half of South America, West Indies.

Spotted Sandpiper

Actitis macularia

(L-8 W-13)

Grayish brown above, white with dark spots below; white eyebrow; bill pinkish with dark tip; legs pinkish. *Winter:* Lacks spots; white below extending toward back at shoulder; grayish smear on side of breast; legs yellowish.

Habits Teeters almost continually while foraging; flight peculiar, with stiff-winged bursts and brief glides.

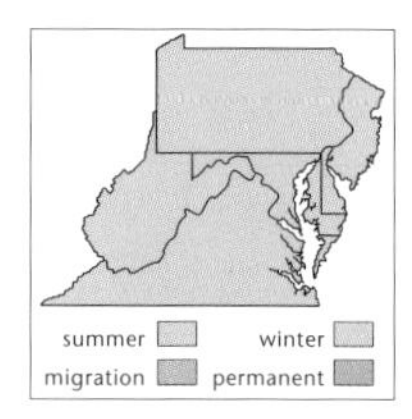

Voice "Peet weet."
Similar Species The Solitary Sandpiper is found in similar habitats; however, it pauses between bobs, shows no white on wings in flight (white on wings in Spotted), has eyering and no pale eyebrow (Spotted has no eyering and a pale eyebrow), and has white spotting on folded wings (absent in Spotted); tail barring is much less prominent in the Spotted than the Solitary.
Habitat Streams, ponds, rivers, lakes, beaches.
Abundance and Distribution Common transient (Aug–Sep, May) and uncommon and local summer resident* (Jun–Aug) throughout; rare to casual in winter along the coast.
Where to Find Beech Fork State Park, West Virginia; Presque Isle State Park, Pennsylvania; C&O Canal, Maryland.
Range Breeds throughout temperate and boreal North America; winters from the southern United States south to northern Argentina, West Indies.

Upland Sandpiper
Bartramia longicauda
(L-12 W-22)

The shape of this bird is unique—a relatively large, heavy body with long legs, long neck, small head, and short bill. Plumage is light brown with dark brown streaks on back and wings; buff head, neck, and breast mottled with brown; large brown eye.

Voice Call—"kip-ip," repeated.
Similar Species The Buff-breasted Sandpiper is found in similar habitats; however, it has clear, unstreaked, buffy underparts contrasting with white wing linings in flight.

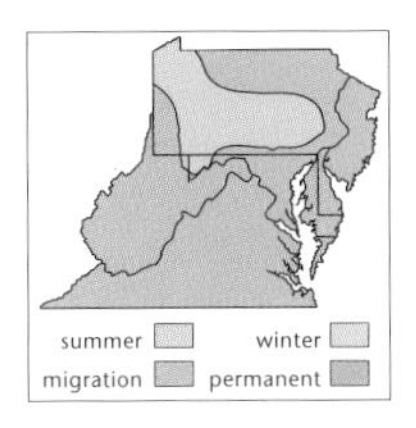

Habitat Prairie, pastures, plowed fields.
Abundance and Distribution Uncommon fall transient (Aug–Sep), rare spring transient (Apr–May), and rare and local summer resident* (May–Aug), mainly in Pennsylvania, although there are scattered, recent breeding records from elsewhere throughout the region.
Where to Find Lucketts, Hwy 661, Virginia; Erie National Wildlife Refuge, Pennsylvania; New Castle County Airport, Delaware.
Range Breeds across northeast and north central United States and in Great Plains of Canada; winters South America.

Whimbrel

Numenius phaeopus
(L-18 W-33)

Mottled gray and brown above; grayish below with dark flecks; crown striped with dark brown and gray.

Voice Call—"kee kee kee kee."
Similar Species Long-billed Curlew is larger, buffy, has a very long, decurved bill, and is not striped prominently on crown.
Habitat Tall grasslands, estuaries, oyster reefs.

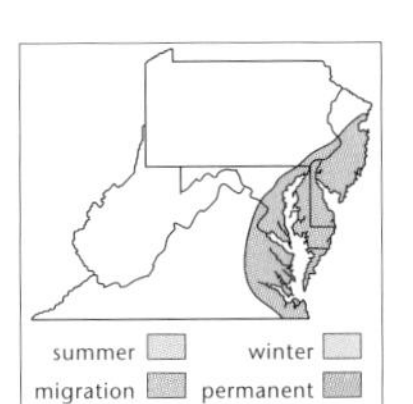

Abundance and Distribution Common transient (Jul–Sep, Apr–May) along the immediate coast; rare in summer and winter.
Where to Find Barnegat National Wildlife Refuge, New Jersey; Assateague Island National Seashore, Maryland; Chincoteague National Wildlife Refuge, Virginia.
Range Breeds in high Arctic of Old and New World; winters in subtropical and tropical regions.

Long-billed Curlew

Numenius americanus

(L-23 W-38)

A large, heavy-bodied bird with very long, decurved bill; mottled brown and buff above; uniformly buffy below; rusty wing linings visible in flight.

Voice Call—a strident "per-leee," repeated.
Similar Species Long-billed Curlew is larger than the Whimbrel, buffy, has a very long, decurved bill, and is not striped prominently on crown (as is the Whimbrel).
Habitat Grasslands, pastures, lawns, golf courses.
Abundance and Distribution Rare to casual visitor along the coast of Virginia, mainly fall and winter.
Where to Find Chincoteague National Wildlife Refuge, Virginia.
Range Breeds in prairie regions of western United States and southwestern Canada; winters from central and southern California, Arizona, Texas, and Louisiana south through southern Mexico.

Hudsonian Godwit

Limosa haemastica

(L-16 W-27)

Mottled dark and light brown and white above; rufous below; white eyebrow; long, upturned bill russet at base, dark at tip; white rump; dark wings with white wing stripe visible in flight. *Winter:* Gray above, whitish below, with white eyebrow.

Voice Call—"quee quee."
Similar Species The Marbled Godwit is mottled brown and buff at all seasons (not gray or cinnamon).

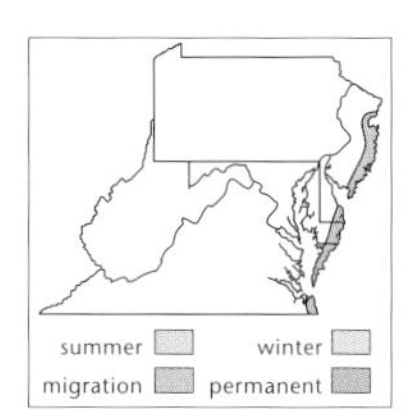

Habitat Ponds, swales, marshes, beaches, mud flats.
Abundance and Distribution Rare fall transient (Aug–Oct) along the immediate coast.
Where to Find Chincoteague National Wildlife Refuge, Virginia; Bombay Hook National Wildlife Refuge, Delaware.
Range Breeds in northern North America; winters in southern South America.

Marbled Godwit

Limosa fedoa

(L-18 W-31)

Mottled buff and dark brown above; buffy below barred with brown; long, upturned bill pink at base, dark at tip; rusty wing linings. *Winter:* Barring on belly faint or lacking.

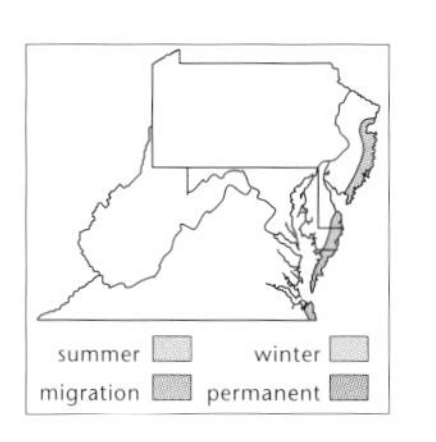

Voice Call—"koo-wik" repeated.
Similar Species The Marbled Godwit is mottled buff and brown at all seasons (not gray and white like the rare Hudsonian Godwit); Marbled Godwit rump is mottled buff (not white like Hudsonian), and underwing is cinnamon (not dark gray like Hudsonian).
Habitat Bay shores, mud flats, estuaries, flooded grasslands.
Abundance and Distribution Rare transient (Aug–Oct, May), mainly in fall along the coast.
Where to Find Chincoteague National Wildlife Refuge, Virginia; Bombay Hook National Wildlife Refuge, Delaware; Back Bay National Wildlife Refuge, Virginia.
Range Breeds northern Great Plains of Canada and extreme northern United States; winters coastal United States from California and South Carolina south to northern South America.

Ruddy Turnstone

Arenaria interpres

(L-10 W-19)

A plump bird with slightly upturned bill; rufous above, white below with black breast; black-and-white facial pattern; 2 white stripes on wings; white band across tail; orange legs. *Winter:* Grayish above; white below, with varying amounts of black on breast and face.

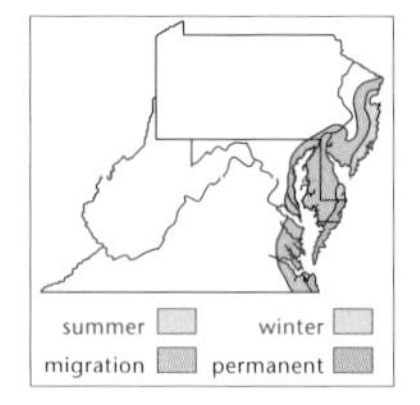

Voice Calls—a single, whistled "tew"; also a low chatter.
Habitat Beaches, mud flats.
Abundance and Distribution Common summer resident (Apr–Sep, nonbreeding) along coast; common to uncommon transient (Jul–Sep, Apr–May) along the Coastal Plain; uncommon to rare winter resident in coastal regions north to Delaware; rare to casual transient elsewhere in the region.
Where to Find South Cape May Meadows, New Jersey; Indian River Inlet, Maryland; Chesapeake Bay Bridge-Tunnel, Virginia.
Range Breeds in high Arctic of Old and New World; winters in south temperate and tropical regions.

Red Knot

Calidris canutus

(L-11 W-21)

A plump bird, mottled brown and russet above and on crown; rusty below. *Winter:* Pale gray above; grayish breast; whitish belly; white eyebrow; white, finely barred rump.

Voice Call—a hoarse "kew kew."

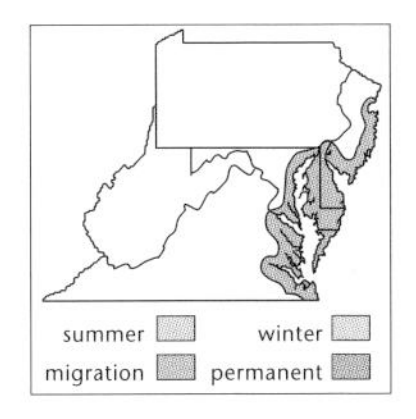

Similar Species Dowitchers are white up the back; Sanderling is smaller and lacks white rump; other peeps of similar plumage are sparrow sized.
Habitat Beaches, bay shorelines, mud flats.
Abundance and Distribution Common transient (Jul–Sep, May) along the immediate coast; rare at other seasons or elsewhere in the region.
Where to Find Delaware Bay, New Jersey; Virginia Coast Reserve, Virginia; Indian River Inlet, Maryland.
Range Breeds in Old and New World high Arctic; winters in temperate and tropical coastal regions.

Sanderling

Calidris alba

(L-8 W-16)

A chunky, feisty peep of the beaches; dappled brown, white, and black on back and crown; rusty flecked with white on face and breast; short, black legs and bill; lacks hind toe (hallux); white wing stripe. *Winter:* Mottled gray and white above; white below; white eyebrow; grayish shoulder patch extends to side of breast.

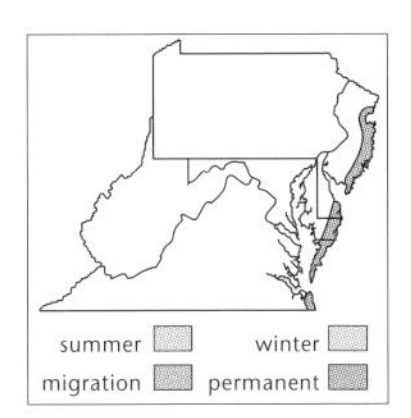

Habits Spends considerable time in chases and fights with other Sanderlings.
Voice Call—“kwit.”
Similar Species Other winter peeps are more or less streaked with grayish on throat and breast (throat and breast are white on winter Sanderling); Sanderling has palest back.
Habitat Beaches.
Abundance and Distribution Common transient and winter resident (Aug–May) and uncommon summer resident (Jun–Jul, nonbreeding) along the coast; rare to casual inland.

Where to Find Cape Henlopen, Delaware; Barnegat National Wildlife Refuge, New Jersey; Back Bay National Wildlife Refuge, Virginia.
Range Breeds in high Arctic of Old and New World; winters on temperate and tropical beaches.

Semipalmated Sandpiper

Calidris pusilla

(L-6 W-12)

Dark brown, russet, and white above; finely streaked brown and white on head and breast; pale eyebrow with some rust on crown and cheek; white belly; partial webbing between toes (hence "semipalmated"). *Winter:* Grayish above with grayish wash on head and breast; white belly.

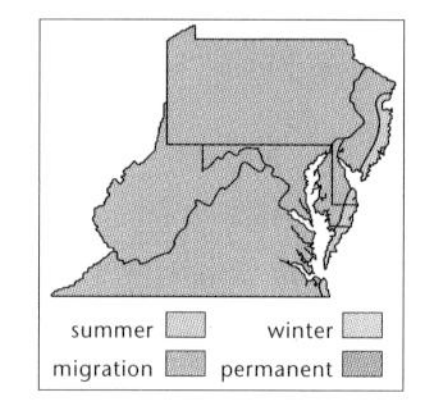

Voice Call—"chit" or "chek."
Similar Species Winter-plumaged Western Sandpiper is essentially identical to winter Semipalmated. They can be separated by voice ("chek" in Semipalmated, "zheet" in Western). The bill of the Western averages longer than that of Semipalmated, and in longer-billed birds there is a noticeable droop at the tip. However, there is considerable overlap in this and all other characters used to separate the winter-plumaged birds in the field, except voice (Phillips 1975).
Habitat Mud flats, ponds, lakes.
Abundance and Distribution Common to uncommon transient (Jul–Oct, Apr–May) throughout; common to uncommon in summer (Jun–Jul, nonbreeding) along the immediate coast.
Where to Find Flat Top Lake, West Virginia; Prince Gallitzin State Park, Pennsylvania; Sandy Point State Park, Maryland.
Range Breeds in high Arctic of North America;

winters along both coasts of Middle and South America to Paraguay (east) and northern Chile (west), West Indies.

Western Sandpiper

Calidris mauri

(L-6 W-12)

Dark brown, russet, and white above; finely streaked brown and white on head and breast; pale eyebrow; partial webbing on toes; similar to Semipalmated Sandpiper but whiter on back and with distinct rusty cheek and supraorbital patches. *Winter:* Grayish above with grayish wash on head and breast; white belly.

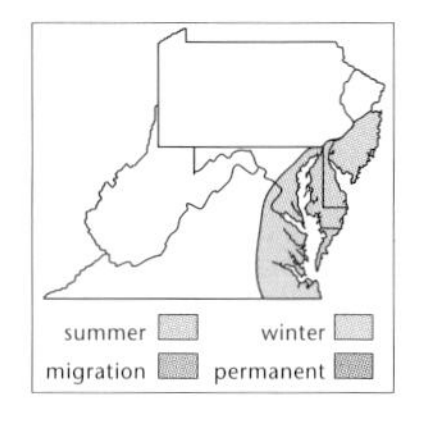

Voice Call—"zheet."

Similar Species Winter-plumaged Western Sandpiper is essentially identical to winter Semipalmated. They can be separated by voice ("chek" in Semipalmated, "zheet" in Western). The bill of the Western averages longer than that of Semipalmated, and in longer-billed birds there is a noticeable droop at the tip. However, there is considerable overlap in this and all other characters used to separate the winter-plumaged birds in the field, except voice (Phillips 1975).

Habitat Beaches, mudflats, ponds, lakes, estuaries, swales.

Abundance and Distribution Common fall transient (Jul–Oct), uncommon winter resident (Nov–Mar), and rare spring transient (Apr–May) along the Coastal Plain; rare to casual transient inland.

Where to Find South Cape May Meadows, New Jersey; Bombay Hook National Wildlife Refuge, Delaware; Chincoteague National Wildlife Refuge, Virginia.

Range Breeds in high Arctic of northern Alaska and northeastern Siberia; winters along both coasts of United States from California and North Carolina south to northern South America, West Indies.

Least Sandpiper

Calidris minutilla

(L-6 W-12)

Dark brown and buff on back and wings; head and breast streaked with brown and white; relatively short, thin bill; white belly; yellowish legs (sometimes brown with caked mud). *Winter:* Brownish gray above; buffy wash on breast.

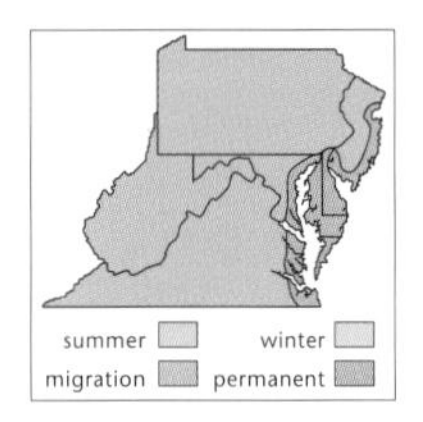

Voice Call—"preep."
Similar Species Yellowish legs are distinctive (other small peeps have dark legs), as is call note.
Habitat Ponds, lakes, swales, mud flats.
Abundance and Distribution Common to uncommon transient (Jul–Sep, Apr–May) throughout; uncommon to rare resident (nonbreeding) along the coast.
Where to Find Lily Pons Water Gardens, Maryland; Roanoke Sewage Treatment Plant, Virginia; John Heinz National Wildlife Refuge, Pennsylvania.
Range Breeds across northern North America; winters from coastal (Oregon, North Carolina) and southern United States south to northern half of South America, West Indies.

White-rumped Sandpiper

Calidris fuscicollis

(L-8 W-15)

Dark and tan on back and wings; finely streaked brown and white on head, breast, and flanks; white rump; dark legs; wings extend beyond tail. *Winter:* Gray above, white below with gray wash on breast and white eyebrow.

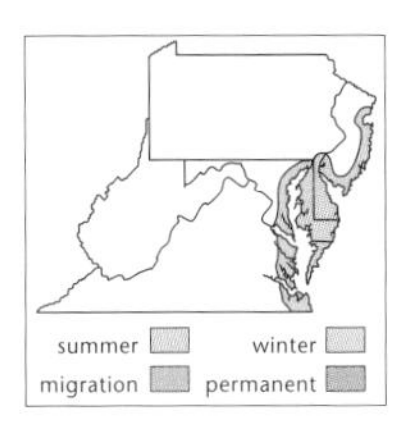

Voice Call—"teet."
Similar Species The rarer Curlew Sandpiper is also white rumped but has long, decurved bill.
Habitat Swales, mud flats, lakes, ponds.
Abundance and Distribution Uncommon transient (Aug–Oct, May–Jun) and uncommon to rare in summer (Jun–Aug, nonbreeding) along the coast; rare to casual elsewhere.
Where to Find Brigantine National Wildlife Refuge, New Jersey; Bombay Hook National Wildlife Refuge, Delaware; Chincoteague National Wildlife Refuge, Virginia.
Range Breeds in high Arctic of North America; winters in South America east of the Andes.

Baird's Sandpiper

Calidris bairdii

(L-8 W-15)

A medium-sized sandpiper, dark brown and buff on back and wings; buffy head and breast flecked with brown; wings extend beyond tail.

Voice Call—"chureep."
Similar Species Brownish (rather than grayish) winter plumage makes this bird look like a large Least Sandpiper. Long wings, black bill and legs,

and dark rump separate it from the Least and other peeps.

Habitat Ponds, lakes, swales.

Abundance and Distribution Rare fall transient (Aug–Oct), mainly along the coast.

Where to Find Assateague Island National Seashore, Maryland; Chincoteague National Wildlife Refuge, Virginia.

Range Breeds in high Arctic of North America and northeastern Siberia; winters in western and southern South America.

Pectoral Sandpiper

Calidris melanotos

(L-9 W-17)

Dark brown, tan, and white on back and wings; head and breast whitish, densely flecked with brown; belly white, distinct dividing line between mottled brown breast and white belly; dark bill; yellow legs.

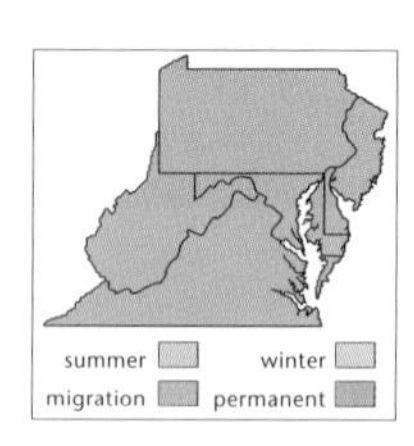

Voice Call—"prip."

Similar Species No other sandpiper shows sharp contrast of breast and belly markings.

Habitat Wet prairies, ponds, lakes, swales, agricultural fields.

Abundance and Distribution Common to uncommon transient (Apr–May, Aug–Oct) throughout; more numerous in fall; rare to casual in winter along the coast.

Where to Find Cheat Lake, West Virginia; Nockamixon State Park, Pennsylvania; Lily Pons Water Gardens, Maryland.

Range Breeds in the high Arctic of northern North America; winters in southern half of South America.

Purple Sandpiper

Calidris maritima

(L-9 W-15)

Dark brown back and wings; head finely streaked with brown and white; belly white; breast and flanks whitish spotted with brown; bill yellow-orange at base, dark at tip; yellow-orange legs; white eyering. *Winter:* Back marked with gray and buff; head and breast slate gray with gray streaking on flanks.

Voice Call—"we-it."
Habitat Rocky shores and jetties.
Abundance and Distribution Uncommon to rare and local winter resident (Oct–May) along the immediate coast.
Where to Find Chesapeake Bay Bridge-Tunnel, Virginia; Cape May Point, New Jersey; Assateague Island National Seashore, Maryland.
Range Breeds in Arctic of northeastern North America, Greenland, Iceland, northern Scandinavia, and Siberia; winters along Atlantic Coast of North America from New Brunswick to Virginia; also shores of North and Baltic seas in Old World.

Dunlin

Calidris alpina

(L-9 W-16)

Rusty and dark brown on back and wings; white streaked with black on neck and breast; crown streaked with russet and black; large, black smudge on belly; long bill droops at tip. *Winter:* Dark gray back; head and breast pale gray with some streaking; white eyebrow; white belly.

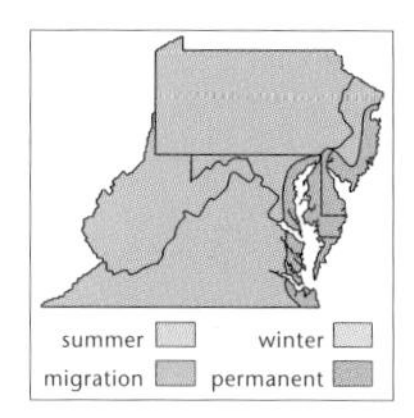

Voice Call—a hoarse "zheet."
Similar Species Dark gray back and long, drooping bill distinguish the winter Dunlin from other gray-breasted peeps.
Habitat Ponds, lakes, beaches, mud flats.
Abundance and Distribution Common (along coast) to uncommon or rare (inland) transient (Oct, Apr–May); common in winter (Nov–Mar) along coast, uncommon to rare in summer (Jun–Sep, nonbreeding).
Where to Find Bombay Hook National Wildlife Refuge, Delaware; Chincoteague National Wildlife Refuge, Virginia; Hereford Inlet, New Jersey.
Range Breeds in Arctic of Old and New World; winters in temperate and northern tropical regions.

Curlew Sandpiper

Calidris ferruginea
(L-8 W-16)

Rusty above and below with long, decurved bill. *Winter:* Dark gray back; head and breast pale gray; white eyebrow; white belly.

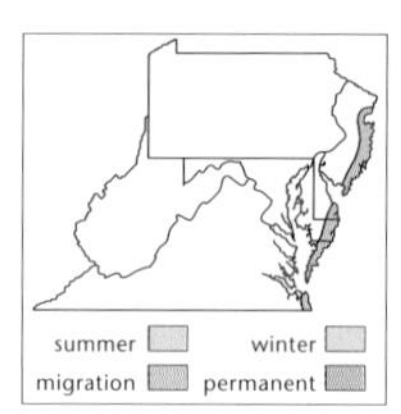

Voice Call—"si-rip."
Similar Species Winter Dunlin is similar, but winter Curlew Sandpiper has white rump.
Habitat Mud flats, beaches.
Abundance and Distribution Rare transient (Jul–Oct, May) along the immediate coast.
Where to Find Bombay Hook National Wildlife Refuge, Delaware; Chincoteague National Wildlife Refuge, Virginia; Little Creek Wildlife Area, Delaware.
Range Breeds in Old World Arctic; winters in Old World temperate and tropical regions.

Stilt Sandpiper

Calidris himantopus

(L-9 W-17)

A trim, long-necked, long-legged sandpiper, dark brown and buff on back and wings; crown streaked with dark brown and white; white eyebrow; chestnut pre- and postorbital stripe; white streaked with brown on neck; white barred with brown on breast and belly; long, yellow-green legs; white rump; long, straight bill (twice length of head). *Winter:* Gray and white mottling on back; gray head with prominent eyebrow; gray throat and breast; whitish belly with gray flecks.

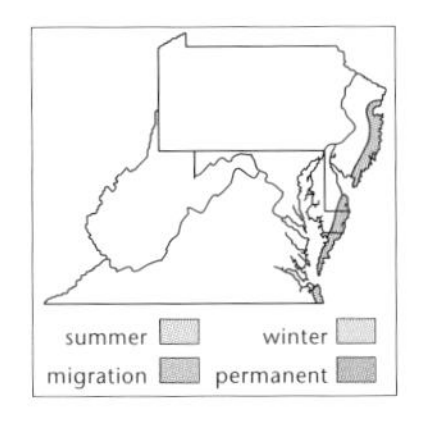

Habits Often feeds with rapid up-and-down motion in breast-deep water.
Voice Call—a low "whurp."
Similar Species Winter Lesser Yellowlegs has barred (not gray) tail; winter dowitchers have longer bill (3 times length of head) and flanks with dark bars or spots and show white on lower back in flight.
Habitat Mud flats, flooded pastures, lakes, ponds.
Abundance and Distribution Uncommon fall transient (Jul–Oct) and rare spring transient (Apr–May) along the coast; rare to casual elsewhere.
Where to Find South Cape May Meadows, New Jersey; Assateague Island National Seashore, Maryland; Craney Island Landfill, Virginia.
Range Breeds in Arctic of north central Canada; winters in central South America.

Buff-breasted Sandpiper

Tryngites subruficollis

(L-8 W-17)

Dark brown feathering edged in buff or whitish on back; buffy head and underparts, spotted with brown on breast; streaked crown; large, dark eye; pale legs; wing linings appear white in flight.

Habits Often seen in small flocks that twist and turn erratically in flight over freshly plowed fields.
Voice Call—a low trill.
Similar Species The Buff-breasted Sandpiper and Upland Sandpiper are found in similar habitats; however, the buffbreasted has unstreaked, buffy underparts contrasting with white wing linings in flight. The Upland Sandpiper has streaked underparts and buffy, streaked wing linings.
Habitat Grassland, stubble fields, overgrazed pastures.
Abundance and Distribution Rare fall transient along the immediate coast (Aug–Oct), casual in spring and elsewhere in the region.
Where to Find Chincoteague National Wildlife Refuge, Virginia; Bombay Hook National Wildlife Refuge, Delaware; South Cape May Meadows, New Jersey.
Range Breeds in Arctic of northern Alaska and northwestern Canada; winters in Paraguay, Uruguay, and northern Argentina.

Ruff

Philomachus pugnax

(Male: L-12 W-24; Female: L-10 W-21)

Dark brown, tan, and white on back; spectacular collar of ruffled feathers around neck and breast—

Sandpipers

black, chestnut, or white; black below; bill and legs yellowish; rump white except for dark central line. *Female* (called a Reeve): Dark gray above and below; white belly; white rump with dark central line; yellow legs and base of bill. *Winter*: Dark gray and white on back; grayish head and breast; whitish belly; bill yellowish at base; legs yellowish.

Voice Call—"tew-ee."
Similar Species Lesser Yellowlegs has white rump with no dark, central line.
Habitat Beaches, estuaries, mud flats.
Abundance and Distribution Rare transient (Aug–Oct, Apr–Jun) along the immediate coast; casual elsewhere and at other seasons.
Where to Find Chincoteague National Wildlife Refuge, Virginia; Bombay Hook National Wildlife Refuge, Delaware; South Cape May Meadows, New Jersey.
Range Breeds in Old World Arctic; winters in Old World temperate and tropical regions.

Short-billed Dowitcher
Limnodromus griseus
(L-11 W-19)

A rather squat, long-billed sandpiper (bill is 3 times length of head); mottled black, white, and buff on back, wings, and crown; cinnamon buff on neck and underparts, barred and spotted with dark brown; belly white; white line above eye with dark line through eye; white tail barred with black; white on rump extending in a wedge up back; bill dark; legs greenish. *Winter:* Grayish crown, nape, and back; white line above eye and dark line through eye; grayish neck and breast fading to whitish on belly; flanks and undertail coverts spotted and barred with dark brown. *Juvenile:* Patterned like adult but paler brown above and on head and with buffy neck and breast.

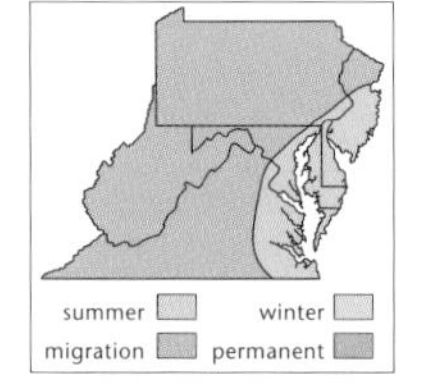

Habits Often feeds with rapid up-and-down motion, breast-deep in water, usually in small flocks.
Voice Call—"tu tu" or "tu tu tu," repeated.
Similar Species Short-billed has a 2- or 3-note whistle ("tu tu tu"), while Long-billed has a thin "eek."
Habitat Mud flats, flooded pastures, ponds, lakes, estuaries.
Abundance and Distribution Common (coast) to rare or casual (inland) transient (Jul–Oct, Apr–May); common in summer, uncommon to rare in winter along the coast.
Where to Find Assateague Island National Seashore, Maryland; Brigantine National Wildlife Refuge, New Jersey; Chincoteague National Wildlife Refuge, Virginia.
Range Breeds across central Canada and southern Alaska; winters coastal United States (California, South Carolina) south to northern South America, West Indies.

Long-billed Dowitcher

Limnodromus scolopaceus
(L-11 W-19)

A squat, long-billed sandpiper (bill is 3 times length of head); mottled black, white, and buff on back, wings, and crown; cinnamon buff on neck and underparts, barred and spotted with dark brown; belly white; white line above eye with dark line through eye; white tail barred with black;

white on rump extending in a wedge up back; bill dark; legs greenish. *Winter:* Grayish crown, nape, and back; white line above eye and dark line through eye; grayish neck and breast fading to whitish on belly; flanks and undertail coverts spotted and barred with dark brown. *Juvenile:* Patterned like adult in winter.

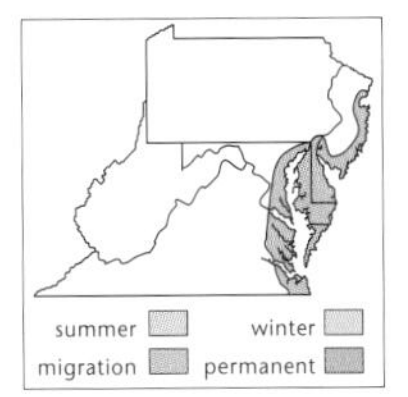

Voice Call—a thin "eek," often repeated.
Similar Species The Long-billed Dowitcher has a thin "eek" call, whereas the Short-billed has a 2- or 3-note whistle ("tu tu tu").
Habitat Mud flats, flooded pastures.
Abundance and Distribution Uncommon fall transient (Aug–Oct), rare winter resident (Nov–Mar) and spring transient (Apr–May) along the coast; casual inland.
Where to Find Assateague Island National Seashore, Maryland; Brigantine National Wildlife Refuge, New Jersey; Chincoteague National Wildlife Refuge, Virginia.
Range Breeds in coastal Alaska, northwestern Canada, and northeastern Siberia; winters southern United States to Guatemala.

Common Snipe

Gallinago gallinago

(L-11 W-17)

Dark brown back with white stripes; crown striped with dark brown and gray; neck and breast streaked grayish brown and white; whitish belly; tan rump; tail banded with rust and black; very long bill (3 times length of head).

Habits Normally solitary, retiring, and wary; gives an explosive "zhrrt" when flushed and flies in erratic swoops and dips.

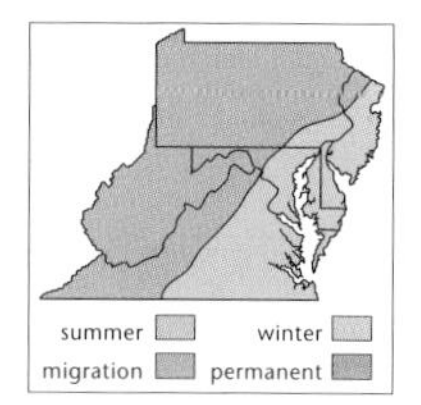

Voice Song given on wing from high in the air, "cheek cheek cheek cheek," plus a winnowing sound (made by air passing through the tail feathers).

Habitat Marshes, bogs, flooded pastures, wet ditches.

Abundance and Distribution Common to uncommon transient (Sep–May) throughout; uncommon to rare winter resident (Oct–Mar) in the Coastal Plain, Piedmont, and lowland, open marshes elsewhere in the region; uncommon to rare and local summer resident* (Apr–Sep) in northern Pennsylvania, West Virginia's Canaan Valley, and northwestern New Jersey.

Where to Find Pymatuning Wildlife Management Area, Pennsylvania (breeding); Dyke Marsh, Virginia; Blackwater National Wildlife Refuge, Maryland.

Range Breeds in north temperate and boreal regions of Old and New World; winters in south temperate and tropical regions.

American Woodcock

Scolopax minor

(L-11 W-17)

A fat, short-legged, long-billed bird with no apparent neck and hardly any tail; dark and grayish brown above; gray-brown neck and breast becoming rusty on belly and flanks; dark brown crown with thin, transverse gray stripes; large, brown eye; overall impression of flushed bird is cinnamon buff because of underparts and wing linings.

Habits Nocturnal; feeds, sings, and displays at night.

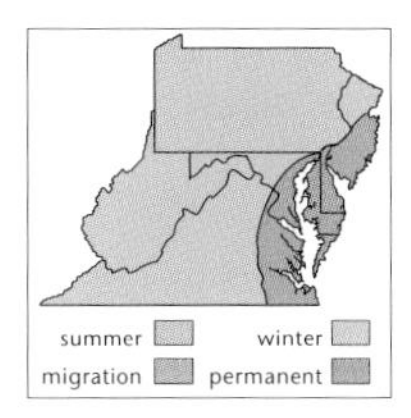

Voice A buzzy, nasal "peent," also a flight song composed of a long trill followed by various "chik," "cheek," and "cherk" sounds made by air passing through the bird's feathers as it descends; wings make a whistling sound when bird is flushed.
Habitat Moist woodlands, swamp borders.
Abundance and Distribution Common to uncommon transient and summer resident* (Feb–Nov) throughout; uncommon to rare and local winter resident (Dec–Jan) along the coast.
Where to Find Great Swamp National Wildlife Refuge, New Jersey; Blackwater National Wildlife Refuge, Maryland; Eastern Shore National Wildlife Refuge, Virginia.
Range Breeds across eastern half of the United States and southeastern Canada; winters from southern half of the breeding range into south Texas and southern Florida.

Wilson's Phalarope
Phalaropus tricolor
(L-9 W-17)

Female: A trim, thin-billed, long-necked bird; chestnut and gray on the back; gray crown and nape; black extending from eye down side of neck; white chin; chestnut throat; white breast and belly. *Male:* Similar but paler. *Winter:* Grayish brown above, white below; gray head with white eyebrow; white chin; rusty throat.

Habits Unlike other shorebirds, phalaropes often swim, whirling in tight circles, while foraging.
Voice Call—a nasal "wak," repeated.
Similar Species Other phalaropes have white foreheads and black-and-white facial pattern in winter.

Habitat Ponds, lakes, mud flats, flooded pastures.
Abundance and Distribution Uncommon fall transient (Aug–Sep) and rare spring transient (May) along the coast.
Where to Find Back Bay National Wildlife Refuge, Virginia; Little Creek Wildlife Area, Delaware; Chincoteague Island National Wildlife Refuge, Virginia.
Range Breeds in western United States, southwestern Canada, and Great Lakes region; winters in western and southern South America.

Red-necked Phalarope
Phalaropus lobatus
(L-8 W-14)

Female: Black with rusty striping on back and flanks; white below; neck and nape orange; head black; chin white. *Male:* Similar in pattern but paler. *Winter:* Black back with white striping; white below; white forehead; gray-black crown; white eyebrow; black postorbital stripe.

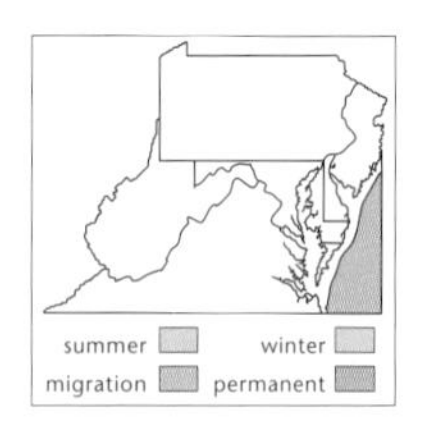

Habits Unlike other shorebirds, phalaropes often swim, whirling in tight circles, while foraging.
Voice Call—a hoarse "chik" or "ker-chik."
Similar Species Winter Red Phalarope has a pearl gray, unstreaked back; Red-necked is almost black on back, striped with white.
Habitat Mainly pelagic in winter but also found occasionally at marshes, ponds, or lakes.
Abundance and Distribution Uncommon to rare transient (Sep–Oct, May), mainly offshore.
Where to Find Pelagic birding trips.
Range Breeds in Arctic regions of the Old and New World; winters at sea in southern Pacific and Indian oceans.

Red Phalarope

Phalaropus fulicaria

(L-9 W-17)

Female: Black back striped with buff; neck and underparts rusty orange; black cap; white cheek; relatively short, heavy bill is yellow at base, black at tip. *Male:* Similar in pattern but paler. *Winter:* Uniform pearl gray on back; white below; white forehead and front half of crown; gray rear half of crown and nape line; dark postorbital stripe; bill blackish.

Habits Unlike other shorebirds, phalaropes often swim, whirling in tight circles, while foraging.

Similar Species Winter Red Phalarope has a pearl gray, unstreaked back; Red-necked is almost black on back, striped with white.

Habitat Mainly pelagic in winter but also found occasionally on beaches and mud flats.

Abundance and Distribution Rare transient and winter visitor, mainly offshore (Sep–May).

Where to Find Pelagic birding trips.

Range Breeds in Old and New World Arctic; winters in southern Pacific and Atlantic oceans.

Jaegers, Skuas, Gulls, Terns, and Skimmers

(Family Laridae)

This family is subdivided into 4 distinct groupings: jaegers and skuas, gulls, terns, and skimmers. The jaegers and skuas are dark, heavy-bodied, Herring Gull–sized birds with hooked, predatory beaks and protruding central tail feathers. The wings are long, pointed, and sharply angled at the wrist.

Three to 4 years are usually spent in subadult and immature plumages before full adult plumage, with the characteristic central tail feathers, is achieved. Until that time, the central tail feathers are only slightly or moderately protruding. These are the hawks of the oceans. Jaegers and skuas feed on stolen fish and the eggs and young of other seabirds. Characteristic behavior is chasing gulls and terns in attempts to rob them of captured fish. Jaegers and skuas are pelagic except during the breeding season. The most probable locality for locating a jaeger is along the immediate coast, on offshore islands, or at sea.

Gulls are variable in size; most are various patterns of gray and white as adults, although several species are black headed. Like jaegers, they have several intermediate plumages between hatching and attainment of full adult plumage at 2 to 4 years of age. Only a few of these plumages are described here: adult breeding plumage, winter plumage (where applicable), and a "typical" immature plumage.

Terns are thin-billed, trim-bodied seabirds. Most species forage by plummeting into the water from a height of 10 to 30 feet to catch fish.

Skimmers are long-winged birds with long bills shaped like a straight razor, in which the lower mandible protrudes beyond the upper. They feed by flying above the water with the lower mandible dipped just below the surface to capture fish.

Identification to species of the immatures for some groups within the Laridae can be difficult or impossible in the field. Harrison (1991) and Grant (1982) provide detailed descriptions and keys for advanced students of these birds.

Great Skua

Stercorarius skua

(L-23 W-59)

A thick-bodied, Herring Gull–sized bird; dark brown throughout with an overall rusty tint and white patch at base of primaries; light ruff at nape; thick scapulars give the bird a definite "hunched" appearance in flight. *Immature:* Generally paler in color with some streaking on breast.

Habits Chases other seabirds to rob food.
Habitat Pelagic, beaches.
Abundance and Distribution Rare offshore in winter (Nov–Mar).
Where to Find Pelagic birding trips.
Range Breeds in Iceland, Faeroe Islands, Orkney Islands, and mainland northern Scotland. Wanders during nonbreeding season over the Norwegian Sea and North Atlantic Ocean.

Pomarine Jaeger

Stercorarius pomarinus

(L-22 W-48)

Protruding central tail feathers are twisted; dark brown above; white collar tinged with yellow; whitish below with dark breast band and undertail coverts. *Dark Phase:* Completely dark brown with white at base of primaries; cheeks tinged with yellow; dark cap. *Immature:* Barred brown or brown and white below.

Similar Species The Pomarine Jaeger has twisted tail feathers; Parasitic Jaeger has pointed tail feathers; Long-tailed Jaeger has very long central tail feathers (half of body length).

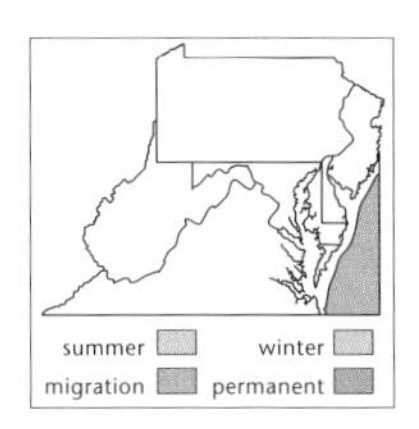

Habitat Pelagic, beaches.
Abundance and Distribution Uncommon to rare transient (Sep–Nov, May) and rare visitor at other times of the year offshore.
Where to Find Pelagic birding trips; Chesapeake Bay Bridge-Tunnel; coastal beaches.
Range Breeds in Old and New World Arctic; winters in temperate and tropical oceans.

Parasitic Jaeger

Stercorarius parasiticus
(L-19 W-42)

Dark brown above; white collar and white below with faint brown breast band; pointed, protruding central tail feathers; wings dark but whitish at base of primaries. *Dark Phase:* Completely dark brown. *Immature:* Dark brown above; barred reddish brown or brown and white below.

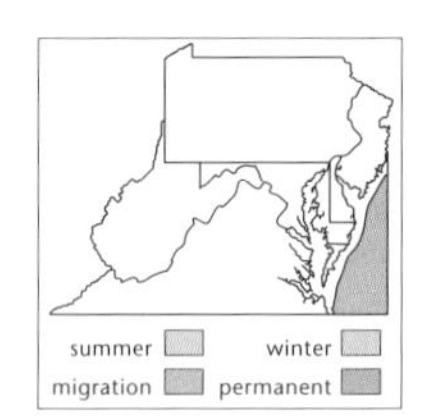

Similar Species Parasitic Jaeger has pointed tail feathers; the Pomarine Jaeger has twisted tail feathers; Long-tailed Jaeger has very long central tail feathers (half of body length).
Habitat Pelagic; beaches.
Abundance and Distribution Uncommon transient (Sep–Nov, May) and rare summer and winter visitor, mainly offshore but occasionally in lower Chesapeake Bay.
Where to Find Pelagic birding trips; Chesapeake Bay Bridge-Tunnel; coastal beaches.
Range Breeds in Old and New World Arctic; winters in temperate and tropical oceans.

Long-tailed Jaeger

Stercorarius longicaudus

(L-32 W-40)

Gray-brown above; white collar and white below with gray-brown under-tail coverts; very long central tail feathers. *Dark Phase:* Completely dark gray-brown (extremely rare). *Immature:* Dark throat; white breast and belly.

Similar Species Parasitic Jaeger has pointed tail feathers; Pomarine Jaeger has twisted tail feathers; Long-tailed Jaeger has very long central tail feathers (half of body length).
Habitat Pelagic, beaches.
Abundance and Distribution Rare transient (Sep–Nov, Apr–May) offshore.
Where to Find Pelagic birding trips.
Range Breeds in Old and New World Arctic; winters mainly in tropical oceans.

Laughing Gull

Larus atricilla

(L-17 W-40)

Black head; gray back and wings with black wingtips; white collar, underparts, and tail; scarlet bill and black legs; partial white eyering. *Winter:* Head whitish with dark gray ear patch; black bill. *Immature:* Brownish above and on breast; white belly and tail with black terminal bar.

Voice A raucous, derisive “kaah kah-kah-kah-kah kaah kaah”; also a single “keeyah.”
Similar Species Laughing Gull has black wingtips and whitish head with dark ear patches in winter;

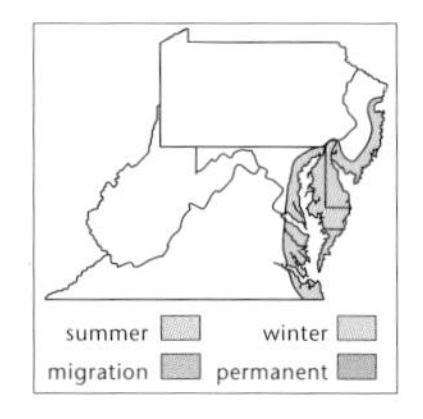

Franklin's Gull shows white tips on black primaries and a darkish hood in winter.
Habitat Beaches, bays, lakes, agricultural areas.
Abundance and Distribution Common transient and summer resident* (Apr–Nov) along the coast; rare to casual in winter mainly along the Virginia coast.
Where to Find Ocean City, Maryland; Hunting Bay, Virginia; Stone Harbor, New Jersey.
Range Breeds along both coasts in North America from New Brunswick on the east and northern Mexico on the west, south to northern South America, West Indies.

Franklin's Gull

Larus pipixcan

(L-15 W-36)

Black head; gray back and wings; white bar bordering black wingtips spotted terminally with white; white collar, underparts, and tail; underparts variously tinged with rose; scarlet bill and legs; partial white eyering. *Winter:* Head with a partial, dark hood; black bill. *Immature:* Similar to winter adult but with black terminal tail band.

Voice "Ayah."
Similar Species Franklin's Gull shows white tips on black primaries and a darkish hood in winter. Laughing Gull has black wingtips and whitish head with dark ear patches in winter.
Habitat Pastures, flooded fields, bays.
Abundance and Distribution Rare summer and fall visitor to the Virginia coast (Jun–Oct); casual elsewhere.
Where to Find Chesapeake Bay-Bridge Tunnel, Virginia; Hunting Bay, Virginia.
Range Breeds north central United States and

south central Canada; winters along Pacific coast of Middle and South America.

Little Gull

Larus minutus

(L-11 W-22)

A small gull; black head; gray back; wings dark below and gray above with white border; white collar, underparts, and tail; black bill; red legs. *Winter:* Light forehead, incomplete gray hood, black ear patch. *Immature:* Back tail band and black stripe running diagonally across gray upper wing; pinkish legs.

Voice "Ki-ki-ki-ki."

Similar Species Other black-headed gulls have black or black-and-white wingtips; those of the adult Little Gull are white; immature Little Gull has dark primaries but lacks the dark subterminal band on secondaries of Franklin's, Black-headed, and Bonaparte's gulls.

Habitat Marshes, bays, rivers, estuaries.

Abundance and Distribution Rare visitor (Dec–Apr) to the immediate coast, most numerous in spring (Apr).

Where to Find Indian River Inlet, Maryland; Back River Waste Water Treatment Plant, Maryland.

Range Breeds in central and western Asia, northern Europe, and locally in the Great Lakes region of North America; winters in coastal regions of the Baltic, Mediterranean, Black, and Caspian seas and rare in North America in the Great Lakes and Atlantic coastal regions from Maine south to Virginia.

Black-headed Gull

Larus ridibundus

(L-16 W-40)

White with gray back and wings; front two-thirds of head with dark brown hood; red bill and legs; white outer primaries with black tips. *Winter:* White head with black ear patch. *Immature:* Like winter adult but with black tail band and brown stripe running diagonally across gray wing; bill flesh-colored with dark tip; legs flesh-colored; incomplete brown hood in first breeding plumage.

Voice "Keerrip."

Similar Species Primaries of Bonaparte's Gull appear white or grayish from below; Black-headed primaries appear dark; Black-headed has red bill (black in Bonaparte's).

Habitat Marshes, bays, estuaries.

Abundance and Distribution Rare visitor (Nov–Apr) to the immediate coast, most numerous in spring (Apr).

Where to Find Craney Island Landfill, Virginia; Hereford Inlet, New Jersey; Indian River Inlet, Maryland.

Range Breeds in northern regions of the Old World, casual in Newfoundland; winters in temperate and northern tropical regions of the Old World and along the Atlantic Coast of North America from Labrador to Virginia.

Bonaparte's Gull

Larus philadelphia

(L-14 W-32)

A small gull; black head; gray back; wings gray with white outer primaries tipped in black; white

collar, underparts, and tail; black bill; red legs. *Winter:* White head with black ear patch. *Immature:* Like winter adult but with black tail band and brown stripe running diagonally across gray wing; flesh-colored legs; black bill.

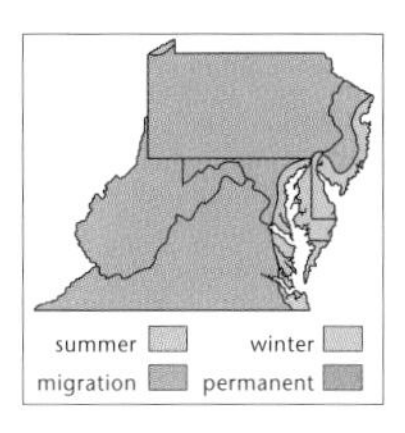

Voice A high-pitched, screechy "aaaanhh."
Similar Species Two other black-headed gull species occur regularly in the region: Laughing Gull and Franklin's Gull. Both are much larger than Bonaparte's Gull. Bonaparte's Gull has red or flesh-colored legs and feet (not dark as in Laughing and Franklin's). Bonaparte's has white outer primaries tipped with black (Laughing Gull has black outer primaries, and Franklin's has black outer primaries tipped with white).
Habitat Marine, bays, estuaries, lakes.
Abundance and Distribution Common winter resident (Oct–Apr) along the Atlantic coast and Lake Erie; uncommon to rare transient (Oct–Nov, Apr–May) elsewhere, more numerous in spring; rare to casual in summer (May–Sep).
Where to Find Presque Isle State Park, Pennsylvania; Cheat Lake, West Virginia; Chesapeake Bay Bridge-Tunnel, Virginia.
Range Breeds across north central and northwestern North America in Canada and Alaska; winters south along both coasts, from Washington on the west and Nova Scotia on the east to central Mexico; also the Great Lakes, Bahamas, and Greater Antilles.

Ring-billed Gull

Larus delawarensis

(L-16 W-48)

White body; gray back; gray wings with black outer primaries tipped with white; bill orange-yellow with black, subterminal ring; legs pale yellow. *Immature:* Mottled with brown or gray; bill pale with black tip; whitish tail with broad, black subterminal band and narrow, white terminal band.

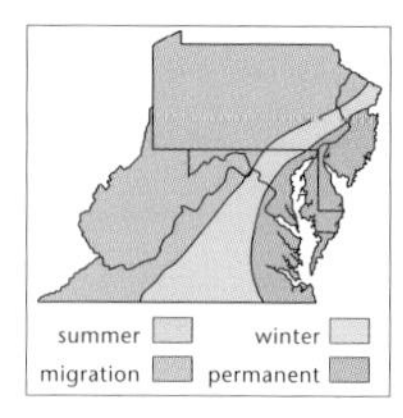

Voice A strident "ayah."

Similar Species Immature Herring Gulls are much larger; tail is mostly dark at tip in immature Herring Gull, lacking the narrow white terminal band of the immature Ring-billed Gull.

Habitat Lakes, bays, beaches, estuaries.

Abundance and Distribution Common to uncommon transient (Aug–Oct, Mar–Apr) throughout; uncommon to rare resident along the Coastal Plain; uncommon winter resident (Nov–Feb) in the Piedmont and at large bodies of open water elsewhere in the region.

Where to Find The Mall, Washington, D.C.; Roanoke Sewage Treatment Plant, Virginia; Shawnee State Park, Pennsylvania.

Range Breeds in the Great Plains region and scattered areas elsewhere in the northern United States and southern Canada; winters across most of United States except northern plains and mountains south to Panama.

Herring Gull

Larus argentatus

(L-24 W-57)

White body; gray back; gray wings with black outer primaries tipped with white; bill yellow with red spot on lower mandible; feet pinkish; yellow eye. *Winter:* Head and breast smudged with brown. *Immature:* Mottled dark brown in first winter; bill black; feet pinkish; tail dark. In second and third winter, grayish above, whitish below variously mottled with brown; tail with broad, dark, terminal band; bill pinkish with black tip; legs pinkish.

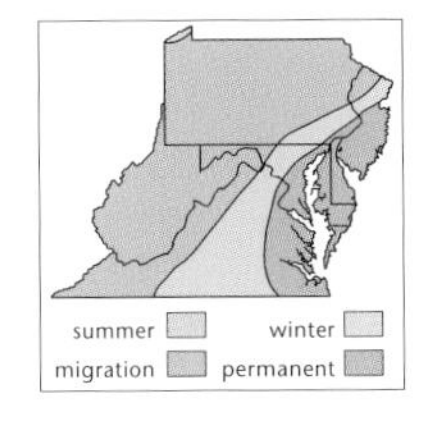

Voice The familiar, rusty gate squawking "Qeeyah kwa kwa kwa kwa" of all Hollywood shows that include an ocean scene.

Similar Species Adult Thayer's Gull can be distinguished (with difficulty) from Herring Gull, with which it used to be considered conspecific, by dark eye, darker pink legs, and outer primary black at base, white at tip (entirely black in adult Herring Gull).

Habitat Beaches, bays, marine, lakes, rivers.

Abundance and Distribution Common to uncommon transient (Oct–Nov, Feb–Apr) throughout; common resident* along the Coastal Plain; common to uncommon and local winter resident (Oct–Apr) in the Piedmont; uncommon and local winter resident inland elsewhere at larger rivers and lakes.

Where to Find Bombay Hook National Wildlife Refuge, Delaware; Ohio River, south of Parkersburg, West Virginia; Delaware Bay, New Jersey.

Range Breeds in boreal and north temperate regions of both Old and New World; winters along coasts of far north southward into temperate and north tropical regions.

Iceland Gull

Larus glaucoides

(L-23 W-42)

White body; gray back and upper wing coverts; white primaries and terminal edge of secondaries; yellow eye; bill pinkish or yellowish with red spot on lower mandible; feet pinkish; long wingtips extend beyond tail in sitting bird. *Winter:* Head and breast smudged with brown; wingtips with some gray. *First Winter:* White with light brown markings; bill black; legs pinkish; tail white marked with brown. *Second and Third Winters:* Paler than adult; some brown smudging; bill pinkish with black tip.

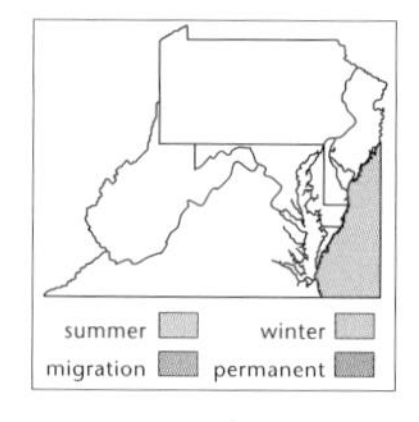

Similar Species Glaucous Gull is larger and has more massive bill and proportionately shorter wings (wings extend well beyond tail in standing Iceland Gull). Glaucous Gull lacks contrasting darker gray wingtips of winter Iceland Gull.

Habitat Seacoasts.

Abundance and Distribution Rare transient and winter visitor (Dec–Apr) mainly along the coast and offshore.

Where to Find Cape May Point, New Jersey; Ocean City, Maryland; Chincoteague National Wildlife Refuge, Virginia.

Range Breeds on islands of the far north in the North Atlantic; winters along northern coasts of the North Atlantic and Baltic Sea.

Lesser Black-backed Gull

Larus fuscus

(L-21 W-45)

White body; dark gray back and wings; outer primaries black sparsely tipped with white, appearing

an almost uniform gray below; bill yellow with red spot on lower mandible; eyes and legs yellow. *Winter:* Head and breast streaked with brown. *First Winter:* White heavily marked with brown; bill black; feet pale pinkish. *Second Winter:* Body white marked with brown; back and wings dark gray marked with brown; bill pinkish with black tip; legs pale yellow or pink.

Similar Species First winter birds very similar to first winter Herring Gull but have paler rump; older birds are darker on the back and have yellow legs (pinkish in Great Black-backed Gull).
Habitat Beaches, marine.
Abundance and Distribution Rare transient and winter visitor (Oct–Apr) mainly along the coast and offshore.
Where to Find Back River Sewage Treatment Plant, Maryland; Seashore State Park, Virginia.
Range Breeds in Iceland and coastal northern Europe; winters along coasts of Europe and eastern North America.

Glaucous Gull

Larus hyperboreus
(L-27 W-60)

A very large gull; white body; pale gray back; yellow eyering and eye; wings pale gray with outer primaries broadly tipped with white above, mostly white from below; bill yellow with red spot on lower mandible; feet pinkish; wingtips barely extend beyond tail in sitting bird. *Winter:* Head and breast smudged with brown. *First Winter:* Mostly white or with light brown markings; bill pinkish with black tip; legs pinkish; tail white marked with brown. *Second and Third Winters:* Back grayish; bill yellowish with dark tip.

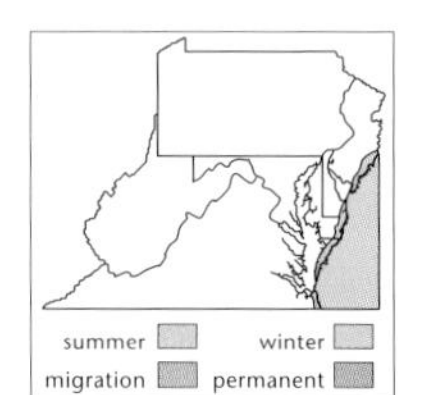

Voice "Usually silent, occasionally utters hoarse, deep, Herring Gull–like scream" (Harrison 1991, 347).
Similar Species The Glaucous Gull is a large, heavy-bodied gull with yellow eyering and eye and white primaries.
Habitat Coastal regions.
Abundance and Distribution Rare winter resident (Dec–Mar) along immediate coast and offshore.
Where to Find Craney Island Landfill, Virginia; Back River Sewage Treatment Plant, Maryland; Cape May Point, New Jersey.
Range Breeds in circumpolar regions of Old and New World; winters coastal regions of northern Eurasia and North America.

Great Black-backed Gull
Larus marinus
(L-30 W-66)

A large gull; white body; black back; yellow eye; wings appear black above with white trailing edge, outer primaries tipped with white; bill very large, yellow with red spot on lower mandible; legs pinkish. *First Winter:* White checked with brown above, paler below; rump paler; bill black; legs pale pink. *Second and Third Winters:* Body white streaked with brown; black mottled with brown on back and wings; bill pinkish or yellowish with black tip; legs pale pinkish.

Voice Hoarse, low squawks and chuckles.
Habitat Seacoasts and large lakes.

Abundance and Distribution Common resident* along the Atlantic coast and common winter resident (Sep–Apr) on Lake Erie coast; scarce elsewhere in the region.

Where to Find Public docks, Erie, Pennsylvania; Delaware Bay, New Jersey; Bombay Hook National Wildlife Refuge, Delaware.

Range Breeds in coastal northeastern North America and northeastern Eurasia; winters in breeding range and south to the southeastern United States and southern Europe.

Black-legged Kittiwake

Rissa tridactyla

(L-17 W-36)

A small gull; white body; gray back; white, slightly forked tail; dark eye; wings gray above, white below tipped with black; bill yellow; legs dark. *Winter:* Nape and back of crown gray with dark ear patch. *First Winter:* White marked with black "half-collar" on nape; gray back; dark diagonal stripe on upper wing; black ear patch; tail white with black terminal band; bill and legs black.

Voice "Kitt-ee-waak."

Similar Species Immature Bonaparte's Gull lacks dark "half-collar" at base of neck.

Habitat Pelagic, coasts.

Abundance and Distribution Uncommon to rare and irregular winter visitor (Nov–Mar), mainly offshore.

Where to Find Pelagic birding trips; Chesapeake Bay Bridge-Tunnel, Virginia.

Range Breeds circumpolar in Old and New World; winters in northern and temperate seas of the Northern Hemisphere.

Gull-billed Tern

Sterna nilotica

(L-15 W-36)

A medium-sized tern with a heavy, black bill; black cap and nape; white face, neck, and underparts; gray back and wings; tail with shallow fork; black legs. *Winter:* Crown white and finely streaked with black.

Habits This bird does not dive like most terns. It swoops and sails over marshland for insects.
Voice A harsh, nasal "kee-yeek" or "ka-wup."
Similar Species Hcavy, black, gull-likc bill and shallow tail fork are distinctive.
Habitat Marshes, wet fields, grasslands, bays.
Abundance and Distribution Uncommon to rare summer resident* (Apr–Aug) along the coast.
Where to Find Eastern Shore National Wildlife Refuge, Virginia; Chincoteague National Wildlife Refuge, Virginia; South Cape May Meadows, New Jersey.
Range Breeds locally in temperate and tropical regions of the world; winters in subtropics and tropics.

Caspian Tern

Sterna caspia

(L-21 W-52)

A large tern with large, heavy, blood-orange bill; black cap and nape; white face, neck, and underparts; gray back and wings; tail moderately forked; black legs. *Winter:* Black cap streaked with white.

Voice A low squawk, "aaaak," repeated.
Similar Species Caspian Tern primaries appear

dark from below, Royal Tern primaries appear whitish; Royal has white forehead (not black flecked with white) most of the year; Royal's call is a screechy "kee-eer," not a squawk like the Caspian. Also, the bill of the Caspian Tern is generally a bright, reddish orange, while that of the Royal Tern is a paler orange.

Habitat Beaches, bays, marine, estuaries, lakes.

Abundance and Distribution Common transient (Aug–Oct, Apr–May) along the coast; uncommon to rare and local summer resident (May–Aug) and scarce breeder,* mainly on coastal islands; rare to casual transient inland.

Where to Find Hunting Bay, Virginia; Assateague Island National Seashore, Maryland; Bombay Hook National Wildlife Refuge, Delaware.

Range Breeds locally inland and along the coast in temperate, tropical, and boreal areas of the world; winters in south temperate and tropical regions.

Royal Tern

Sterna maxima

(L-20 W-45)

A large tern with yellow or yellow-orange bill; black cap with short, ragged crest; black nape; white face, neck, and underparts; gray back and wings; tail forked; black legs. *Winter:* White forehead and crown; black nape.

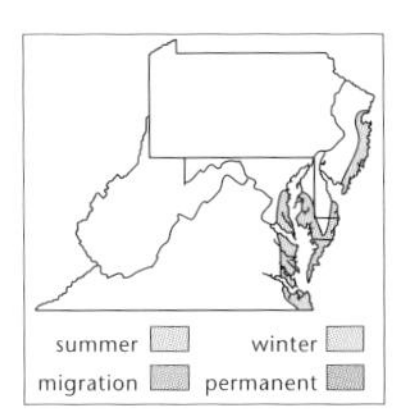

Voice A screechy "keee-eer."

Similar Species Royal Tern primaries appear whitish from below, Caspian Tern primaries appear dark; Royal has white forehead (not black flecked with white as in Caspian Tern) most of the year; Royal's call is a screechy "kee-eer," not a squawk like the Caspian. The bill of the Caspian

Tern is generally a bright, reddish orange, whereas that of the Royal Tern is a paler orange.
Habitat Beaches, bays.
Abundance and Distribution Common summer resident* (Apr–Oct) along the coast.
Where to Find Cape Henlopen, Delaware; South Cape May Meadows, New Jersey; Virginia Coast Reserve, Virginia.
Range Breeds locally in temperate and tropical regions of Western Hemisphere and West Africa; winters in warmer portions of breeding range.

Sandwich Tern

Sterna sandvicensis
(L-15 W-34)

A trim, medium-sized tern with long, pointed, black bill tipped with yellow; black cap and short, ragged crest; black nape; white face, neck, and underparts; gray back and wings; tail deeply forked; black legs. *Winter:* White forehead; crown white, finely streaked with black; black nape.

Voice "Keerr-rik."
Similar Species The combination of slim, yellow-tipped, black bill and black legs is distinctive.
Habitat Marine, bays, often associated with Royal Tern.
Abundance and Distribution Uncommon summer resident* (Apr–Sep), mainly on Virginia coastal islands, although found along bays and inlets during the postbreeding period (Jul–Sep).
Where to Find Virginia Coast Reserve, Virginia; Chesapeake Bay Bridge-Tunnel, Virginia; Chincoteague National Wildlife Refuge, Virginia.
Range Breeds locally along Atlantic, Gulf, and Caribbean coasts of North and South America

from Virginia to Argentina, West Indies; winters through subtropical and tropical portions of breeding range and on the Pacific Coast of Middle and South America from southern Mexico to Peru.

Roseate Tern

Sterna dougallii

(L-16 W-28)

A medium-sized tern; black bill reddish at base; black cap and nape; white face, neck; underparts white tinged with pink; gray back and wings; black outer primaries; tail very long, white, and deeply forked, extends well beyond folded wings in sitting bird; red legs. *Winter:* White forehead and crown; back of crown and nape black; bill black. *First Breeding:* Like adult but with white forehead.

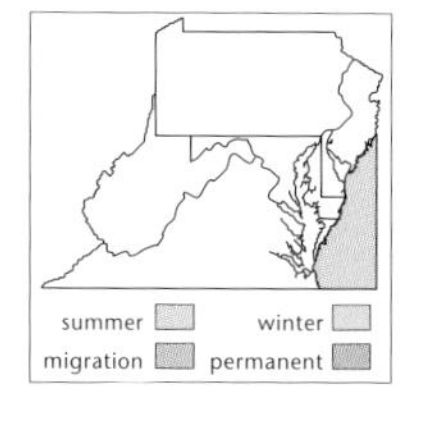

Habits Flies with distinctive rapid, shallow wing beats.

Voice "Chewee," mellow (for a tern).

Similar Species Roseate Tern has completely white tail; Common Tern has gray outer edgings and white inner edgings; Forster's Tern has white outer edgings and gray inner edgings; tail of Common Tern does not extend beyond folded wings of sitting bird, as it does in the Roseate; Roseate Tern "chewee" call is distinctive ("kee-arr" in Forster's, "keeeyaak" or "keyew" in Common).

Habitat Beaches, marine, bays.

Abundance and Distribution Rare to casual transient and summer visitor (May–Jul) along the coast and offshore; has bred.*

Where to Find Chesapeake Bay Bridge-Tunnel, Virginia; Cape Henlopen, Delaware.

Range Breeds locally along coasts of temperate and tropical regions of the world (in United States from Maine to North Carolina); winters mainly in tropical portions of breeding range.

Common Tern

Sterna hirundo

(L-14 W-31)

A medium-sized tern; red bill, black at tip; black cap and nape; white face, neck, and underparts; gray back and wings; entire outer primary, tips, and basal portions of other primaries are black (gray from below); forked tail with gray outer edgings, white inner edgings; red legs. *Winter:* White forehead and crown; back of crown and nape black; bill black; sitting bird shows a dark bar at the shoulder (actually upper wing coverts). *First Breeding:* Similar to adult winter.

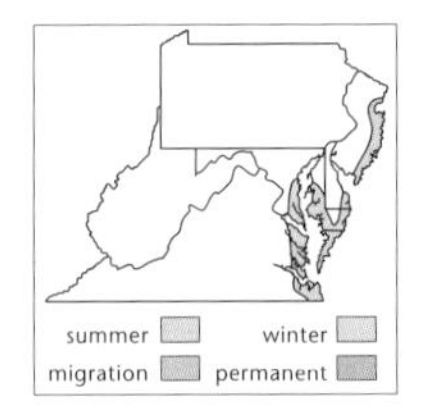

Voice A harsh "keeeyaak," also "keyew keyew keyew."

Similar Species Winter Arctic Tern has white (not dark) basal portions of primaries from below. Roseate has completely white tail; Common Tern has gray outer edgings and white inner edgings; Forster's Tern has white outer edgings and gray inner edgings; Common Tern tail does not extend beyond the folded wings of sitting bird, as it does in the Roseate; Roseate Tern "chewee" call is distinctive ("kee-arr" in Forster's, "keeeyaak" or "keyew" in Common).

Habitat Beaches, bays, marshes, lakes, rivers.

Abundance and Distribution Common transient and summer resident* (Apr–Sep) along the immediate coast; rare to casual transient (Aug–Oct, Apr–May) inland.

Where to Find Holgate, New Jersey; Indian River Inlet, Maryland; Chincoteague National Wildlife Refuge, Virginia.
Range Breeds locally in boreal and temperate regions of Old and New World (mostly Canada, northeastern United States, and West Indies); winters in south temperate and tropical regions.

Arctic Tern

Sterna paradisaea
(L-16 W-31)

A medium-sized tern; red bill; black cap and nape; white face; grayish neck, breast, and belly; white undertail coverts; gray back; wings gray above with dark outer edge of outer primary, white below with primaries narrowly tipped in black; tail deeply forked; red legs. *Winter:* White forehead and crown; back of crown and nape black; bill black.

Voice A high, nasal "ki-kee-yah"; also "kaaah."
Similar Species Winter Arctic Tern has white basal portions of primaries from below, not dark as in Common Tern.
Habitat Mostly marine, beaches.
Abundance and Distribution Rare offshore spring transient (May), rare to casual along coast.
Where to Find Pelagic birding trips; Chesapeake Bay Bridge-Tunnel, Virginia.
Range Breeds in the Arctic; winters in the Antarctic.

Forster's Tern

Sterna forsteri

(L-15 W-31)

A medium-sized tern; orange or yellow bill with black tip; black cap and nape; white face and underparts; gray back and wings; forked tail, white on outer edgings, gray on inner; orange legs. *Winter:* White head with blackish ear patch; bill black.

Voice "Chew-ik."
Similar Species Winter Common and Arctic terns have back of crown and nape black; crown and nape arc usually white in Forster's Tern.
Habitat Marshes, bays, beaches, marine.
Abundance and Distribution Common transient and summer resident* (Apr–Nov) along the coast; rare to casual transient inland; uncommon in winter, mainly in southeastern Virginia.
Where to Find Brigantine National Wildlife Refuge, New Jersey; Bombay Hook National Wildlife Refuge, Delaware; Assateague Island National Seashore, Maryland.
Range Breeds locally across northern United States, south central and southwestern Canada, and Atlantic and Gulf coasts; winters coastally from southern United States to Guatemala and Greater Antilles.

Least Tern

Sterna antillarum

(L-9 W-20)

ENDANGERED A small tern; yellow bill with black tip; black cap and nape with white forehead; white face and underparts; gray back and wings; tail deeply forked; yellow legs. *Winter:* White head with blackish postorbital stripe and nape; bill brown; legs dull yellow.

Voice A high-pitched "ki-teek ki-teek ki-teek."
Similar Species Least Tern has black nape, not white as in winter Forster's Tern; Forster's Tern is much larger.
Habitat Bays, beaches, rivers, lakes, ponds.
Abundance and Distribution Common transient and summer resident* (May–Sep) along the coast.
Where to Find Kiptopeke State Park, Virginia; South Cape May Meadows, New Jersey; Assateague Island National Seashore, Maryland.
Range Breeds along both coasts from central California (west) and Maine (east) south to southern Mexico; endangered inland population breeds along rivers of Mississippi drainage; breeds in West Indies; winters along coast of northern South America.

Bridled Tern

Sterna anaethetus
(L-15 W-30)

Dark gray above, white below; black cap with white forehead and eyebrow; white collar; black bill and feet; deeply forked, brown tail edged in white. *Immature:* Similar in pattern to adult, but paler above with grayish brown cap.

Similar Species Sooty Tern lacks collar; its white stripe extends to, but not behind, eye (extends behind eye in Bridled Tern).
Habitat Marine, bays, beaches.
Abundance and Distribution Uncommon to rare summer and fall visitor (Jul–Oct) offshore.

Where to Find Pelagic birding trips.
Range Breeds locally on islands in tropical seas (including West Indies).

Sooty Tern

Sterna fuscata

(L-17 W-32)

Black above, white below; black cap with white forehead; black bill and feet; tail deeply forked, black with white edging. *Immature:* Dark brown throughout with white spotting.

Habits Does not dive for fish as other terns do; plucks fish from surface.
Voice A creaky "waky wak."
Similar Species Sooty Tern lacks collar, and its white stripe extends to, but not behind, eye (extends behind eye in Bridled Tern).
Habitat Marine, bays, beaches.
Abundance and Distribution Rare to casual summer and fall visitor (Jul–Oct) offshore.
Where to Find Pelagic birding trips.
Range Breeds locally on tropical and subtropical coasts and islands throughout the world.

Black Tern

Chlidonias niger

(L-10 W-24)

A small tern; dark gray throughout except white undertail coverts; slightly forked tail. *Winter:* Dark gray above, white below with dark smudge at shoulder; white forehead; white crown streaked with gray; gray nape; white collar.

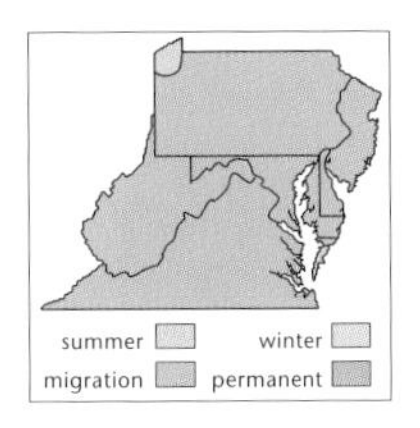

Voice A high-pitched "kik kik kik."
Habits Swallowlike flight in pursuit of insects.
Habitat Marshes, bays, estuaries, lakes, ponds.
Abundance and Distribution Common fall transient (Jul–Sep) and uncommon to rare spring transient (May) along the coast. Uncommon to rare transient (Jul–Sep, Apr–May) inland, more numerous in spring in West Virginia. Summer resident* (May–Jul) in western Pennsylvania (Crawford and Erie counties) and as a nonbreeder at Chincoteague.
Where to Find Presque Isle State Park, Pennsylvania (breeding); Chincoteague National Wildlife Refuge, Virginia; Little Creek Wildlife Area, Delaware.
Range Breeds in temperate and boreal regions of Northern Hemisphere; winters in tropics.

Black Skimmer

Rynchops niger
(L-18 W-45)

Black above; white below; long, straight razor–shaped bill, red with black tip; lower mandible is longer than upper. *Immature:* Mottled brown and white above.

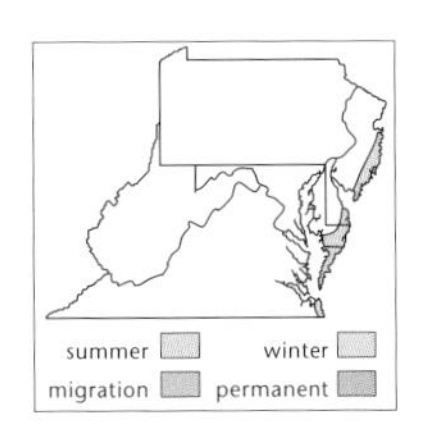

Habits Flies over the water with the lower mandible dipped below the surface to capture fish.
Voice A mellow "kip," "kee-yip," or "kee-kee-yup."
Habitat Beaches, bays, estuaries.
Abundance and Distribution Common transient and summer resident* (Apr–Oct) along the coast; rare in winter; rare to casual inland.
Where to Find Holgate, New Jersey; Port Mahon Impoundment, Delaware; Grandview Beach, Virginia.

Range Breeds from southern California and New York south to southern South America along coasts and on major river systems; winters in south temperate and tropical portions of breeding range, West Indies.

Auks, Guillemots, and Murres
(Family Alcidae)

Aquatic, penguinlike birds streamlined for swimming and diving in pursuit of fish.

Dovekie
Alle alle
(L-9 W-12)

A small, chunky, short-necked seabird; black above; black breast; white belly. *Winter:* Breast and throat white.

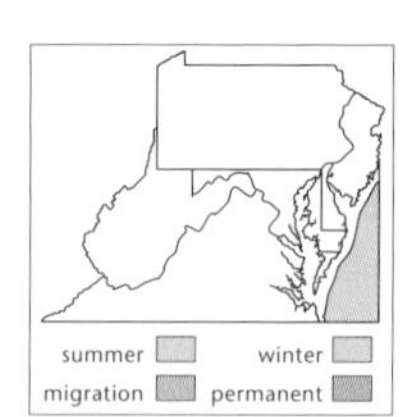

Similar Species The combination of small size, chubby, thick-necked shape, and extremely rapid wing beats is distinctive.
Habitat Pelagic.
Abundance and Distribution Common to uncommon in winter (Dec–Feb) offshore.
Where to Find Pelagic birding trips.
Range Breeds in the high Arctic—Greenland, Iceland, Novaya, Zemlya, Spitsbergen, Jan Mayen, and perhaps Ellesmere. Winters at sea in the Arctic and North Atlantic oceans.

Razorbill

Alca torda

(L-16 W-25)

A heavy-bodied, thick-necked seabird with distinctive broad bill with whitish vertical stripe; black above with black throat and white line from top of bill to eye; white below; wings black above with white wing linings. *Winter:* White throat; no white eyeline. *Immature:* Dark gray above; white below; dark bill is thick but much smaller than that of adults.

Similar Species The bill of the adult is unmistakable.
Habitat Pelagic.
Abundance and Distribution Common to uncommon in winter (Dec–Mar) offshore.
Where to Find Pelagic birding trips.
Range Breeds along northeastern coast of North America (Maine–Labrador), southwestern Greenland, Iceland, and northern Europe (northern France to northern Russia). Winters at sea in the North Atlantic.

Doves and Pigeons

(Order Columbiformes, Family Columbidae)

These are small to medium-sized, chunky birds with relatively small heads. Larger species are generally referred to as "pigeons," while smaller ones are "doves." They feed on fruits and seeds and produce "crop milk" (specialized cells sloughed from the esophagus) to feed their young. Clutch size is normally 2 throughout the group.

Rock Dove

Columba livia

(L-13 W-23)

The Rock Dove, or domestic pigeon, is highly variable in color, including various mixtures of brown, white, gray, and dark blue; the "average" bird shows gray above and below with head and neck iridescent purplish green; white rump; dark terminal band on tail.

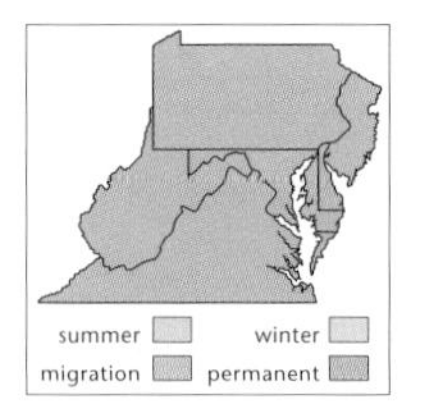

Voice A series of low "coo"s.
Habitat Cities, towns, agricultural areas.
Abundance and Distribution Common permanent resident* throughout.
Range Resident of Eurasia and North Africa; introduced into Western Hemisphere, where resident nearly throughout at farms, towns, and cities.

Mourning Dove

Zenaida macroura

(L-12 W-18)

Tan above, orange-buff below; brownish wings spotted with dark brown; tail long, pointed, and edged in white; gray cap; black whisker; purplish bronze iridescence on side of neck; blue eyering; brown eye and bill; pink feet.

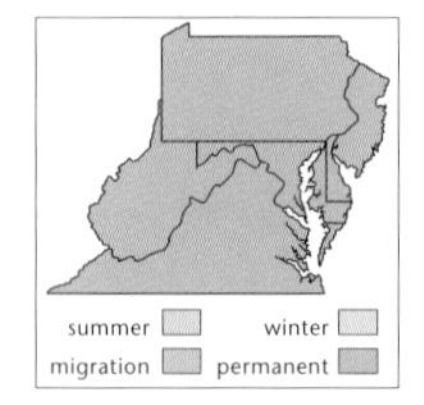

Voice A low, hoarse "hoo-wooo hooo hooo."
Similar Species The long, pointed tail separates Mourning Dove from other doves in the region.
Habitat Woodlands, pastures, old fields, second growth, scrub, residential areas.
Abundance and Distribution Common summer resident* (Apr–Oct) throughout; common winter

resident in lowlands and foothills; uncommon to rare or absent in highlands.
Where to Find Catoctin Mountain Park, Maryland; Ohiopyle State Park, Pennsylvania; National Zoological Park, Washington, D.C.
Range Breeds from southern Canada south through United States to highlands of central Mexico, Costa Rica, Panama, Bahamas, Greater Antilles; winters in south temperate and tropical portions of breeding range.

Common Ground-Dove

Columbina passerina

(L-7 W-11)

A very small dove; grayish brown above with dark brown spotting on wing; gray tinged with pink below, with scaly brown markings on breast; mottled gray crown; red eye; pink bill; flesh-colored legs; rusty wing patches show in flight; black outer tail feathers with white edgings. *Female:* Like male but grayish below (not pinkish), and gray crown is less distinct.

Habits Thrusts head forward while walking.
Voice "Hoo-wih," repeated.
Habitat Sandy or grassy shrublands.
Abundance and Distribution Rare summer visitor (Apr–Oct) in extreme southeastern Virginia.
Where to Find Mackay Island National Wildlife Refuge, Virginia.
Range Southern United States south to northern half of South America, West Indies.

Cuckoos
(Order Cuculiformes, Family Cuculidae)

Cuckoos are a family of trim, long-tailed birds found in both the Old and New Worlds; several species lay their eggs in other birds' nests. Like parrots, woodpeckers, and some other avian groups, cuckoos perch with 2 toes forward and 2 toes backward (most birds have 3 toes forward and one toe backward).

Black-billed Cuckoo
Coccyzus erythropthalmus
(L-12 W-16)

A thin, streamlined bird with long, graduated tail; red eyering; black bill; brown above; white below; dark brown tail with white tips.

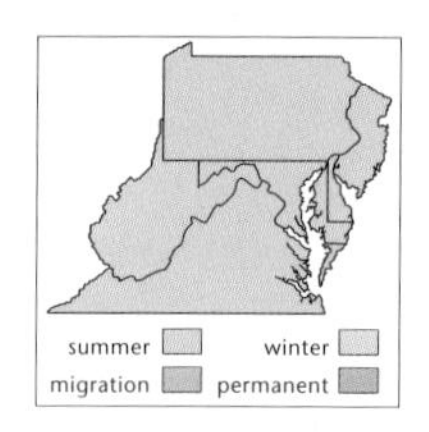

Habits Slow, deliberate, reptilian foraging movements.
Voice A mellow "coo-coo-coo," repeated.
Similar Species Yellow-billed Cuckoo has yellow eyering and lower mandible, more prominent white tail spots, rusty wing patch visible in flight.
Habitat Deciduous and mixed forest, overgrown fields with shrubby second growth.
Abundance and Distribution Common (north and highlands) to uncommon or rare (south and lowlands) transient and summer resident* (May–Sep).
Where to Find Beech Fork State Park, West Virginia; Dickey Ridge Visitor Center, Shenandoah National Park, Virginia; Kahle Lake, Pennsylvania.
Range Breeds in the northeastern quarter of United States and in southeastern Canada; winters in the northern half of South America.

Yellow-billed Cuckoo

Coccyzus americanus

(L-12 W-17)

A thin, streamlined bird with long, graduated tail; dark bill with yellow lower mandible and eyering; brown above; white below; dark brown tail with white tips; rusty wing patches visible in flight.

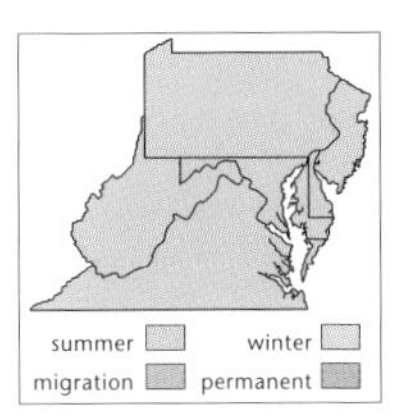

Habits Slow, deliberate, reptilian foraging movements.

Voice A metallic "ka-ka-ka-ka-ka-ka cow cow cow"; also a low, hoarse "cow."

Similar Species Yellow-billed Cuckoo has yellow eyering and lower mandible and rusty wing patches visible in flight (lacking in the Black-billed Cuckoo).

Habitat Deciduous woodlands and second growth.

Abundance and Distribution Common transient and summer resident* (May–Sep) throughout; more numerous in lowlands.

Where to Find Great Swamp National Wildlife Refuge, New Jersey; Gunpowder Falls State Park, Maryland; Bombay Hook National Wildlife Refuge, Delaware.

Range Breeds southeastern Canada and nearly throughout United States into northern Mexico and West Indies; winters north and central South America.

Barn Owls

(Order Strigiformes, Family Tytonidae)

This small, cosmopolitan family comprises 12 species, of which only the Barn Owl occurs regularly in North America.

Barn Owl

Tyto alba

(L-16 W-45)

Tawny and gray above; white sparsely spotted with brown below; white, monkey-like face; large, dark eyes; long legs.

Habits Almost strictly nocturnal.
Voice Eerie screeches and hisses.
Habitat Old fields, pasture, grasslands, farmland; roosts in barns, caves, abandoned mine shafts.

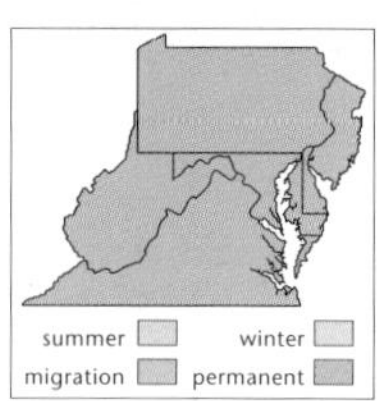

Abundance and Distribution Uncommon to rare permanent resident.* Easily overlooked because of nocturnal habits. Recent breeding bird surveys found the bird to be rare in northern and western Pennsylvania and rare or absent over much of West Virginia.
Where to Find Elliott Island, Maryland; Mud Level Rd, Shippensburg, Pennsylvania; Presquile National Wildlife Refuge, Virginia.
Range Resident in temperate and tropical regions nearly throughout the world.

Typical Owls

(Family Strigidae)

These owls comprise a large family of fluffy-plumaged, mainly nocturnal raptors. Plumage coloration is mostly muted browns and grays; 2 toes forward and 2 backward when perched. Ear structures are asymetrical for ranging distance to prey.

Eastern Screech-Owl

Otus asio

(L-9 W-22)

Red Phase: A small, long-eared, yellow-eyed owl with a pale bill; rufous above; streaked with rust, brown, and white below; rusty facial disk with white eyebrows. *Gray Phase:* Gray rather than rusty.

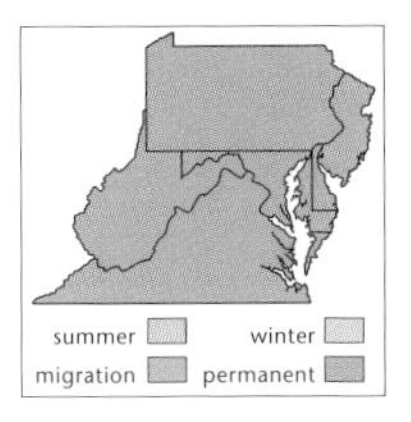

Voice A quavering, descending whistle; also a low, quavering whistle on a single pitch.
Habitat Woodlands, residential areas.
Abundance and Distribution Common resident* throughout.
Where to Find Cranberry Glades Botanical Area, Monongahela National Forest, West Virginia; Great Swamp National Wildlife Refuge, New Jersey; Carvin Cove Reservoir, Virginia.
Range Resident from southeastern Canada and the eastern half of the United States to northeastern Mexico.

Great Horned Owl

Bubo virginianus

(L-22 W-52)

A large owl; mottled brown, gray, buff, and white above; grayish white below barred with brown; yellow eyes; rusty facial disc; white throat (not always visible); long ear tufts.

Habits Eats skunks as well as many other small to medium-sized vertebrates.
Voice A series of low hoots, "ho-hoo hooo hoo"; female's call is higher pitched than male's.

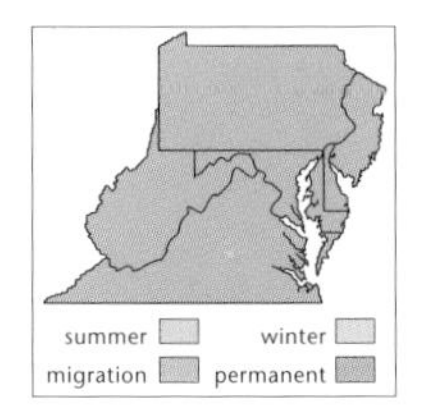

Similar Species The smaller Long-eared Owl is streaked rather than barred below, and ear tufts are placed more centrally; other large owls lack long ear tufts.
Habitat Woodlands.
Abundance and Distribution Uncommon permanent resident* throughout.
Where to Find Woodbourne Sanctuary, Pennsylvania; Bombay Hook National Wildlife Refuge, Delaware; Patuxent Research Refuge, Maryland.
Range Most of New World except polar regions, Amazonia, much of Central America, West Indies, and most other islands.

Snowy Owl

Nyctea scandica
(L-24 W-60)

White with various amounts of gray-and-black barring; yellow eyes. *Immature:* Much more buff-and-brown spotting and barring than adults.

Voice A harsh "ka-ow" but usually silent.
Habitat Tundra, fields, pastures.
Abundance and Distribution Rare to casual and irregular winter visitor (Dec–Feb). This species appears in certain years and is absent in others, at 5–10-year intervals.
Where to Find Brigantine National Wildlife Refuge, New Jersey; Tioga/Hammond Lakes National Recreation Area, Pennsylvania; Chincoteague National Wildlife Refuge, Virginia.
Range Resident in northern polar regions of Old and New World.

Barred Owl

Strix varia

(L-18 W-42)

A large, dark-eyed owl; brown mottled with white above; white barred with brown on throat and breast, streaked with brown on belly.

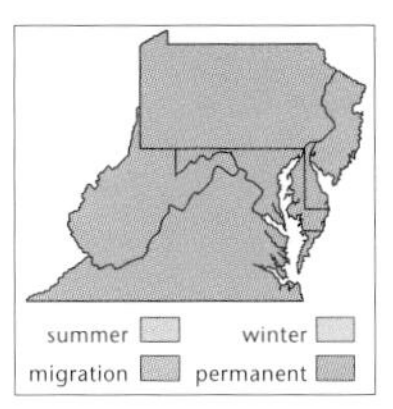

Voice "Haw haw haw ha-hoo-aw" ("who cooks for you all").

Similar Species The Barred Owl has concentric rings of brown and white on the facial disk, a neck ruff of horizontal brown and white bars, and a breast and belly streaked with brown and white. This pattern and the lack of ear tufts are distinctive.

Habitat Deciduous and mixed forest.

Abundance and Distribution Uncommon resident* throughout.

Where to Find Catoctin Mountain State Park, Maryland; Bombay Hook National Wildlife Refuge, Delaware; Sleepy Creek Wildlife Management Area, West Virginia.

Range Resident from southern Canada and eastern half of United States; locally in northwestern United States and in the central plateau of Mexico.

Long-eared Owl

Asio otus

(L-14 W-39)

Medium-sized, trim, yellow-eyed owl with long ear tufts; mottled brown and white above; whitish streaked with brown below; rusty facial disks.

Habits Nocturnal.

Voice "Haaaaaa," like a baby's cry; also 3–4 quick, high-pitched "hoh"s.

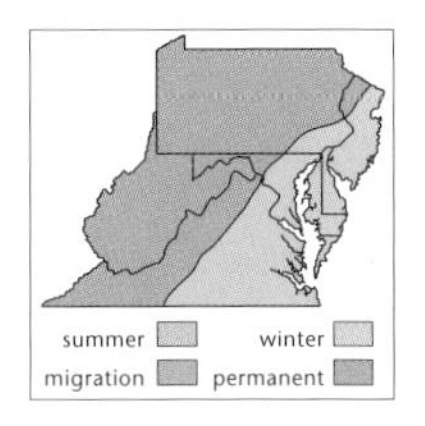

Similar Species Long-eared Owl is much smaller and slimmer than Great Horned Owl and is streaked rather than barred below; ear tufts are placed more centrally; other large owls lack long ear tufts.
Habitat Coniferous and mixed woodlands, especially dense cedar (*Thuja*) thickets in winter, when several birds can be found in communal roosts.
Abundance and Distribution Uncommon to rare and local resident* in Pennsylvania, northwestern New Jersey, West Virginia, and the highlands of Virginia; rare winter visitor (Nov–Mar) elsewhere in the region.
Where to Find Presque Isle State Park, Pennsylvania; Great Swamp National Wildlife Refuge, New Jersey; Natural Chimneys Regional Park, Virginia.
Range Breeds in temperate and boreal regions of the Northern Hemisphere; winters in temperate areas and mountains of northern tropical regions.

Short-eared Owl
Asio flammeus
(L-15 W-42)

A medium-sized owl; brown streaked with buff above; buff streaked with brown below; round facial disk with short ear tufts and yellow eyes; buffy patch shows on upper wing in flight; black patch at wrist visible on wing lining.

Habits Crepuscular. Forages by coursing low over open areas, reminiscent of Northern Harrier.
Voice A sharp "chik-chik" and various squeaks.
Similar Species Tawny color and lack of ear tufts separate this from other owl species.
Habitat Grasslands, marshes, estuaries, agricultural fields.

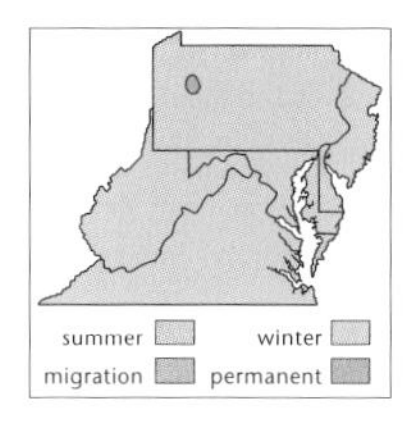

Abundance and Distribution Uncommon to rare winter resident (Nov–Mar) throughout, more numerous in coastal areas; rare to casual summer resident* (Apr–Oct) in New Jersey and Pennsylvania (strip mine areas of Clarion and Jefferson counties and grasslands of the Philadelphia airport).
Where to Find Port Mahon Marsh, Delaware; Eastern Shore of Virginia National Wildlife Refuge, Virginia; Elliott Island, Maryland; Mount Zion Strip Mines, Pennsylvania (resident*).
Range Breeds in tundra, boreal, and north temperate areas of Northern Hemisphere, also in Hawaiian Islands; winters in southern portions of breeding range south to northern tropical regions.

Northern Saw-whet Owl

Aegolius acadicus
(L-8 W-20)

A small owl; dark brown above with white spotting; white below with broad chestnut stripes; facial disk with gray at base of beak, gray eyebrows, rest of face finely streaked brown and white; yellow eyes. *Immature:* Dark brown back spotted with white; solid chestnut below; chestnut facial disk with white forehead and eyebrows.

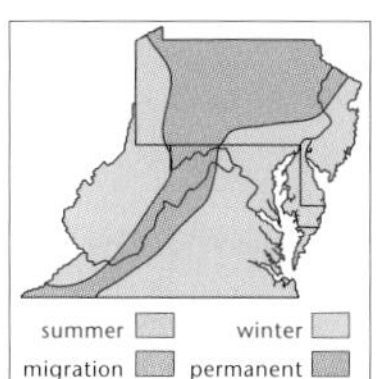

Habits A tame owl, easily approached.
Voice A whistled monotone "hoo hoo hoo hoo"; also a raspy note, given in series of 3 (like noises of saw whetting).
Similar Species Chestnut underparts are distinctive.
Habitat Coniferous forest, especially spruce-fir; dense thickets of pine and cedar *(Thuja)* in winter.
Abundance and Distribution Rare resident* in highland spruce-fir forest of the region; rare and irregular winter resident (Oct–Mar) elsewhere; has

been reported in summer in cedar swamps of the New Jersey pine barrens.
Where to Find Mount Rogers, Virginia (resident); Lebanon State Forest, New Jersey; Allegheny National Forest, Pennsylvania (resident).
Range From central Canada south to northern United States and in western mountains south to southern Mexico; winters in breeding range south and east to southern United States.

Goatsuckers, Nightjars, and Nighthawks
(Order Caprimulgiformes, Family Caprimulgidae)

Most of the species in this family have soft, fluffy plumage of browns, grays, and buff. Long, pointed wings and extremely large, bristle-lined mouth are also distinctive characters of the group. Usually crepuscular or nocturnal, they forage for insects in batlike fashion with characteristic swoops and dives.

Common Nighthawk
Chordeiles minor
(L-9 W-23)

Mottled dark brown, gray, and white above; whitish below with black bars; white throat; white band across primaries and tail; tail slightly forked. *Female:* Buffy rather than white on throat.

Habits Crepuscular. Normally forages high above the ground; has a dive display in which the bird plummets toward the ground, swerving up at the

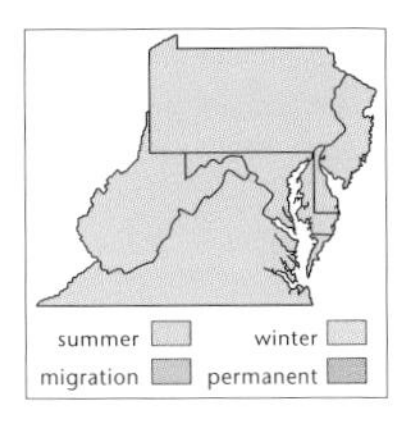

last moment and making a whirring sound with the wings.
Voice "Bezzzt."
Similar Species In flight, the sleek dark form with the prominent white bar across the primaries is distinctive. Sitting, the bird is readily distinguished from other goatsuckers by the white throat and white bar visible even on the folded primaries.
Habitat A variety of open and semi-open situations in both urban and rural environments.
Abundance and Distribution Common to uncommon transient and summer resident* (May–Sep) throughout.
Where to Find Cape Henlopen State Park, Delaware; Red Creek Recreation Area, Monongahela National Forest, West Virginia; The Mall, Washington, D.C.
Range Breeds locally from central and southern Canada south through United States south to Panama; winters in South America.

Chuck-will's-widow

Caprimulgus carolinensis
(L-12 W-25)

A large nightjar; tawny, buff, and dark brown above; chestnut and buff barred with dark brown below; white throat; long, rounded tail with white inner webbing on outer 3 feathers; rounded wings. *Female:* Has buffy tips rather than white inner webbing on outer tail feathers.

Habits Nocturnal.
Voice "Chuck-wills-widow"; also moaning growls.
Similar Species Lacks white wing patches of nighthawks; smaller, darker Whip-poor-will has extensive white on tail in male, buff in female.

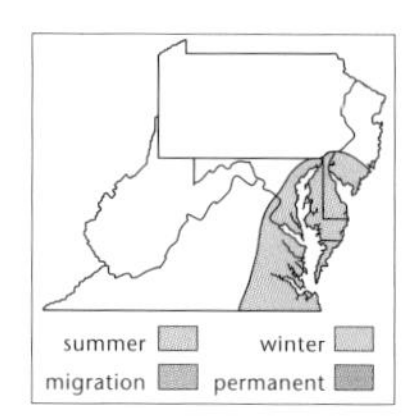

Habitat Open deciduous, mixed and pine woodlands.
Abundance and Distribution Uncommon summer resident* (May–Sep) of the Coastal Plain and southern Virginia Piedmont; has been recorded in recent years as a rare and irregular summer resident in southwestern Pennsylvania, West Virginia, and the mountains of Virginia.
Where to Find Brigantine National Wildlife Refuge, New Jersey; Piscataway National Park, Maryland; Pocahontas State Park, Virginia.
Range Breeds across the eastern half of the United States, mainly in the southeast; permanent resident in parts of the West Indies; winters from eastern Mexico south through Central America to Colombia; also winters in south Florida and the West Indies.

Whip-poor-will

Caprimulgus vociferous
(L-10 W-19)

Dark brown mottled with buff above; grayish barred with dark brown below; dark throat and breast with white patch; rounded tail with white outer 3 feathers; rounded wings. *Female:* Has buffy tips rather than white on outer tail feathers.

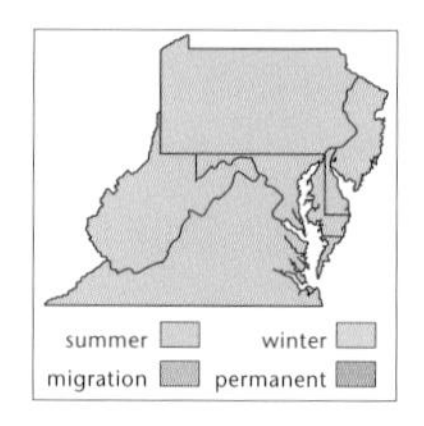

Habits Nocturnal.
Voice A rapid, whistled "whip-poor-will," monotonously repeated.
Habitat Deciduous, mixed, and coniferous forest; also second-growth forest and woodlots.
Abundance and Distribution Common to uncommon summer resident* (Apr–Sep) throughout.
Where to Find Fowler's Hollow State Park, Pennsylvania; Cypress Swamp Conservation Area,

Delaware; Stonecoal Lake Wildlife Management Area, West Virginia.
Range Breeds across southeastern and south central Canada and eastern United States, also in southwestern United States through the highlands of Mexico and Central America to Honduras; winters from northern Mexico to Panama, in Cuba, also rarely along Gulf and Atlantic coasts of southeastern United States.

Swifts
(Order Apodiformes, Family Apodidae)

Swifts are small to medium-sized, stubby-tailed birds with long, pointed wings. In flight, an extremely rapid, shallow wing beat is characteristic.

Chimney Swift
Chaetura pelagica
(L-5 W-12)

Dark throughout; stubby, square tail; long, narrow wing.

Voice Rapid series of "chip"s.
Similar Species Cigar shape (stub tail), rapid wing beat, and constant twittering during flight separate this bird from swallows.
Habitat Widely distributed over most habitat types wherever appropriate nesting and roosting sites are available (chimneys, cliffs, caves, crevices, hollow trees). Often seen over towns and cities where chimneys are available for roosting.
Abundance and Distribution Common transient and summer resident* (May–Sep) throughout.

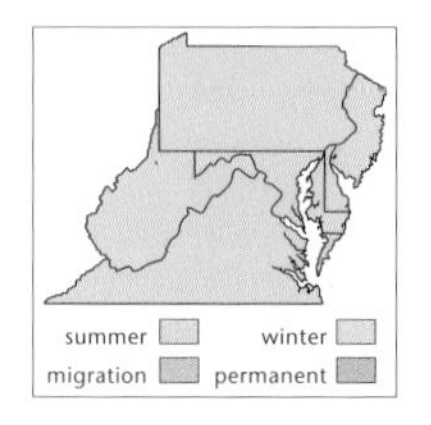

Where to Find Northern Virginia Regional Park, Virginia; Inner Harbor, Baltimore, Maryland; Great Swamp National Wildlife Refuge, New Jersey.
Range Breeds throughout eastern North America; winters mainly in Peru.

Hummingbirds
(Family Trochilidae)

Tiny, brilliantly colored birds that feed on nectar and insects. The only bird species in our region that can fly backward.

Ruby-throated Hummingbird
Archilochus colubris
(L-4 W-4)

Green above; whitish below; red throat. *Female and Immature:* Green above, whitish below.

Voice Song—"chip chipit chip chipit chipit."

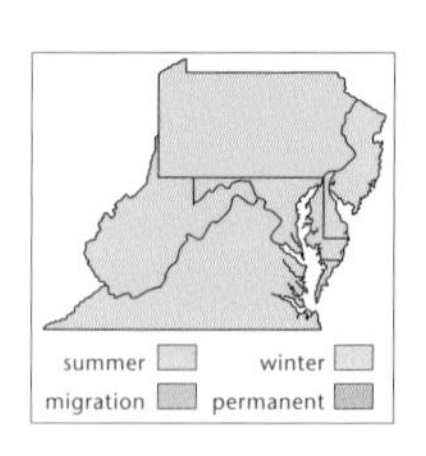

Similar Species The rubythroat is the only hummingbird found regularly in the eastern United States.
Habitat Woodlands, second growth, brushy pastures and fields, gardens.
Abundance and Distribution Common transient and summer resident* (May–Sep) throughout.
Where to Find Berwind Lake Wildlife Management Area, West Virginia; Bowman's Hill Wildflower Preserve, Pennsylvania; State Arboretum, Virginia.
Range Breeds in southern Canada and the eastern half of the United States; winters from southern Mexico to Costa Rica; also south Florida and Cuba.

Kingfishers
(Order Coraciiformes, Family Alcedinidae)

Stocky, vocal birds; predominantly slate blue or green with long, chisel-shaped bills and shaggy crests; generally found near water, where they dive from a perch or from the air for fish.

Belted Kingfisher
Ceryle alcyon
(L-13 W-22)

Blue-gray above with ragged crest; white collar, throat, and belly; blue-gray breast band. *Female:* Like male but with chestnut band across belly in addition to blue-gray breast band.

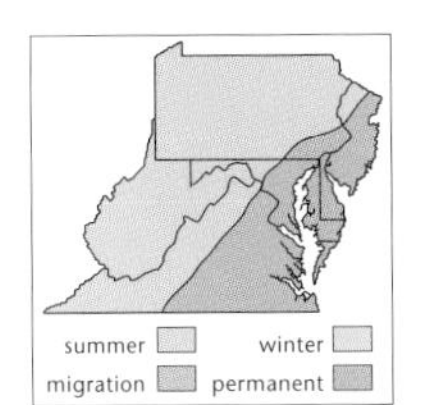

Voice A loud rattle.
Habitat Rivers, lakes, ponds, bays.
Abundance and Distribution Common summer resident* (Mar–Oct) throughout; withdraws from highlands and northern parts of region in winter.
Where to Find Seneca Creek State Park, Maryland; Wharton State Forest, New Jersey; Otter Creek, Virginia.
Range Breeds throughout most of temperate and boreal North America, excluding arid southwest; winters from southern portion of breeding range south through Mexico and Central America to northern South America, also winters in West Indies and Bermuda.

Woodpeckers
(Order Piciformes, Family Picidae)

A distinctive family of birds with plumages of mainly black and white. Long claws, stubby, strong feet, and stiff tail feathers enable woodpeckers to forage by clambering over tree trunks. The long, heavy, pointed bill is used for chiseling, probing, flicking, or hammering for arthropods. Most species fly in a distinct, undulating manner.

Red-headed Woodpecker
Melanerpes erythrocephalus
(L-10 W-18)

Black above, white below; scarlet head. *Immature:* Head brownish.

Voice A loud "keeeer."
Habitat Open oak woodlands, riparian stands.

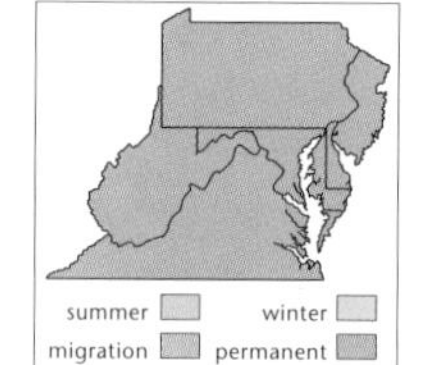

Abundance and Distribution Uncommon to rare and local resident* throughout; withdraws in winter from highland areas.
Where to Find Colonel Denning State Park, Pennsylvania; Beech Fork State Park, West Virginia; Mason Neck National Wildlife Refuge, Virginia.
Range Eastern North America from central plains and southern Canada to Texas and Florida.

Red-bellied Woodpecker
Melanerpes carolinus
(L-10 W-16)

Nape and crown red; barred black and white on back and tail; white rump; dirty white below with reddish tinge on belly. *Female:* Crown whitish.

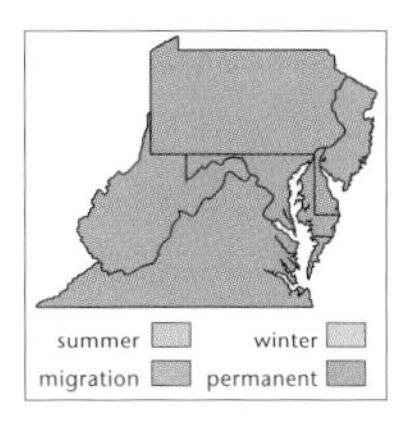

Voice A repeated "churr."
Similar Species Only the rare Red-cockaded, among other woodpeckers found in the region, has a back barred black and white, and it does not have the red cap and nape and white rump of the Red-bellied Woodpecker.
Habitat Riparian forests and mixed woodlands.
Abundance and Distribution Common resident* nearly throughout except in highlands; scarce or absent from parts of northern Pennsylvania.
Where to Find McKee-Beshers Wildlife Management Area, Maryland; Great Swamp National Wildlife Refuge, New Jersey; Bombay Hook National Wildlife Refuge, Delaware.
Range Eastern United States.

Yellow-bellied Sapsucker
Sphyrapicus varius
(L-8 W-15)

Black and white above; creamy yellow below with dark flecks on sides; breast black; throat, forehead, and crown red; black and white facial pattern. *Female:* White throat. *Immature:* Black wings with white patch, white rump, and checked black-and-white tail of adults, but barred brownish and cream on head, back, breast, and belly.

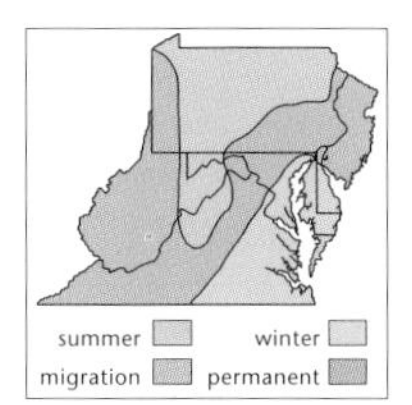

Voice A thin, nasal "cheerr."
Similar Species The yellow belly serves to distinguish this bird from other woodpeckers in the region.
Habitat Deciduous and mixed forest.
Abundance and Distribution Uncommon summer resident* (May–Sep) in northern Pennsylvania; rare and local summer resident in the mountains (above 3,000 feet) in southern Pennsylvania,

Virginia, and Pocahontas, Randolph, and Barbour counties in eastern West Virginia. Uncommon transient (Sep–Oct, Apr–May) elsewhere. Winter resident (Sep–May) in the Piedmont and Coastal Plain of Maryland, Virginia, and Delaware.
Where to Find Woodbourne Sanctuary, Pennsylvania (breeding); James River Park, Richmond, Virginia (winter); Seneca Creek State Park, Maryland (winter).
Range Breeds in temperate and boreal wooded regions of Canada and the United States; winters from the southern United States south to Panama and the West Indies.

Downy Woodpecker

Picoides pubescens

(L-7 W-12)

Black and white above; white below; red cap on back portion of crown. *Female:* Black cap.

Voice A sharp "peek"; a rapid, descending series of "pik"s.

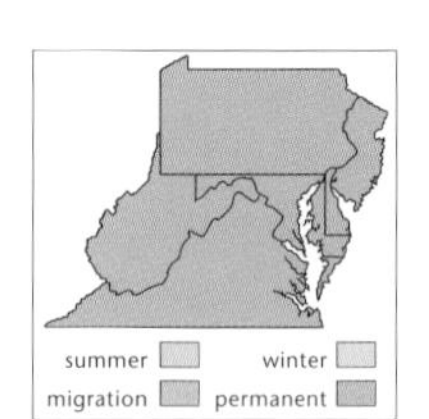

Similar Species The white back separates this sparrow-sized woodpecker from all others except the thrush-sized Hairy Woodpecker, which is much larger and has a lower, usually single, call note.
Habitat Forests, second growth, and park lands.
Abundance and Distribution Common resident* throughout.
Where to Find Lums Pond State Park, Delaware; Tomlinson Run State Park, West Virginia; Wawayanda State Park, New Jersey; Rock Creek Park, Washington, D.C.
Range Temperate and boreal North America south to southern California, Arizona, New Mexico, and Texas.

Hairy Woodpecker

Picoides villosus

(L-9 W-15)

Black and white above; white below; red cap on back portion of crown. *Female:* Black cap.

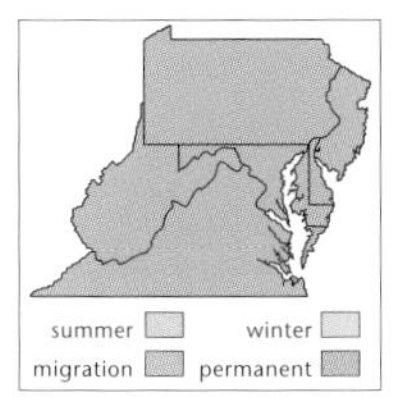

Voice A sharp "pick"; a descending series of "pick"s.

Similar Species The white back separates this thrush-sized woodpecker from all others except the sparrow-sized Downy Woodpecker, which is much smaller and has a higher-pitched call note and rattle.

Habitat Coniferous, mixed, deciduous, and riparian woodlands.

Abundance and Distribution Uncommon resident* throughout.

Where to Find Patuxent Research Refuge, Maryland; Ridley Creek State Park, Pennsylvania; Staunton River State Park, Virginia.

Range Temperate and boreal North America south through the mountains of western Mexico and Central America to Panama, Bahamas.

Red-cockaded Woodpecker

Picoides borealis

(L-8 W-15)

ENDANGERED Barred black and white above; white below flecked with black along sides; white cheek bordered below by black stripe; red spot behind eye. *Female:* No red spot behind eye.

Voice A "srip" call is given regularly by family group members as they forage.

Similar Species Only the Red-bellied, among other woodpeckers found in the region, also has a back barred black and white, and it has a red cap and nape and white rump, which the Red-cockaded lacks.
Habitat Pine savanna, especially mature long leaf and loblolly pine stands.
Abundance and Distribution Rare resident* of extreme southeastern Virginia.
Where to Find Pine forests in rural Sussex and Southampton counties of Virginia.
Range Southeastern United States.

Northern Flicker

Colaptes auratus
(L-12 W-20)

Barred brown and black above; tan with black spots below; black breast; white rump. Eastern forms have yellow underwings and tail linings, gray cap and nape with red back of crown, tan face and throat with black mustache. *Female:* Lacks mustache.

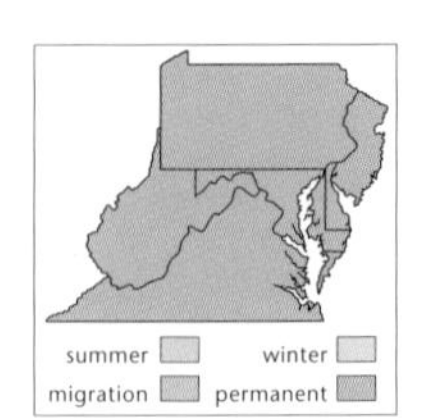

Voice A long series of "kek"s, first rising, then falling; a "kleer" followed by several "wika"s.
Habitat Forests, parklands, orchards, woodlots, and residential areas.
Abundance and Distribution Common transient and summer resident* (Mar–Oct) throughout; uncommon (south, Piedmont, Coastal Plain) to rare (north, mountains) winter resident (Nov–Feb).
Where to Find C&O Canal, Maryland; Bombay Hook National Wildlife Refuge, Delaware; Canaan Valley State Park, West Virginia.
Range Nearly throughout temperate and boreal North America south in highlands of Mexico to

Oaxaca. Leaves northern portion of breeding range in winter.

Pileated Woodpecker

Dryocopus pileatus

(L-17 W-27)

Black, crow-sized bird; black-and-white facial pattern; red mustache and crest; white wing lining. *Female:* Black mustache.

Voice Loud "kuk kuk kuk kuk kuk kuk" stops abruptly, does not trail off like flicker. The drum is a loud rattle on a large log, first increasing, then decreasing in pace, and lasting 2–3 seconds.

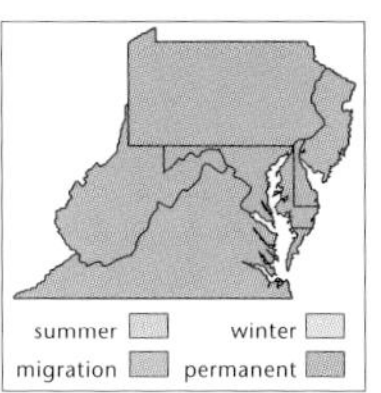

Habitat Deciduous and mixed forest.

Where to Find Glover-Archbold Park, Washington, D.C.; Reed's Gap State Park, Pennsylvania; Staunton River State Park, Virginia.

Range Much of temperate and boreal North America, excluding Great Plains and Rocky Mountain regions.

Flycatchers and Kingbirds

(Order Passeriformes, Family Tyrannidae)

A large group of mostly tropical songbirds, robin-sized or smaller. They are characterized by erect posture, muted plumage coloration (olives, browns, and grays), a broad bill, hooked at the tip, and bristles around the mouth. Flycatchers are mostly insectivorous, often capturing their prey on the wing.

Olive-sided Flycatcher

Contopus cooperi

(L-8 W-13)

Olive above; white below with olive on sides of breast and belly; white tufts on lower back sometimes difficult to see.

Voice "Whit whew whew" (Quick, three beers).

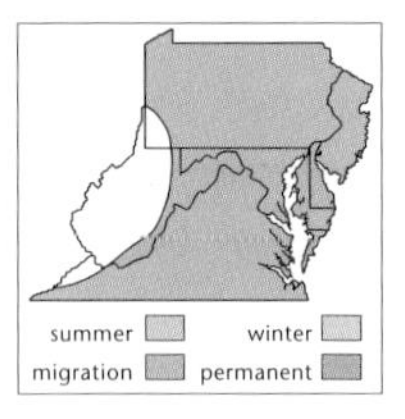

Similar Species Eastern Wood-Pewee is smaller, has wing bars, and lacks dark olive along sides of breast and belly.

Habitat Breeds in hemlock and spruce-fir forests, where it often perches on lone, tall, dead snags; found in a variety of forested habitats during migration.

Abundance and Distribution Uncommon to rare transient (Sep, May) nearly throughout; casual in western West Virginia. Rare and local summer resident* (May–Sep) in the mountains of Pennsylvania, eastern West Virginia (Randolph and Pocahontas counties), and Virginia (Mount Rogers).

Where to Find Cranberry Glades Botanical Area, West Virginia (breeding); Buzzard Swamp Wildlife Management Area, Pennsylvania; Grayson Highlands State Park, Virginia.

Range Breeds in boreal forest across the northern tier of North America and through mountains in west, south to Texas and northern Baja California; winters in mountains of South America from Colombia to Peru.

Eastern Wood-Pewee

Contopus virens

(L-6 W-10)

Dark olive above; greenish wash on breast; belly whitish; white wing bars.

Voice Plaintive "peeeaweee, peeeaweee, peeyer" song; "peeyer" call.

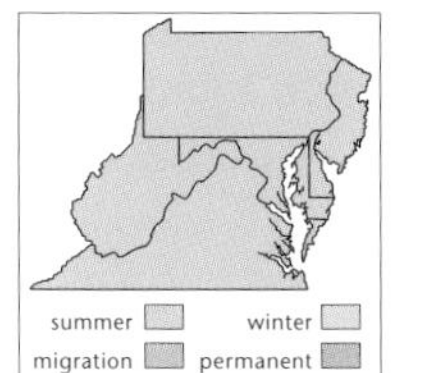

Similar Species Pewees lack the eyering of *Empidonax* species (flycatchers) and show distinct wing bars, unlike phoebes.
Habitat Deciduous and mixed woodlands.
Abundance and Distribution Common transient and summer resident* (May–Sep) throughout.
Where to Find Rock Creek Park, Washington, D.C.; High Point State Park, New Jersey; Green Ridge State Forest, Maryland.
Range Breeds in eastern United States and southeastern Canada; winters in northern South America.

Yellow-bellied Flycatcher

Empidonax flaviventris

(L-5 W-8)

Greenish above; yellowish below; white wing bars and eyering.

Voice Song—a whistled "perweee"; call—a distinctive "squeeup." Both sexes sing and call during migration and on wintering ground.

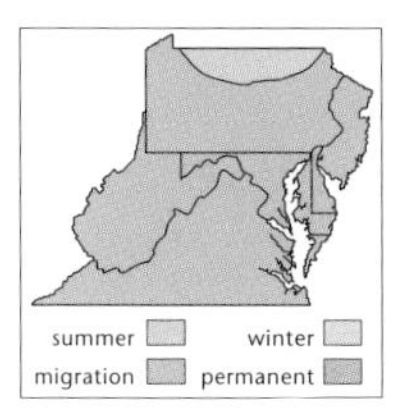

Similar Species All *Empidonax* flycatchers are similar. The 2 yellowish species in this region are Yellow-bellied and Acadian. Calls are distinctive for each. Acadian usually has white throat, is not as yellow below as the other 2, and has bluish legs. Yellow-bellied legs are usually dark brown or black.

Habitat Breeds in highland bogs in hemlock and spruce-fir forests; on migration in a variety of forest and shrublands.
Abundance and Distribution Rare and local summer resident* (May–Sep) in northern Pennsylvania; scattered breeding records for West Virginia (Cranberry Back Country Wilderness, Pocahontas County) and Virginia (Mount Rogers); rare transient (Sep, May) elsewhere.
Where to Find Buzzard Swamp Wildlife Management Area, Pennsylvania; Cranberry Glades Botanical Area, West Virginia; Mount Rogers, Virginia.
Range Breeds northern United States and southern Canada west to British Columbia; winters from central Mexico south to Panama.

Acadian Flycatcher

Empidonax virescens

(L-6 W-9)

Green above; whitish below; yellow on flanks; white wing bars; whitish eyering; bluish legs.

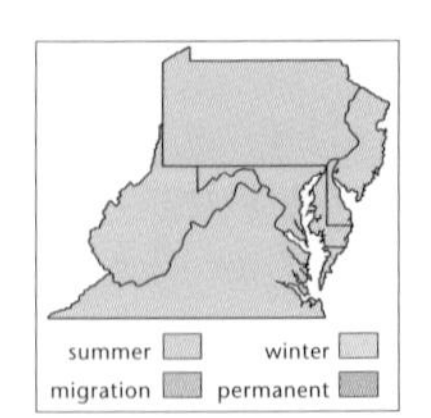

Voice Song—a repeated "pit see"; call—a thin "peet."
Similar Species All *Empidonax* are similar. The 2 yellowish species in this region are Yellow-bellied and Acadian. Calls are distinctive for each. Acadian usually has white throat, is not as yellow below as the others, and has bluish legs. Yellow-bellied legs are usually dark brown or black.
Habitat Moist forest, riparian forest.
Abundance and Distribution Common to uncommon transient and summer resident* (May–Sep) throughout; somewhat scarcer as a breeder in northern Pennsylvania and northern New Jersey.
Where to Find Blackwater National Wildlife

Refuge, Maryland; Shenandoah Mountain Recreation Area, West Virginia; Trap Pond State Park, Delaware.
Range Breeds across the eastern United States; winters from Nicaragua south through Central America to northern South America.

Alder Flycatcher

Empidonax alnorum
(L-6 W-9)

Brownish olive above, whitish below; greenish on flanks; white wing bars and eyering.

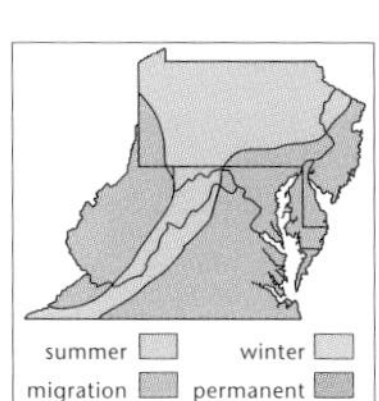

Voice Song—"fee-bee-o," call—"pep."
Similar Species All of the whitish *Empidonax* are similar but distinguishable by song or in the hand using the key by Phillips et al. (1966).
Habitat Swamp thickets, riparian forest.
Abundance and Distribution Uncommon summer resident* (May–Sep) in northern and western Pennsylvania and New Jersey; rare and local summer resident* (May–Sep) in the highlands of West Virginia, western Maryland, and Virginia; rare transient (Sep, May) elsewhere.
Where to Find Stokes State Forest, New Jersey; Swallow Falls State Park, Maryland; Buzzard Swamp Wildlife Management Area, Pennsylvania.
Range Breeds across the northern tier of the continent in Canada and Alaska and south through the Appalachians to North Carolina; winters in northern South America.

Willow Flycatcher

Empidonax traillii

(L-6 W-9)

Brownish olive above; whitish below, turning to a creamy yellow on belly; greenish on flanks; white wing bars and eyering.

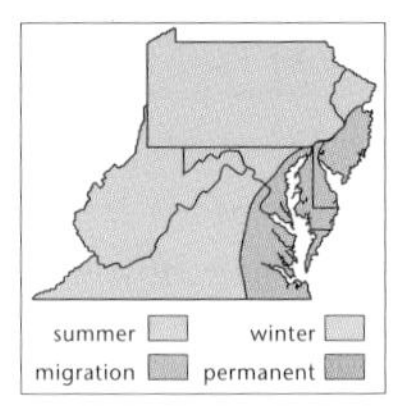

Voice A buzzy, slurred "fitz-bew" song; "wit" call note.

Similar Species A combination of behavior, voice, habitat, and minute details of appearance is often needed to separate species of *Empidonax* flycatchers. The song of the Willow Flycatcher is distinctive, and it is the characteristic *Empidonax* species of willow bogs and swamps. On migration, it is very difficult to separate from other members of the group.

Habitat Willow thickets, shrubby fields, riparian forest.

Abundance and Distribution Uncommon transient and summer resident* (May–Sep) in the Piedmont and highlands; uncommon to rare transient and rare summer resident* (May–Sep) in the Coastal Plain.

Where to Find Lewis Wetzel Wildlife Management Area, West Virginia; Clinch Valley College Campus, Virginia; Erie National Wildlife Refuge, Pennsylvania.

Range Breeds through much of north and central United States; winters in northern South America.

Least Flycatcher

Empidonax minimus

(L-5 W-8)

Brownish above; white below; wing bars and eyering white.

Habits Bobs tail.
Voice Song—"che bec," given by both sexes during migration and on wintering grounds; call—a brief "wit."
Similar Species A combination of behavior, voice, habitat, and minute details of appearance is often needed to separate one species of *Empidonax* flycatcher from another. Tail bobbing habit of the Least can be used to separate it from the Willow Flycatcher.
Habitat Open woodlands.
Abundance and Distribution Uncommon transient and summer resident* (May–Sep) in Pennsylvania and the Piedmont and Appalachian highlands elsewhere; uncommon to rare transient and rare summer resident* (May–Sep) in the Coastal Plain.
Where to Find Grayson Highlands State Park, Virginia; Stokes State Forest, New Jersey; Prince Gallitzin State Park, Pennsylvania.
Range Breeds in northern United States and southern and central Canada west to British Columbia, south in Appalachians to north Georgia; winters from central Mexico to Panama.

Eastern Phoebe

Sayornis phoebe

(L-7 W-11)

Dark brown above; whitish or yellowish below. Dark cap often has crest-like appearance. Bobs tail.

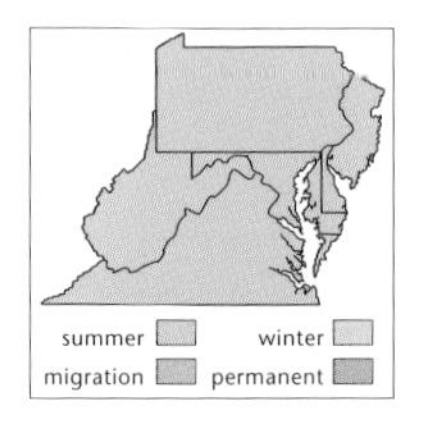

Voice Song—"febezzt feebezzt," repeated; call—a clear "tship."
Similar Species Eastern Wood-Pewee has distinct wing bars, which the phoebe lacks; *Empidonax* flycatchers have wing bars and eyering.
Habitat Riparian forest; deciduous and mixed forest edge; good nest site localities (cliffs, eaves, bridges) seem to determine breeding habitat.
Abundance and Distribution Common transient and summer resident* (Mar–Oct) throughout; rare winter resident (Nov–Feb), mainly along the coast.
Where to Find Bombay Hook National Wildlife Refuge, Delaware; Meadowside Nature Center, Maryland; Natural Chimneys Regional Park, Virginia.
Range Breeds from eastern British Columbia across southern Canada south through eastern United States (except southern Coastal Plain); winters southeastern United States south through eastern Mexico to Oaxaca and Veracruz.

Great Crested Flycatcher

Myiarchus crinitus
(L-9 W-13)

Brown above; throat and breast gray; belly yellow; rufous tail.

Voice "Wheep."
Similar Species The rare Western Kingbird (coast) also has a gray throat. However, it has a gray head (not brown), no crest, a greenish back, and a dark tail with white outer tail feathers.
Habitat Deciduous and mixed forest.
Abundance and Distribution Common transient and summer resident* (May–Sep) throughout.
Where to Find Cowan's Gap State Park, Pennsylvania; Great Swamp National Wildlife Refuge, New

Jersey; Falls of Hills Creek, West Virginia; Scott's Run Nature Preserve, Virginia.

Range Breeds in the eastern United States, central and southeastern Canada; winters in southern Florida, Cuba, and southern Mexico through Middle America to Colombia and Venezuela.

Western Kingbird

Tyrannus verticalis

(L-9 W-15)

Pearl gray head and breast; back and belly yellowish; wings dark brown; tail brown with white outer tail feathers; male's red crown patch usually concealed.

Voice Various harsh twitters; call—a sharp "whit."

Similar Species In contrast to the Great Crested Flycatcher, the Western Kingbird has a gray throat (not brown), no crest, a greenish (not brown) back, and a dark tail with white (not rusty) outer tail feathers.

Habitat Grasslands, pastures, old fields.

Abundance and Distribution Rare fall transient (Sep–Oct) along the coast.

Where to Find Chincoteague National Wildlife Refuge, Virginia; Higbee Beach Wildlife Management Area, New Jersey; Assateague Island National Seashore, Maryland.

Range Breeds in western North America from southern British Columbia and Manitoba to southern Mexico; winters from southern Mexico to Costa Rica.

Eastern Kingbird

Tyrannus tyrannus

(L-9 W-15)

Blackish above; white below; terminal white band on tail; red crest is usually invisible.

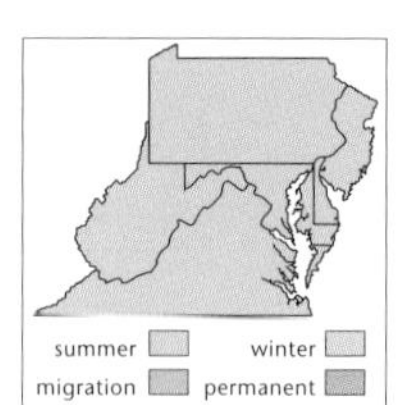

Habits The male is very aggressive to intruders on breeding territory, dive bombing any large bird that happens to cross his airspace and harassing mammals, including humans, that trespass.

Voice A harsh, high-pitched rattle—"kit kit kitty kitty," or "tshee" repeated.

Habitat Wood margins, open farmland, hedgerows.

Abundance and Distribution Common transient and summer resident* (May–Aug) throughout.

Where to Find Meadowside Nature Center, Maryland; Hog Island Waterfowl Management Area, Virginia; Bombay Hook National Wildlife Refuge, Delaware.

Range Breeds from central Canada south through eastern and central United States to eastern Texas and Florida; winters in central and northern South America.

Shrikes

(Family Laniidae)

The New World species in this group are medium-sized, stocky, predominantly gray-and-white birds with hooked beaks and strong feet for grasping and killing insects, small mammals, birds, reptiles, and amphibians.

Loggerhead Shrike

Lanius ludovicianus

(L-9 W-13)

Gray above, paler below; black mask, wings, and tail; white outer tail feathers and wing patch.

Habits Hunts from exposed perches; wary, flees observer with rapid wing beats; larders prey on thorns and barbs.

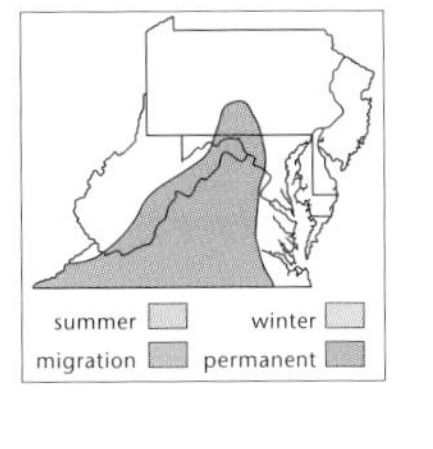

Voice Song—a series of weak warbles and squeaks; call—a harsh "chaaa," often repeated 3–4 times.
Similar Species The Northern Mockingbird is slighter, lacks heavy, hooked bill, has longer tail, and has no black mask. Unlike the Northern Shrike, the Loggerhead Shrike does not hover or pump tail, is smaller and grayer with black rather than whitish forehead, grayish rather than whitish rump, and black rather than whitish lower mandible.
Habitat Old fields, farmlands, hedgerows; prefers open, barren areas where perches (such as fenceposts or telephone poles) are available for sallying.
Abundance and Distribution Uncommon to rare and local resident* in Virginia west of the Coastal Plain, Maryland (Washington and Frederick counties), south central Pennsylvania, and eastern and southern West Virginia; rare to casual visitor* elsewhere in the region.
Where to Find Shenandoah Valley farmlands, Virginia.
Range Breeds locally over most of the United States and southern Canada south through highlands of Mexico to Oaxaca; winters in all but north portions of breeding range.

Northern Shrike

Lanius excubitor

(L-10 W-14)

Pearly gray above, faintly barred below; black mask, primaries, and tail; white wing patch and outer tail feathers; heavy, hooked bill.

Habits Sits on high, exposed perches; pumps tail; sometimes hovers while hunting.

Voice Song of trills and buzzes; harsh chatter call is given infrequently.

Similar Species Loggerhead Shrike doesn't hover or pump tail, is smaller and grayer with black rather than whitish forehead, grayish rather than whitish rump, and black rather than whitish lower mandible.

Habitat Open woodlands and farmland.

Abundance and Distribution Rare and irregular winter visitor (Nov–Feb) to New Jersey; casual elsewhere; more numerous in some years than others.

Where to Find Rosedale Park, Princeton, New Jersey.

Range Breeds across boreal regions of the Old and New World; winters in north temperate and southern boreal regions.

Vireos

(Family Vireonidae)

A predominantly tropical family of small, feisty birds, most of which are patterned in greens, yellows, grays, and browns. The relatively thick bill has a small, terminal hook. Most are woodland species that deliberately glean twigs, leaves, and branches for insect larvae.

White-eyed Vireo

Vireo griseus

(L-5 W-8)

Greenish gray above, whitish washed with yellow below; white eye; yellowish eyering and forehead ("spectacles"); white wing bars. *Immature:* Brown eye until Aug–Sep.

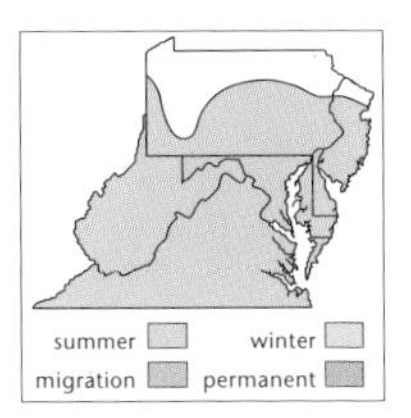

Habits Solitary and retiring; both sexes sing occasionally in winter.

Voice Song—"tstlik tstlit-a-lur tslik" or variations; call—a whiney "churr" repeated.

Similar Species *Empidonax* flycatchers have whiskers and flattened bill; lack "spectacles" and white eye.

Habitat Riparian woodland, thickets, swamp borders, overgrown fields.

Abundance and Distribution Common to uncommon summer resident* (May–Sep) except in highlands, northern Pennsylvania, and northern New Jersey, where scarce.

Where to Find Seneca Creek State Park, Maryland; Norman G. Wilder Wildlife Area, Delaware; Ridley Creek State Park, Pennsylvania.

Range Breeds in the eastern United States from southern Minnesota to Massachusetts southward to Florida and eastern Mexico to Veracruz. Winters from the southeastern United States south through eastern Mexico and Central America to northern Nicaragua, winters in Bahamas and Greater Antilles.

Yellow-throated Vireo

Vireo flavifrons

(L-6 W-10)

Greenish above with gray rump; yellow throat, breast, and "spectacles" (lores, forehead, eyering); white belly and wing bars.

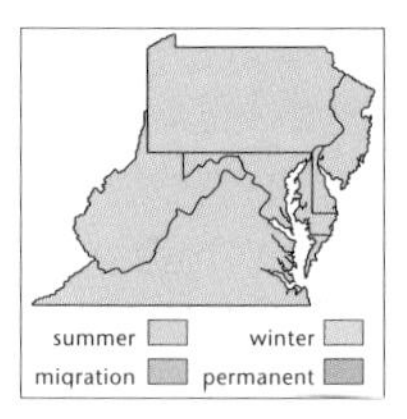

Voice Like a hoarse, slow Red-eyed Vireo, "chearee, . . . chewia, . . . tsuweet."

Similar Species The smaller, slimmer, more active Pine Warbler has yellow-green rump (not gray), white tail spots, and thin bill.

Habitat Deciduous forest, riparian woodland.

Abundance and Distribution Common to uncommon transient and summer resident* (May–Sep) nearly throughout (scarce in swamplands of southern New Jersey and highlands of central Pennsylvania and southeastern West Virginia).

Where to Find Violette's Lock, Maryland; Nanticoke Wildlife Area, Delaware; Swift Creek Lake, Virginia.

Range Breeds in the eastern United States. Winters from southern Mexico south through Central and northern South America; Bahamas; Greater Antilles.

Blue-headed Vireo

Vireo solitarius

(L-6 W-10)

Greenish on the back with a gray rump; whitish below with yellow flanks; 2 prominent, white wing bars; gray head and white "spectacles" (lores, forehead, eyering).

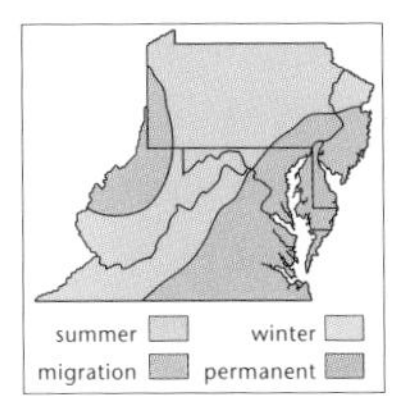

Habits A deliberate forager like most vireos but in contrast to most warblers, which forage in much more active fashion.
Voice Song—a rich series of "chu wit, . . . chu wee, . . . cheerio . . . ," similar to Red-eyed Vireo but with longer pauses between phrases. Call—a whiney "cheeer."
Similar Species The gray head and white spectacles of the Blue-headed Vireo are distinctive in this region.
Habitat Coniferous and mixed woodlands.
Abundance and Distribution Common transient and summer resident* (Apr–Oct) in highlands and northern Pennsylvania; uncommon to rare transient (Sep–Oct, Apr) elsewhere; rare winter visitor (Nov–Mar), mainly along the coast.
Where to Find Hemlock Springs Overlook, Shenandoah National Park, Virginia; Spruce Knob, West Virginia; Erie National Wildlife Refuge, Pennsylvania.
Range Breeds across central and southern Canada, northern and western United States. Winters from the southern United States through Mexico and Central America to Costa Rica; Cuba.

Warbling Vireo
Vireo gilvus
(L-6 W-9)

Grayish green above, pale yellow below; white eye stripe; no wing bars.

Voice Song—a long, wandering series of warbles, almost always ending with an upward inflection. Call—a hoarse "tswee."
Similar Species Can be difficult to separate from some pale Philadelphia Vireos, most of which are yellower below and have a dark loral spot.

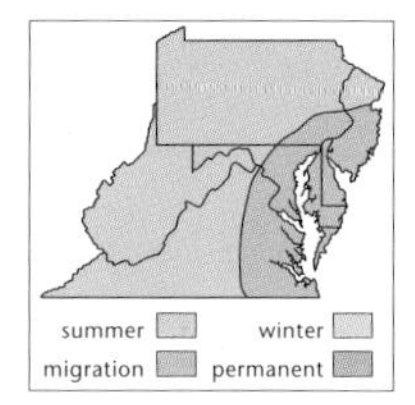

Habitat Riparian woodlands and floodplains; often frequents tall elms and sycamores.
Abundance and Distribution Uncommon transient and local summer resident* (May–Aug) in Piedmont and highlands; rare transient and rare to casual summer resident* along Coastal Plain.
Where to Find Beech Fork State Park, West Virginia; Ridley Creek State Park, Pennsylvania; Shenandoah River State Park, Virginia.
Range Breeds across North America south of the Arctic region to central Mexico. Winters from Guatemala to Panama.

Philadelphia Vireo

Vireo philadelphicus
(L-5 W-8)

Grayish green above, variably yellowish below; white eye stripe and dark loral spot; no wing bars.

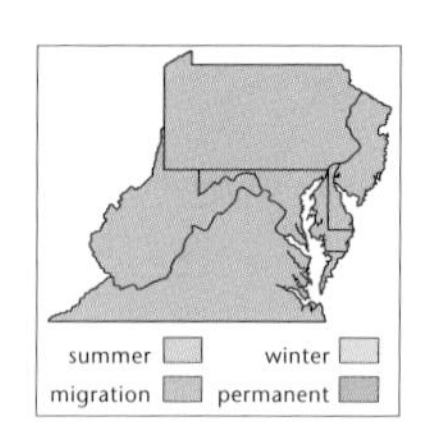

Habits Deliberate foraging movements with much peering and poking in outer clumps of leaves and twigs; even hanging occasionally like a chickadee.
Voice Song is similar to that of the Red-eyed Vireo but higher and slower.
Similar Species Some Warbling Vireos can be difficult to separate from pale Philadelphia Vireos, which are yellower below and have a dark loral spot. Could be mistaken for immature Tennessee Warbler or Orange-crowned Warbler except for thick vireo bill and deliberate foraging behavior.
Habitat Deciduous and mixed forest.
Abundance and Distribution Uncommon fall transient (Sep) and rare spring transient (May) throughout.

Where to Find Violette's Lock, Maryland; Island Beach State Park, New Jersey; Ohiopyle State Park, Pennsylvania.
Range Breeds eastern and central boreal North America in Canada and northern United States. Winters Guatemala to Panama.

Red-eyed Vireo

Vireo olivaceus
(L-6 W-10)

Olive above, whitish below; gray cap; white eyeline; red eye (brown in juvenile).

summer winter
migration permanent

Voice Song—a leisurely, seemingly endless series of phrases: "cheerup, cherio, chewit, chewee," through the middle of long summer days. Male occasionally sings from nest while incubating. Call—a harsh "cheear."
Similar Species Red-eyed Vireo is the only gray-capped vireo without wing bars.
Habitat Deciduous and mixed woodlands.
Abundance and Distribution Common transient and summer resident* (May–Sep) throughout.
Where to Find Lebanon State Forest, New Jersey; Green Ridge State Forest, Maryland; Canaan Valley State Park, West Virginia.
Range Breeds over much of North America (except western United States, Alaska, and northern Canada). Subspecies breed in Central and South America. United States subspecies winters in northern South America.

Crows and Jays
(Family Corvidae)

Medium-sized to large songbirds; crows and ravens have generally dark plumages; jays are blues, greens, and grays. Most members are social, aggressive, and highly vocal.

Blue Jay
Cyanocitta cristata
(L-11 W-16)

Blue above; a dirty white below; blue crest; white wing bars and tip of tail; black necklace.

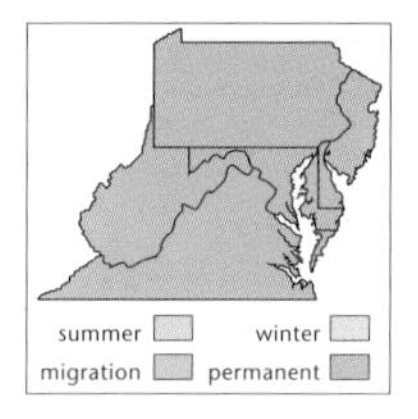

Voice A loud, harsh cry, "jaaaay," repeated; a liquid gurgle; many other cries and calls; mimics hawks.
Habitat Deciduous forest, parklands, residential areas.
Abundance and Distribution Common resident* throughout.
Where to Find National Zoo, Washington, D.C.; East Lynn Lake Wildlife Management Area, West Virginia; Blackiston Wildlife Area, Delaware.
Range Temperate and boreal North America east of the Rockies.

American Crow
Corvus brachyrhynchos
(L-18 W-36)

Black throughout.

Voice A series of variations on "caw"—fast, slow, high, low, and various combinations; nestlings and

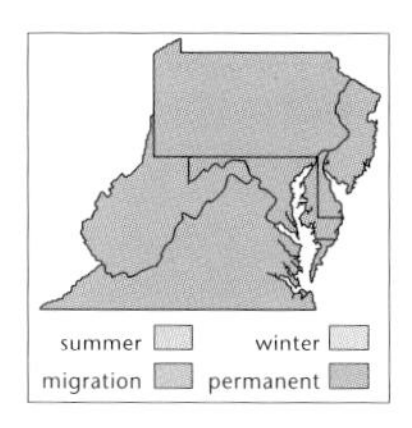

fledglings use an incessant, nasal "caah."
Habitat Deciduous and mixed woodlands, grasslands, farmlands, and residential areas in winter.
Similar Species Ravens croak and have wedge-shaped rather than square tail of crows. American Crow is distinguished from the smaller Fish Crow mainly on voice: American Crows usually give the familiar "caw," whereas Fish Crows have a high-pitched, nasal "caah" (although juvenile American Crows have a similar call). All 3 species can occur simultaneously along the east slope of the Blue Ridge.
Abundance and Distribution Common resident* throughout.
Where to Find Gunpowder Falls State Park, Maryland; Shenandoah River State Park, Virginia; Prince Gallitzin State Park, Pennsylvania.
Range Temperate and boreal North America; migratory in northern portion of range.

Fish Crow

Corvus ossifragus
(L-15 W-33)

Black throughout.

Voice A high, nasal "cah."
Habitat Floodplain forests, bayous, and coastal waterways.

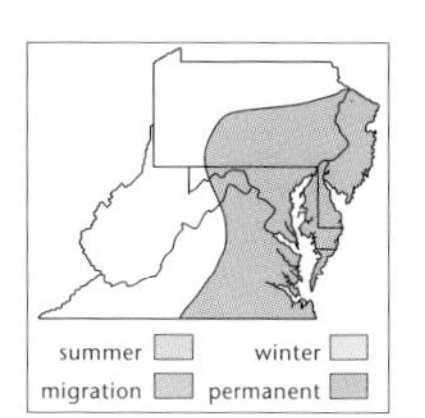

Similar Species Ravens croak and have wedge-shaped rather than square tail of crows. American Crows are best distinguished from the smaller Fish Crow by voice: American Crows usually give the familiar "caw," whereas Fish Crows have a high-pitched, nasal "caah" (although juvenile American Crows have a similar call). All 3 species can occur simultaneously along the east slope of the Blue Ridge.

Abundance and Distribution Common (Coastal Plain) to uncommon or rare (Piedmont) resident.* Withdraws toward the coast and southward in winter.
Where to Find The Mall, Washington, D.C.; Brigantine National Wildlife Refuge, New Jersey; Prime Hook National Wildlife Refuge, Delaware; Tinicum National Wildlife Refuge, Pennsylvania.
Range Coastal Plain and parts of the Piedmont of the eastern United States from Maine to eastern Texas.

Common Raven

Corvus corax

(L-24 W-48)

Black throughout; large, heavy bill; shaggy appearance in facial region; wedge-shaped tail.

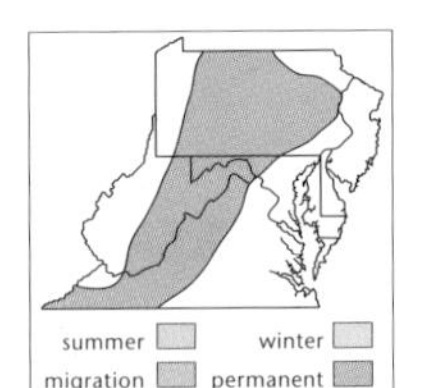

Voice A variety of harsh, low croaks.
Habitat Rugged crags, cliffs, and canyons, as well as a variety of forested and open lands.
Similar Species Ravens croak and have wedge-shaped rather than square tail of crows.
Abundance and Distribution Uncommon resident* of highlands in the Appalachian and Blue Ridge mountains throughout.
Where to Find Dickey Ridge Visitor Center, Shenandoah National Park, Virginia; Spruce Knob, West Virginia; Cook Forest State Park, Pennsylvania.
Range Arctic and boreal regions of the Northern Hemisphere south through mountainous regions of the Western Hemisphere to Nicaragua and in the Eastern Hemisphere to North Africa, Iran, the Himalayas, Manchuria, and Japan.

Larks
(Family Alaudidae)

Small, terrestrial songbirds with extremely long hind claws. The breeding song of these open country inhabitants is often given in flight.

Horned Lark
Eremophila alpestris
(L-7 W-13)

Brown above; white below with black bib; face and throat whitish or yellowish with black forehead and eyeline; black horns raised when singing. *Female:* Similar pattern but paler. *Immature:* Nondescript brownish above; whitish below with light streaking; has dark tail with white edging and long hind claw like adult.

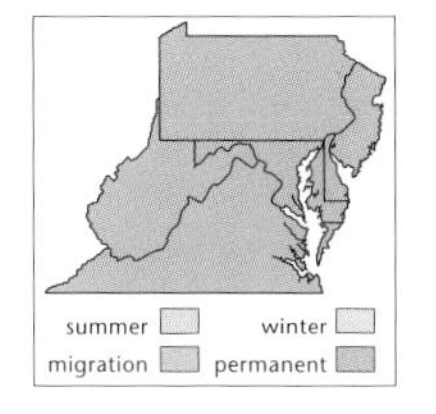

Voice Song—series of high-pitched notes, often given in flight; call—a thin "tseet."
Similar Species Immature lark is similar to pipits but is grayish white rather than buffy on face and underparts.
Habitat Overgrazed pasture, airports, golf courses, plowed fields, surface mines, roadsides, sand flats.
Abundance and Distribution Locally common to uncommon resident;* scarce in winter in highlands.
Where to Find Cape Henlopen State Park, Delaware; Curllsville Strip Mines, Pennsylvania; Grayson Highlands State Park, Virginia.
Range Breeds in North America south to southern Mexico; winters in southern portion of breeding range.

Swallows and Martins
(Family Hirundinidae)

Mostly small, sleek birds with pointed wings. Swallows forage on flying insects caught on the wing.

Purple Martin
Progne subis
(L-8 W-16)

Iridescent midnight blue throughout. *Female and Immature:* Dark, iridescent blue above; dirty white, occasionally mottled with blue below; grayish collar.

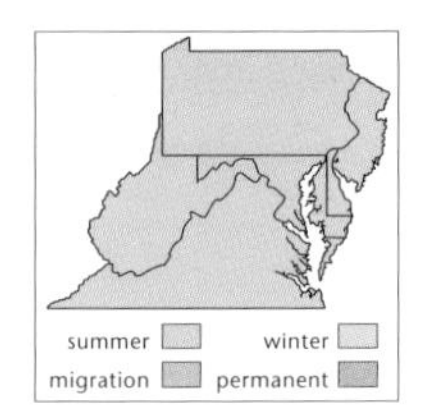

Voice Various squeaky "quik"s and "querks"s repeated in a short, harsh series.
Similar Species The male martin is the only entirely dark swallow in the region. The female and immature are variously whitish below but not pure white, as in the Tree Swallow. The Bank Swallow is white below with a brown breast band, and the Northern Rough-winged Swallow, although dingy white below, is brown, not iridescent blue as in martins. Martins are much larger than any other swallows found in the area.
Habitat Open areas, limited by nesting sites. Nests colonially, originally in hollow trees, now often in specially constructed martin houses.
Abundance and Distribution Locally common transient and summer resident* (Mar–Sep) nearly throughout; scarcer in highlands.
Where to Find Violette's Lock, Maryland; Bombay Hook National Wildlife Refuge, Delaware; Pocahontas State Park, Virginia.
Range Breeds in open areas over most of temperate North America south to the highlands of southern Mexico; winters in South America.

Tree Swallow

Tachycineta bicolor

(L-6 W-13)

Iridescent blue above; pure white below with slightly forked tail. *Female:* Similar in pattern to male but duller. *Immature:* Grayish brown above; whitish below.

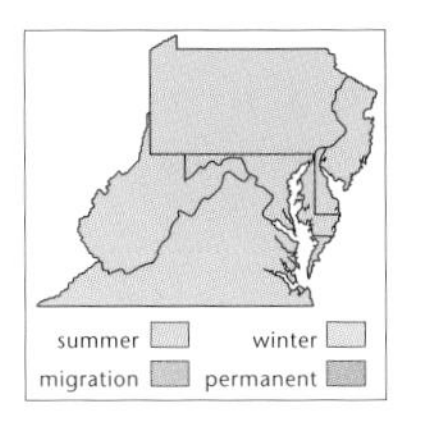

Voice Series of "weet tuwit tuweet" with twitters.
Similar Species The stark contrast between dark blue back and white underparts is distinctive among the region's swallows.
Habitat Lakes, ponds, marshes, open fields—any open area during migration. Limited in summer by breeding sites (holes in hollow trees or nest boxes).
Abundance and Distribution Common transient (Aug–Oct, Apr) and common to uncommon and local summer resident* (May–Aug) throughout.
Where to Find Canaan Valley State Park, West Virginia; Lily Pons Water Gardens, Maryland; Prince Gallitzin State Park, Pennsylvania.
Range Breeds over most of boreal and temperate North America south to the southern United States; winters from southern United States south to Costa Rica and Greater Antilles.

Northern Rough-winged Swallow

Stelgidopteryx serripennis

(L-6 W-12)

Grayish brown above; grayish throat and breast becoming whitish on belly.

Voice A harsh "treet," repeated.
Similar Species Of the 2 brown-backed swallows, the Bank Swallow has a distinct dark breast band

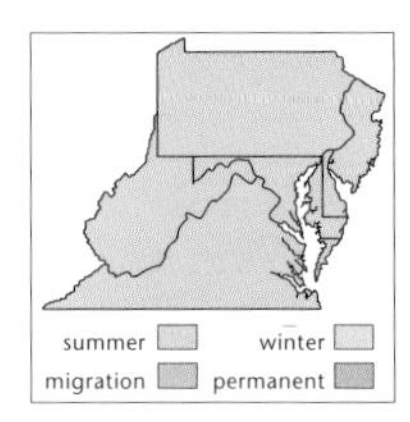

set off by white throat and belly; the roughwing lacks a breast band.

Habitat Lakes, rivers, ponds, streams, most open areas during migration. Nests in holes in banks, drain pipes, and abandoned rodent burrows, normally in single pairs.

Abundance and Distribution Common to uncommon transient and summer resident* (Apr–Sep) throughout.

Where to Find Mason Neck National Wildlife Refuge, Virginia; Cape May Point, New Jersey; Bluestone Wildlife Management Area, West Virginia.

Range Breeds locally over most of temperate North America; winters from southern Texas and northern Mexico south to Panama.

Bank Swallow

Riparia riparia

(L-6 W-12)

Brown above; white below with brown collar.

Voice "Bzzt," repeated in a harsh rattle.

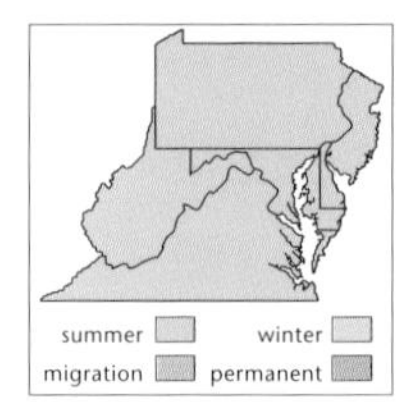

Similar Species Among the brown-backed swallows of the region, the Bank Swallow has a distinct dark breast band set off by white throat and belly; Northern Rough-winged Swallow lacks breast band.

Habitat Lakes, rivers, ponds, most open areas during migration. Nests in holes in banks, normally in colonies.

Abundance and Distribution Common to uncommon transient (Jul–Aug, May) throughout; uncommon summer resident* (May–Aug) in Pennsylvania, Maryland (mostly in Chesapeake Bay area),

and the Coastal Plain of Virginia; uncommon to rare and local summer resident elsewhere in the region.
Where to Find Lily Pons Water Gardens, Maryland; Tioga/Hammond Lakes National Recreation Area, Pennsylvania; Bombay Hook National Wildlife Refuge, Delaware.
Range Cosmopolitan, breeding over much of continental Northern Hemisphere, wintering in the tropics of Asia, Africa, and the New World.

Cliff Swallow

Hirundo pyrrhonota

(L-6 W-12)

Dark above; whitish below with dark orange or blackish throat; orange cheek; pale forehead; orange rump; square tail. *Immature:* Similar to adults but duller.

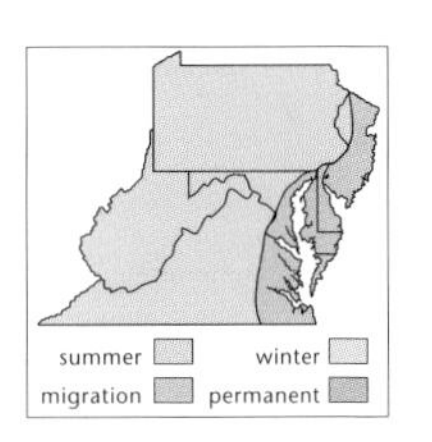

Voice Song—a series of harsh twitters.
Similar Species The Barn Swallow is somewhat similar in pattern, but has a very long, deeply forked tail. The mud nest of the Cliff Swallow is gourdlike, whereas that of the Barn Swallow is open-topped.
Habitat Most open areas during migration; agricultural areas, near water. This species seems to be limited by nesting locations during the breeding season (farm buildings, culverts, bridges, cliffs—near potential mud source).
Abundance and Distribution Uncommon to rare and local transient and summer resident* (Apr–Sep) nearly throughout; scarce or absent as a breeder on the Coastal Plain and most of New Jersey, where found only at a few colonies.
Where to Find Bull's Island State Park, New Jersey;

Little Creek Wildlife Area, Delaware; Kerr Reservoir, Virginia.
Range Breeds over most of North America to central Mexico; winters central and southern South America.

Barn Swallow
Hirundo rustica
(L-7 W-13)

Dark blue above; orange below with deeply forked tail. *Immature:* Paler with shorter tail.

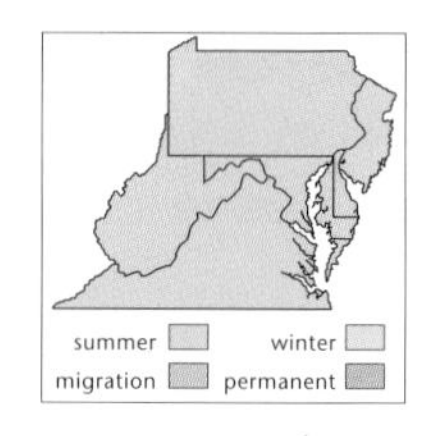

Voice Song is a series of squeaks and twitters, some harsh, some melodic.
Habitat Open areas near water; like most swallows, it requires special sites for nesting, such as bridges, culverts, buildings.
Similar Species The Barn Swallow is somewhat similar in pattern to the Cliff Swallow but has a very long, deeply forked tail. The mud nest of the Cliff Swallow is gourdlike, while that of the Barn Swallow is open-topped; both are placed under eaves, bridges, culverts, and dams and against similar protected walls. Barn Swallows prefer barns, garages, and similar roofed structures.
Abundance and Distribution Common transient and summer resident* throughout.
Where to Find Reed's Gap State Park, Pennsylvania; Summersville Lake Wildlife Management Area, West Virginia; Meadowside Nature Center, Maryland.
Range Cosmopolitan—breeds over much of the Northern Hemisphere; winters in South America, Africa, northern Australia, Micronesia.

Chickadees and Titmice
(Family Paridae)

Small, perky, active birds with plumages of gray and white; generally, these birds forage in small groups, feeding on arthropods and seeds.

Carolina Chickadee
Poecile carolinensis
(L-5 W-8)

Black cap and throat contrasting with white cheek, dark gray back, and grayish white breast and belly.

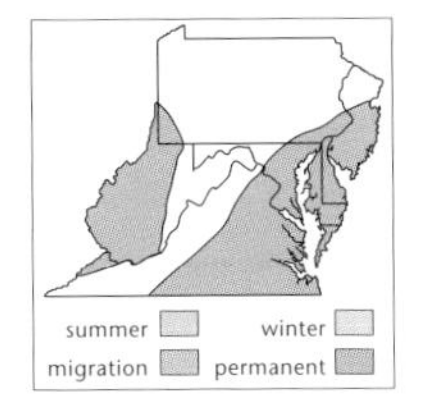

Habits Hangs upside down from branches and twigs while foraging.
Voice "See dee see doo"; "chick a dee dee dee."
Habitat Deciduous and mixed forest, residential areas.
Similar Species The Black-capped Chickadee has a 2-note song, "fee-bee," while the smaller Carolina Chickadee has a 4-note song, "see-dee see doo." The blackcap wing coverts and secondary outer edgings are whiter than those of the Carolina.
Abundance and Distribution Common resident* of the region except in northern Pennsylvania and Appalachian highlands; overlaps with the Black-capped Chickadee in the foothills of the Appalachians.
Where to Find Glover-Archbold Park, Washington, D.C.; Caledonia State Park, Pennsylvania; Norman G. Wilder Wildlife Area, Delaware.
Range Eastern United States from Kansas east to New Jersey and south to Florida and east Texas.

Black-capped Chickadee

Poecile atricapilla

(L-5 W-8)

Black cap, throat, and bib contrasting with white cheek; dark gray back; grayish white breast and belly; white edging to gray secondaries.

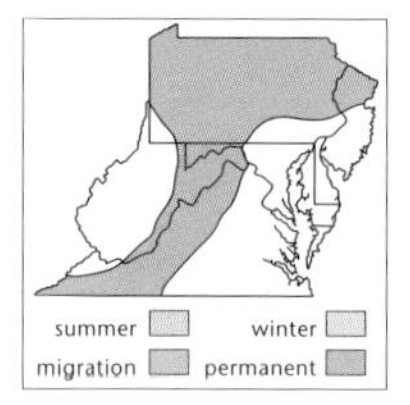

Habits Hangs upside down from branches and twigs while foraging.
Voice "Fee bee"; "chicka dee dee."
Habitat Woodlands, residential areas.
Similar Species The Black-capped Chickadee has a 2-note song, "fee-bee," while the smaller Carolina Chickadee has a 4-note song, "see-dee see doo." The blackcap wing coverts and secondary outer edgings are whiter than those of the Carolina.
Abundance and Distribution Common resident* in Pennsylvania (except extreme southwest and southeast corners), northern New Jersey, and the highlands of the Appalachian Mountains in Virginia, West Virginia, and western Maryland; overlaps with the Carolina Chickadee in the foothills of the Appalachians; strays into neighboring lowlands in winter.
Where to Find Byrd Visitor Center, Shenandoah National Park, Virginia; Caledonia State Park, Pennsylvania; Canaan Valley State Park, West Virginia.
Range Temperate and boreal North America south to California, New Mexico, Oklahoma, and New Jersey; south in Appalachians to North Carolina.

Tufted Titmouse

Baeolophus bicolor

(L-6 W-9)

Gray above; white below with buffy sides; gray crest; black forehead.

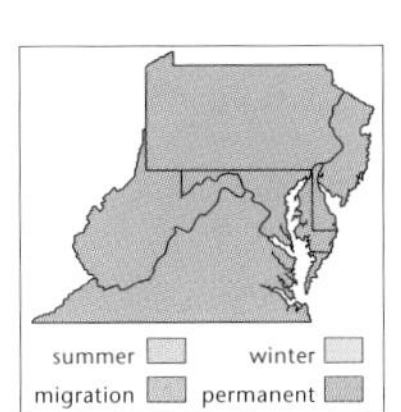

Habits Hangs upside down in chickadee fashion from branches and twigs while foraging.
Voice "Peter peter peter"; also "peet peet peet" on occasion; various chickadee-like calls.
Habitat Deciduous forest.
Similar Species Chickadee-like habits and crest are distinctive.
Abundance and Distribution Common resident* nearly throughout except at high elevations in the Appalachian Mountains; somewhat scarcer in northern Pennsylvania.
Where to Find Rock Creek Park, Washington, D.C.; Great Swamp National Wildlife Refuge, New Jersey; Berwind Lake Wildlife Management Area, West Virginia.
Range Eastern United States west to Nebraska, Iowa, Oklahoma, and west Texas and in Mexico south to Hidalgo and northern Veracruz.

Nuthatches

(Family Sittidae)

Nuthatches are peculiar birds in both appearance and behavior. Long-billed, hunched little beasts, built for probing bark on tree trunks. They creep along the trunks, usually from the top down (not bottom up like Brown Creepers).

Red-breasted Nuthatch

Sitta canadensis

(L-5 W-8)

Gray above; orange buff below; black cap with white line over eye. *Female and Immature:* Paler buff below.

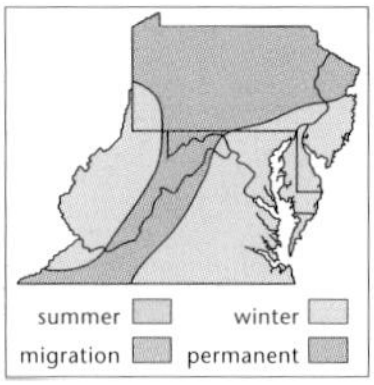

summer winter
migration permanent

Habits Inches along, often upside down, probing bark for invertebrates on branches and trunks.
Voice A nasal, high pitched series of "anh"s.
Habitat Coniferous forest.
Abundance and Distribution Uncommon resident* in northern Pennsylvania and highlands of the Appalachians in southern Pennsylvania, western Maryland, Virginia, and West Virginia; uncommon to rare and irregular winter visitor elsewhere.
Where to Find Cook Forest State Park, Pennsylvania; Dark Hollow Falls, Shenandoah National Park, Virginia; Woodbine Recreation Area, West Virginia.
Range Breeds across the northern tier of Canada and the United States south to North Carolina in the Appalachians and to New Mexico in the Rockies. Winters throughout breeding range and most of the United States to northern Mexico.

White-breasted Nuthatch

Sitta carolinensis

(L-6 W-11)

Gray above; white below with black cap; buffy flanks. *Female:* Cap is gray or dull black.

Habits Forages on tree trunks, generally working down the trunk.
Voice A series of "yank"s.

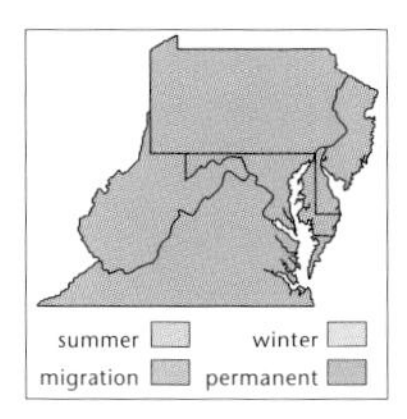

Habitat Deciduous and mixed forest, parklands, residential areas.
Abundance and Distribution Common (inland) to uncommon (along coast) resident* throughout.
Where to Find High Point State Park, New Jersey; Green Ridge State Forest, Maryland; Babcock State Park, West Virginia.
Range Southern tier of Canada and most of the United States; local in Great Plains region; highlands of Mexico south to Oaxaca.

Brown-headed Nuthatch

Sitta pusilla

(L-5 W-8)

Gray above; buff below; brown cap; white cheeks, throat, and nape.

Habits Inches along, often upside down, probing bark for invertebrates on branches and trunks.
Voice "Ki tee"; also various "kit" calls.
Habitat Pine forest.
Similar Species The bird is clearly a nuthatch from its foraging behavior, and the brown cap and white nape distinguish it from other nuthatches in the region.
Abundance and Distribution Common to uncommon resident in the pine forests of southern Delaware, coastal Maryland, and southeastern Virginia.
Where to Find Assawoman Wildlife Management Area, Delaware; Blackwater National Wildlife Refuge, Maryland; Chincoteague National Wildlife Refuge, Virginia.
Range Southeastern United States from Delaware to east Texas.

Creepers
(Family Certhiidae)

This is a small family with only 5 representatives worldwide, one of which, the Brown Creeper, occurs in the New World. Creepers, like nuthatches, are built for probing tree trunks. These no-necked bits of brown-and-white fluff have long, decurved bills and stiff tails for foraging over tree trunks, picking arthropod larvae from bark. They usually forage beginning from the bottom of a tree trunk and moving up, opposite from the way a nuthatch works a trunk.

Brown Creeper
Certhia americana
(L-5 W-8)

Brown streaked and mottled with white above; white below; white eyeline; decurved bill.

Habits Creeps nuthatchlike over trunks but usually begins at the base of the trunk and works up.

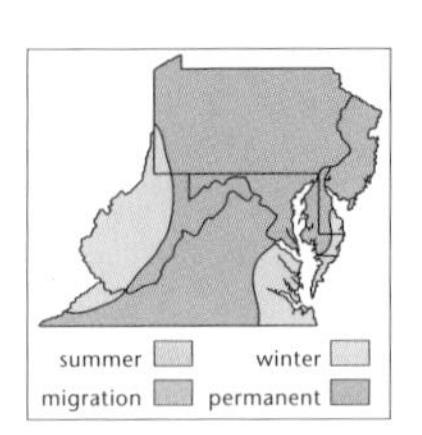

Voice A high-pitched "tseeee."
Habitat Breeds in a variety of forest types, seemingly wherever loose bark provides suitable nesting sites; most woodlands in winter.
Similar Species No other small, brown-and-white bird forages on tree trunks in the region.
Abundance and Distribution Uncommon to rare and local summer resident* (Apr–Sep) in Pennsylvania, northern New Jersey, and the highlands of Virginia, West Virginia, and Maryland. Also breeds locally in the Piedmont and coastal regions of Maryland and New Jersey wherever dead trees with loose, hanging bark provide nesting sites.

Uncommon transient and winter resident (Oct–Mar) throughout.

Where to Find Patuxent Research Refuge, Maryland; R. B. Winter State Park, Pennsylvania; Grayson Highlands State Park, Virginia.

Range Breeds in boreal, montane, and transitional zones of North America south through the highlands of Mexico and Central America to Nicaragua; winters nearly throughout in temperate, boreal, and montane regions of the continent.

Wrens
(Family Troglodytidae)

Wrens are a large, mostly tropical family of superb songsters. Most species have plumages of brown and white, sharp, decurved bills, and short, cocked tails. Nests are often built in a cavity or a completely closed structure of grasses.

Carolina Wren
Thryothorus ludovicianus

(L-6 W-8)

A rich brown above; buff below with prominent white eyeline and whitish throat.

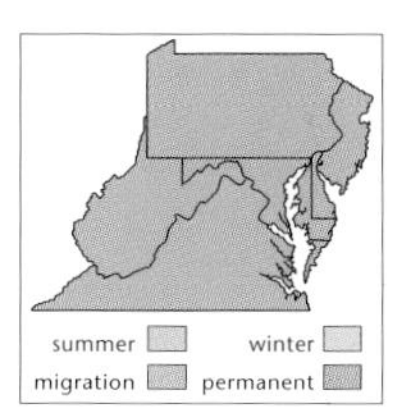

Voice Song is a loud, clearly whistled "tea kettle tea kettle tea kettle" or variations on this theme; also a sharp trill call.

Habitat Thickets, tangles, and undergrowth of moist woodlands, riparian forest, swamps.

Abundance and Distribution Common resident* in lowlands and mid-elevations throughout, except northern Pennsylvania, where uncommon to rare.

Where to Find Dyke Marsh, Virginia; Wharton State Forest, New Jersey; Panther State Forest, West Virginia.
Range Eastern North America from Iowa, Minnesota, New York, and Massachusetts south through the southeastern states, eastern Mexico, and Central America along the Caribbean slope to Nicaragua.

Bewick's Wren

Thryomanes bewickii
(L-5 W-7)

Brown above; whitish below with prominent white eyeline; long, active tail, barred black, brown, and white below.

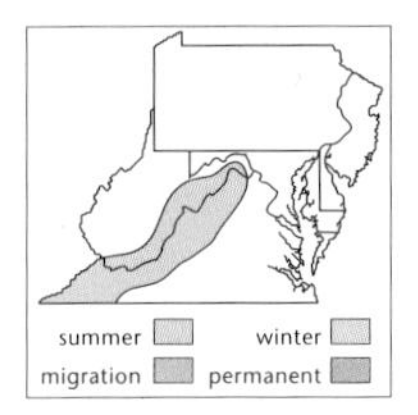

Voice A series of buzzy whistles and warbles reminiscent of the Song Sparrow; also a harsh "churr."
Habitat Riparian thickets, shrubby fields.
Abundance and Distribution Rare and local summer resident* (May–Oct) in Appalachian highlands of Virginia and West Virginia; formerly in Maryland.
Where to Find Formerly along Route 601, Highland County, Virginia. Also, 3 recent confirmed breeding records for West Virginia in Jefferson, Monroe, and Mingo counties.
Range Breeds in most of temperate North America (except Atlantic coastal and southern portions of Gulf states) south in highlands to southern Mexico; populations in eastern United States have become scarce in recent years; winters in southern portion of breeding range.

House Wren

Troglodytes aedon

(L-5 W-7)

Brown above, buff below; buff eyeline; barred flanks.

Voice A rapid, descending trill, "chipy-chipy-chipy-chipy," with associated buzzes and churrs.

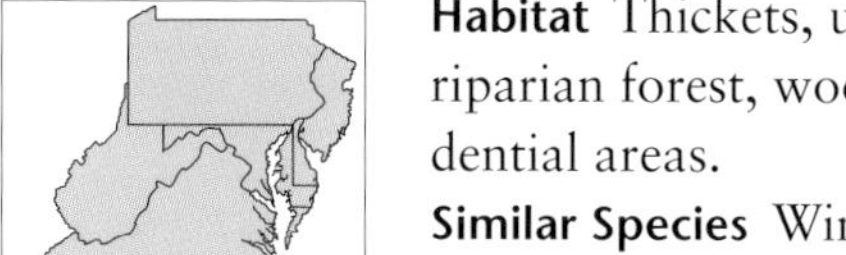

Habitat Thickets, undergrowth, and tangles in riparian forest, woodlands, hedgerows, and residential areas.

Similar Species Winter Wren is a richer brown below and has shorter tail and more prominent barring on flanks and belly. Songs and calls are very different for the 2 species.

Abundance and Distribution Common transient and summer resident* (Apr–Sep) throughout; rare winter resident (Oct–Apr) along the coast.

Where to Find Bombay Hook National Wildlife Refuge, Delaware; Tobyhanna State Park, Pennsylvania; Great Swamp, New Jersey.

Range Breeds in temperate North America from southern Canada south to south central United States; winters in southern United States south to southern Mexico. The Southern House Wren, now considered to be conspecific with the Northern, is resident over much of Mexico, Central and South America, and the Lesser Antilles.

Winter Wren

Troglodytes troglodytes

(L-4 W-6)

Brown above; brownish buff below; buffy eyeline; barred flanks and belly; short tail.

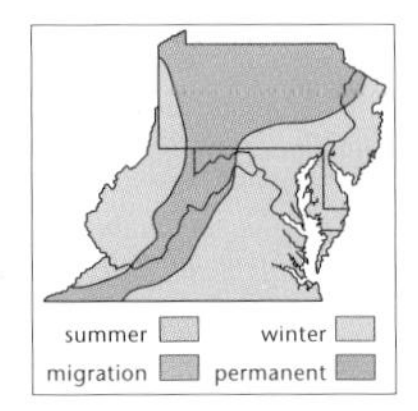

Voice A rich, cascading series of trills and warbles; calls are staccato "chuck"s, "kip"s, "churr"s.
Habitat Thickets, tangles, undergrowth of coniferous fens, bogs, and swamps; lowland riparian thickets in winter.
Similar Species Winter Wren is a richer brown below, has shorter tail, and more prominent barring on flanks and belly than House Wren. Songs and calls are very different for the 2 species.
Abundance and Distribution Uncommon summer resident* (Apr–Sep) in northern Pennsylvania and highlands of southern Pennsylvania, Virginia, Maryland, West Virginia, and extreme northwestern New Jersey; uncommon winter resident (Oct–Mar) elsewhere.
Where to Find Mount Rogers, Virginia; Canaan Valley State Park, West Virginia; Savage River State Forest, Maryland.
Range Breeds from Alaska and British Columbia to Labrador, south along the Pacific coast to central California and in the Appalachians to north Georgia; winters from the central and southern United States to northern Mexico.

Sedge Wren

Cistothorus platensis
(L-4 W-6)

Crown and back brown streaked with white; pale buff below; indistinct white eyeline.

Voice Song—a weak series of gradually accelerating "tsip"s; call—a sharp "chip" or "chip chip."
Similar Species Marsh Wren has plain brown (unstreaked) crown, distinct white eyeline, and black back streaked with white.

Habitat Low, wet marshes; grasslands; estuaries.
Abundance and Distribution Uncommon to rare transient and winter resident (Sep–Apr) along the coast of Delaware, Maryland, and Virginia; rare and local summer resident* (Apr–Sep) at scattered marshy sites throughout the region.
Where to Find Saxis Wildlife Management Area, Virginia; Moraine State Park, Pennsylvania (breeding); Jug Bay Natural Area, Maryland (breeding).
Range Breeds across the northeastern United States and southeastern Canada from southeastern Saskatchewan to New Brunswick, south to Virginia and Oklahoma; winters along Atlantic and Gulf Coastal Plain from New Jersey to northern Mexico. Scattered resident populations in Mexico and Central and South America.

Marsh Wren

Cistothorus palustris

(L-5 W-7)

Brown above; black back prominently streaked with white; white eyeline and throat; buff underparts.

Voice Song—a rapid series of dry "tsik"s, like the sound of an old sewing machine; call—a sharp "tsuk."

summer
winter
migration
permanent

Habitat Cattail and bulrush marshes, wet grasslands; saltwater marshes.
Similar Species Marsh Wren has plain brown (unstreaked) crown, distinct white eyeline, black back streaked with white. The crown and back of the Sedge Wren are streaked with white; Sedge Wren has no black on the back.
Abundance and Distribution Common to uncommon summer resident* (Apr–Sep) along the coast and in northwestern Pennsylvania; rare to casual

winter resident (Oct–Mar) along the coast from Delaware southward; uncommon to rare and local summer resident* (Apr–Sep) at scattered inland localities in Pennsylvania, Maryland, and Virginia.
Where to Find Great Swamp National Wildlife Refuge, New Jersey; Dyke Marsh, Virginia; Bombay Hook National Wildlife Refuge, Delaware.
Range Breeds across central and southern Canada and northern United States, southward along both coasts to northern Baja California and southeastern Texas; rare and local in inland United States; winters along both coasts and in southern United States to southern Mexico.

Kinglets
(Family Regulidae)

A small, New World family of birds containing only 2 species.

Golden-crowned Kinglet
Regulus satrapa
(L-4 W-7)

Greenish above, whitish below; white wing bars, white eyeline; male has orange crown bordered by yellow and black; female has yellow crown bordered in black.

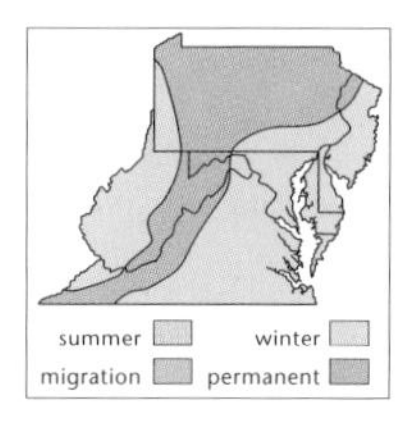

Habits Continually flicks wings while foraging.
Voice Call—a very high-pitched "tse tse tse"; song—a series of "tsee"s first rising, then descending in pitch.
Similar Species Ruby-crowned has broken white eyering, which Golden-crowned lacks; Golden-

crowned has white eyeline, which Ruby-crowned lacks.

Habitat Breeds in mature spruce-fir and pine forest of the highlands; found in a variety of woodland habitats in winter.

Abundance and Distribution Uncommon to rare and local summer resident* (Apr–Oct) in northern Pennsylvania and highlands of southern Pennsylvania, Virginia, West Virginia, and Maryland; common to uncommon winter resident (Oct–Mar) throughout.

Where to Find Mount Rogers, Virginia; Savage River State Forest, Maryland; Big Pocono State Park, Pennsylvania.

Range Breeds in boreal North America (except Great Plains) and in the Appalachians south to North Carolina; south in Rockies to Guatemala; winters from southern Canada and the United States south through highland breeding range in Mexico and Guatemala.

Ruby-crowned Kinglet

Regulus calendula

(L-4 W-7)

Greenish above, whitish below; white wing bars, broken white eyering. Adult male has scarlet crown. *Immature and Female:* Lack scarlet crown.

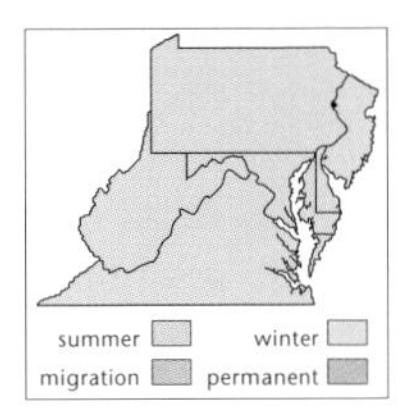

Habits Continually flicks wings while foraging.

Voice Song begins with a series of "tsee" notes followed by lower pitched "tew" notes and terminated by a series of "teedadee"s. Distinctive "tsi tit" call.

Similar Species Ruby-crowned has broken white eyering, which Golden-crowned lacks; Golden-crowned has white eyeline, which Ruby-crowned lacks.

Habitat Coniferous forests of the boreal zone in summer; a variety of woodlands during winter.
Abundance and Distribution Common to uncommon transient (Oct–Nov, Mar–May) and uncommon to rare (highlands) winter resident (Nov–Mar) throughout.
Where to Find John Heinz National Wildlife Refuge, Pennsylvania; Dennis Creek Wildlife Management Area, New Jersey; Meadowside Nature Center, Maryland.
Range Breeds in boreal North America from Alaska to Labrador south to northern New York, Michigan, and Minnesota and in the Rockies south to New Mexico and Arizona; winters over most of the United States, Mexico, and Guatemala; resident populations on Guadalupe Island (Baja California) and in Chiapas.

Gnatcatchers
(Family Sylviidae)

The sylviids are a group of mainly Old World species, most of which are slim, arboreal birds, comparable in size to the New World warblers (Parulidae). The family is represented in the Mid-Atlantic region by one species, the Blue-Gray Gnatcatcher.

Blue-gray Gnatcatcher
Polioptila caerulea
(L-5 W-7)

Bluish gray above, white below; tail black above with white outer feathers (tail appears white from below); white eyering. Adult male in breeding

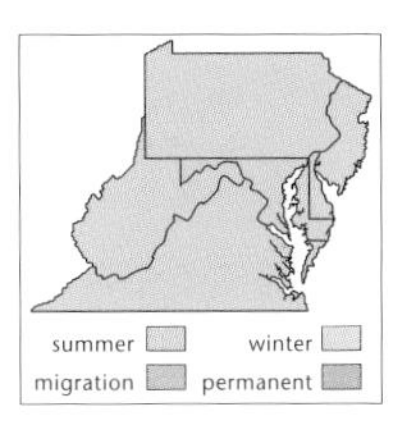

plumage (Apr–Aug) has black forehead and eyeline. Sexes nearly identical in nonbreeding plumage.

Voice Song—a barely audible string of high-pitched warbles and squeaks; call—"tsee."
Habitat Deciduous woodlands and riparian forest.
Abundance and Distribution Common transient and summer resident* (Apr–Sep) nearly throughout (scarce as a breeding bird in highlands and northern portions of region).
Where to Find Mason Neck National Wildlife Refuge, Virginia; Lewis Wetzel Wildlife Management Area, West Virginia; Wharton State Forest, New Jersey.
Range Breeds in the temperate United States south through Mexico to Guatemala. Winters in the southern Atlantic (from Virginia south) and Gulf states and in the west from California, Arizona, New Mexico, and Texas south throughout Mexico and Central America to Honduras. Resident in the Bahamas.

Thrushes and Bluebirds
(Family Turdidae)

A group of medium-sized birds, most of which are cryptically colored in browns and grays. The thrush family is considered by many to contain the most beautiful singers in the bird world. Though you may debate whether the Wood Thrush, Hermit Thrush, Slate-colored Solitaire, or Nightingale holds the top position, few would contest their place in the Top Ten.

Eastern Bluebird

Sialia sialis

(L-7 W-12)

Blue above, brick red below with white belly. *Female:* Similar but paler.

Voice Song—a whistled "cheer cheerful farmer"; call—"tur lee."
Habitat Old fields, orchards, parkland, open woodlands, open country with scattered trees—appears limited by the availability of suitable nest sites (cavities in posts and nest boxes).

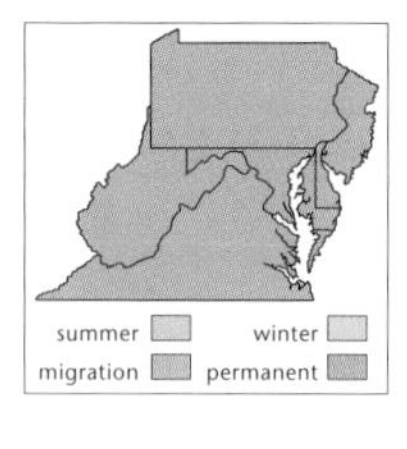

Abundance and Distribution Common summer resident* (Mar–Oct) throughout; common to uncommon winter resident (Nov–Feb) except in highlands and northern portions of the region, where scarce or absent.
Where to Find McKee-Beshers Wildlife Management Area (Hughes Hollow), Maryland; Moraine State Park, Pennsylvania; Bombay Hook National Wildlife Refuge, Delaware.
Range Breeds in eastern North America from southern Saskatchewan to New Brunswick south to Florida and Texas; through highlands of Mexico and Central America to Nicaragua; Bermuda. Winters in southern portion of breeding range (central and eastern United States southward).

Veery

Catharus ustulatus

(L-7 W-12)

Russet above; throat buff with indistinct spotting; whitish belly.

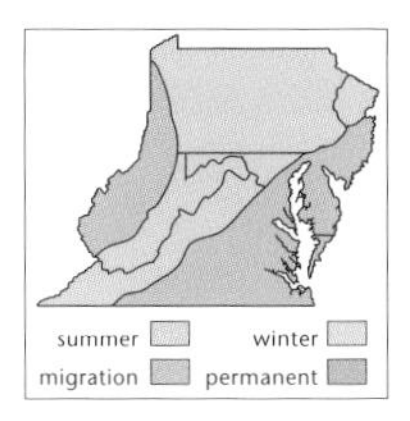

Voice Song—a whistled, wheezy, descending "zheew zheew zhoo zhoo"; call—"zink."
Similar Species Breast spotting is more distinct in Swainson's and Gray-cheeked thrushes; Swainson's has distinct eyering; russet tail of Hermit Thrush contrasts with gray-brown back.
Habitat Mature deciduous and mixed forest.
Abundance and Distribution Uncommon transient (Sep, May) throughout; uncommon and local summer resident* (May–Sep) in the highlands of West Virginia and Virginia and wet woodlands of Pennsylvania, Maryland, and New Jersey.
Where to Find Glover Archbold Park, Washington, D.C.; Assunpink Wildlife Management Area, New Jersey; Dolly Sods Scenic Area, West Virginia.
Range Breeds across southern Canada and northern United States, south in mountains to Georgia in the east and Colorado in the west; winters in northern South America.

Gray-cheeked Thrush
Catharus minimus
(L-8 W-12)

Grayish brown above, whitish below; heavy dark spotting at throat diminishing to smudges at belly and flanks.

Voice Song—"zhee zheeoo titi zhee"; call—"zheep."

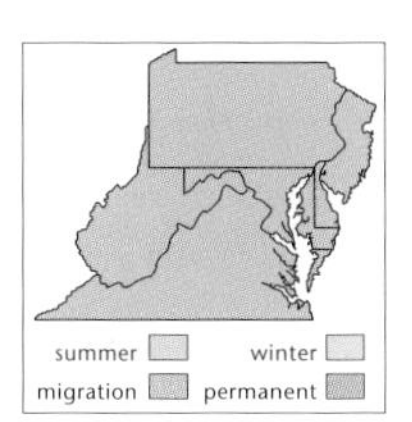

Similar Species The Gray-cheeked Thrush is distinguished from the other *Catharus* thrushes more by what it does not have than by what it does. It does not have a rich buff eyering or buffy underparts like the Swainson's Thrush; it does not have a rusty tail like the Hermit Thrush; and it does not have a rusty back and head like the Veery. In our region, if you see a grayish brown thrush with no eyering, it is likely to be this bird.

Habitat Breeding—coniferous forest and shrubby taiga; migration and winter—riparian forest, deciduous and mixed woodlands and scrub.

Abundance and Distribution Uncommon (fall) to rare (spring) transient throughout (Sep–Oct, May).

Where to Find C&O Canal, Maryland; Assawoman State Wildlife Management Area, Delaware; Ramseys Draft, Virginia.

Range Breeds from northeastern Siberia across Alaska and the northern tier of Canada; winters in northern South America.

Bicknell's Thrush

Catharus bicknelli

(L-8 W-12)

Grayish brown above, whitish below; heavy dark spotting at throat diminishing to smudges at belly and flanks.

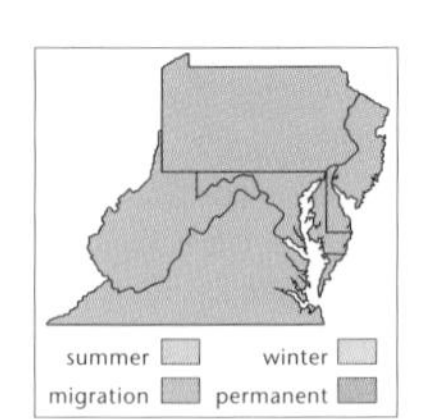

Voice Song—"zhee zheeoo titi zhee"; call—"zheep."

Similar Species Bicknell's Thrush is virtually indistinguishable from the Gray-cheeked Thrush in the field. Both are distinguished from other *Catharus* species by lack of a rich buff eyering and underparts (Swainson's), lack of a rusty tail (Hermit), or lack of a rusty back and head (Veery).

Habitat Breeding—coniferous forest; migration and winter—riparian forest, deciduous and mixed woodlands and scrub.

Abundance and Distribution Uncommon (fall) to rare (spring) transient throughout (Sep–Oct, May).

Where to Find C&O Canal, Maryland; Assawoman State Wildlife Management Area, Delaware; Ramseys Draft, Virginia.

Range Breeds in northeastern United States and southeastern Canada; winters in Greater Antilles.

Swainson's Thrush

Catharus ustulatus

(L-7 W-12)

Brown or grayish brown above, buffy below with dark spotting; lores and eyering buffy.

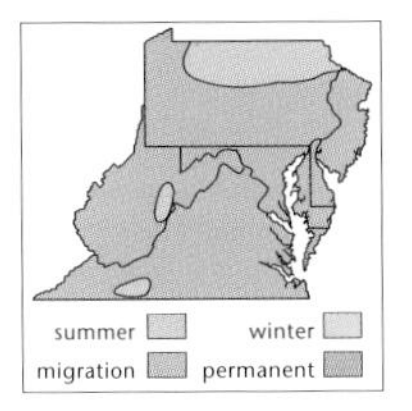

Voice Song—"zhoo zhoo zhee zhee" with rising pitch; call—"zheep."
Similar Species The whitish or buff eyering of the Swainson's Thrush distinguishes this bird from the Gray-cheeked Thrush and western populations of the Veery. Also, flanks of the Swainson's Thrush are buffy, not grayish as in western Veeries.
Habitat Breeding—thickets in coniferous forest, bogs, alder swamps; migration and winter—moist woodlands, riparian thickets.
Abundance and Distribution Common to uncommon transient (Sep–Oct, May) throughout, more numerous in fall than spring; uncommon to rare and local summer resident in highland spruce and hemlock forest of Pennsylvania, West Virginia, and Mount Rogers, Virginia.
Where to Find Cook Forest State Park, Pennsylvania (breeding); Cranberry Glades Botanical Area, West Virginia (breeding); Great Swamp National Wildlife Refuge, New Jersey (migration).
Range Breeds in boreal North America from Alaska across Canada to Labrador south to the northern United States; south in mountains and along coast to southern California and in mountains to northern New Mexico. Winters from southern Mexico to the highlands of South America.

Hermit Thrush

Catharus guttatus

(L-7 W-12)

Grayish brown above; whitish below with dark spotting; whitish eyering; rusty tail, often flicked.

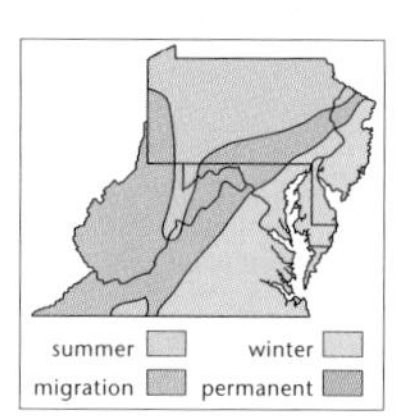

Voice Song—distinct phrases, each beginning with a long, whistled note followed by a trill; trills of different phrases have different inflection. Call—"tuk tuk tuk" or a raspy "zhay."

Similar Species The rusty tail of the Hermit Thrush distinguishes this bird from the other *Catharus* thrush species (Swainson's, Veery, and Gray-cheeked).

Habitat Breeding—Spruce-fir and hemlock forest; migration and winter—riparian thickets, broadleaf woodlands.

Abundance and Distribution Common to uncommon and local summer resident* in northern Pennsylvania and New Jersey and highlands of southern Pennsylvania, West Virginia, western Maryland, and Mount Rogers, Virginia; common to uncommon transient (Sep–Oct, Apr–May) throughout; common to uncommon winter resident along the Coastal Plain and in the Piedmont and lower mountain valleys in the southern portion of the region.

Where to Find Stokes State Forest, New Jersey (breeding); Swallow Falls State Park, Maryland (breeding); Chincoteague National Wildlife Refuge, Virginia (winter).

Range Breeds in boreal Canada and northern United States, south in mountains to southern California, Arizona, New Mexico, and west Texas. Winters from the southern United States north along coasts to southern British Columbia and New Jersey and south through Mexico (excluding

Yucatan Peninsula) to Central America. Resident population in Baja California.

Wood Thrush

Hylocichla mustelina

(L-8 W-13)

Russet above, white below with distinct black spots; reddish brown crown and nape.

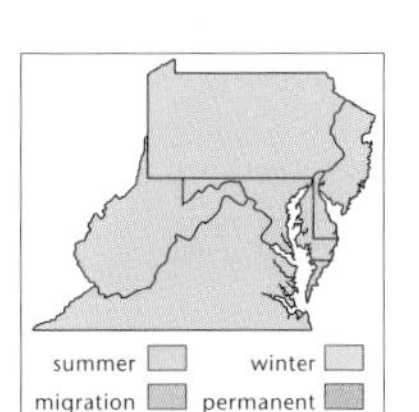

Voice Song—distinct phrases of buzzy trills separated by pauses; call—"bup bup bup."
Similar Species None of the *Catharus* thrushes have the clear white underparts and distinct, black spots.
Habitat Deciduous or mixed forest, riparian woodland.
Abundance and Distribution Common transient and summer resident* (May–Sep) throughout.
Where to Find Trap Pond State Park, Delaware; East Lynn Lake Wildlife Management Area, West Virginia; R. B. Winter State Park, Pennsylvania.
Range Breeds in the eastern United States and southeastern Canada. Winters from southern Mexico to Panama and northwestern Colombia.

American Robin

Turdus migratorius

(L-10 W-17)

Dark gray above, orange-brown below; white lower belly; white throat with dark streaks; partial white eyering.

Voice Song—a varied series of whistled phrases, "cheerily cheerup cheerio"; call—"tut tut."

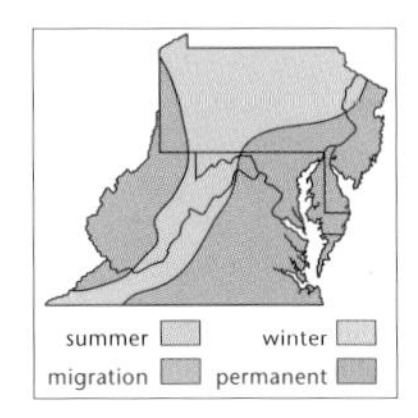

Habitat A wide variety of forest and parkland will serve as breeding habitat so long as there is rich, moist soil containing earthworms for foraging and mud available for nest construction. Deciduous and mixed forest, scrub, parkland, riparian forest, oak woodlands, and residential areas in winter.

Abundance and Distribution Common summer resident* (Mar–Oct) throughout; common to uncommon winter resident (Nov–Feb) along the Coastal Plain and Piedmont; uncommon to rare and irregular in winter elsewhere, generally in flocks.

Where to Find National Zoological Park, Washington, D.C.; Hoffman Sanctuary, New Jersey; Dickey Ridge Visitor Center, Shenandoah National Park, Virginia.

Range Breeds nearly throughout Canada and United States south through central highlands of Mexico. Winters from southern half of breeding range into Guatemala; western Cuba; Bahamas. Resident population in Baja California.

Thrashers and Mockingbirds
(Family Mimidae)

Long-tailed, medium-sized birds, several of which have a somewhat fierce appearance due to their decurved bills and colored eye (orange, yellow, white). Some of the species in the family include phrases from the songs of other birds in their own songs.

Gray Catbird
Dumetella carolinensis
(L-9 W-12)

Slate gray throughout, black cap and tail, rusty undertail coverts.

summer winter
migration permanent

Voice Song—a stream of whistles, mews, squeaks; sometimes mimics other species. Call—a nasal, catlike mew.

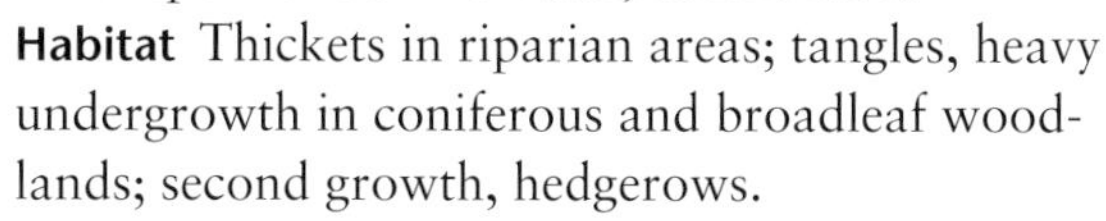

Habitat Thickets in riparian areas; tangles, heavy undergrowth in coniferous and broadleaf woodlands; second growth, hedgerows.

Abundance and Distribution Common summer resident* (Apr–Oct) throughout; uncommon to rare winter resident (Nov–Mar) along the Coastal Plain from Cape May southward; rare elsewhere in the Piedmont and Coastal Plain in winter, often at feeders.

Where to Find Seneca Creek, Maryland; Ridley Creek State Park, Pennsylvania; Presquile National Wildlife Refuge, Virginia.

Range Breeds throughout eastern, central, and northwestern United States and southern Canada. Winters along central and southern Atlantic and Gulf coasts south through the Gulf and Caribbean lowlands of Mexico to Panama; Bahamas, Greater Antilles. Resident in Bermuda.

Northern Mockingbird
Mimus polyglottos
(L-10 W-14)

Gray body, paler below; black tail with white outer tail feathers; white wing bars and white patches on wings.

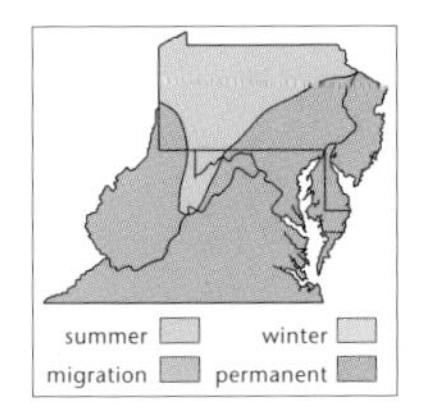

Habits Often flies to the ground and spreads wings and tail in a very mechanical fashion while foraging.
Voice A variety of whistled phrases, repeated several times and including pieces of other bird songs.
Similar Species Loggerhead Shrike has heavy, hooked bill, short tail, and black mask.
Habitat Old fields, hedgerows, agricultural areas, residential areas.
Abundance and Distribution Common permanent resident* nearly throughout, except highlands and northwestern Pennsylvania, where a local summer resident,* withdrawing in winter.
Where to Find National Arboretum, Washington, D.C.; Prime Hook National Wildlife Refuge, Delaware; Great Swamp National Wildlife Refuge, New Jersey.
Range Resident in central and southern United States and Mexico to Oaxaca; Bahamas, Greater Antilles.

Brown Thrasher

Toxostoma rufum

(L-11 W-13)

Rufous above, buff below with dark streaking; yellowish eye.

Voice A long sequence of brief phrases of squeaky warbles, each phrase given twice.

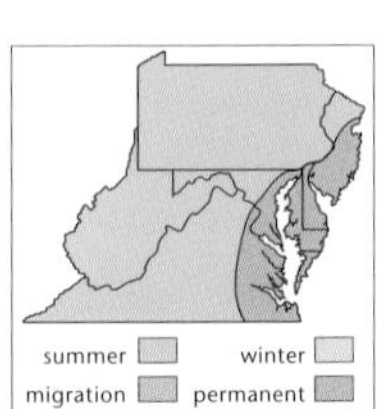

Habitat Tangles, undergrowth, and thickets of forests, old fields, hedgerows.
Abundance and Distribution Common transient and summer resident* throughout (Apr–Sep); uncommon to rare winter resident (Oct–Mar) along the Coastal Plain.
Where to Find Lewis Wetzel Wildlife Management Area, West Virginia; Meadowside Nature Center,

Thrashers and Mockingbirds

Maryland; Moraine State Park, Pennsylvania.
Range Breeds across eastern and central North America from southern Canada west to Alberta and south to east Texas and southern Florida. Winters in southern portion of breeding range.

Starlings
(Family Sturnidae)

Starlings are an Old World family with no members native to the New World. They are medium-sized, relatively long-billed songbirds, often with iridescent plumage. Many of the species mimic the songs of other birds.

European Starling
Sturnus vulgaris
(L-9 W-15)

Plump, short-tailed, and glossy black with purple and green highlights and long, yellow bill in summer. *Winter:* Dark billed and speckled with white.

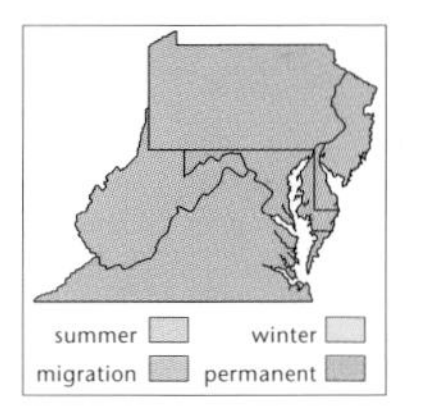

Habits Erect, waddling gait when foraging on ground; forms large roosting flocks during non-breeding season.
Voice Various squeaks, harsh "churr"s, whistles, and imitations of other bird songs, often with repeated phrases; call—a high pitched, rising "tseee."
Similar Species Other black birds are shorter billed, longer tailed; winter bird is speckled with white; summer bird has yellow bill.
Habitat Towns, farmland.
Abundance and Distribution Common resident* of cities, towns, and farmlands throughout.

Where to Find The Mall, Washington, D.C.; Dover, Delaware; Wheeling, West Virginia.
Range New World—resident from southern Canada to northern Mexico, Bahamas, and Greater Antilles; Old World—breeds in temperate and boreal regions of Eurasia; winters in southern portions of breeding range into north Africa, Middle East, and southern Asia.

Pipits
(Family Motacillidae)

Chaff-colored, long-tailed birds of open country. Like larks, they have an extremely long hind claw and often call and sing on the wing. They pump their tails with each step as they walk along the ground in search of seeds and insects.

American Pipit
Anthus rubescens
(L-7 W-11)

Sparrowlike in size and coloration but sleek and erect in posture, walks rather than hops, and has thin bill; grayish brown above; buffy below streaked with brown; whitish throat and eyeline; white outer tail feathers; dark legs; wags tail as it walks; undulating flight.

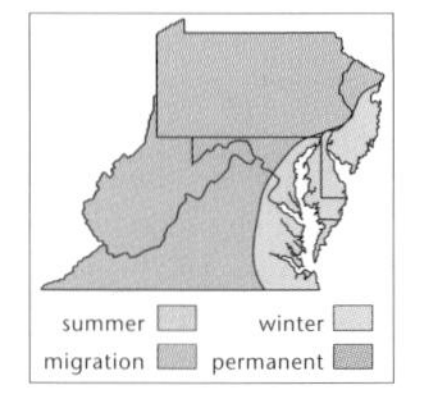

Voice Flight song—a sibilant "chwee" repeated; call—"tsee-eet."
Similar Species American Pipit has unstreaked back and dark legs; Sprague's Pipit has streaked back and flesh-colored legs. Vesper Sparrow has cone-shaped bill (not narrow); hops rather than walks.

Habitat Short grassland, plowed fields, swales, mudflats, roadsides; pond, stream and river margins.
Abundance and Distribution Common to uncommon transient (Oct–Nov, Mar–Apr) throughout; common winter resident (Nov–Mar) along coast; uncommon to rare and irregular winter resident inland.
Where to Find Eastern Shore National Wildlife Refuge, Virginia; Blackwater National Wildlife Refuge, Maryland; Bombay Hook National Wildlife Refuge, Delaware.
Range Breeds in Arctic regions of the Old and New World and in mountainous areas and high plateaus of temperate regions; winters in temperate regions and high, arid portions of the tropics.

Waxwings
(Family Bombycillidae)

Waxwings are a small family of sleek, brown, crested birds with bright waxy tips on their secondaries. They are highly social during the nonbreeding season, feeding mostly in flocks on tree fruits and berries.

Cedar Waxwing
Bombycilla cedrorum
(L-7 W-11)

Natty brown above and below with sharp crest; black face and throat; yellow wash on belly; tail tipped with yellow; red waxy tips to secondaries. *Immature:* Faint streaking below, lacks waxy tips.

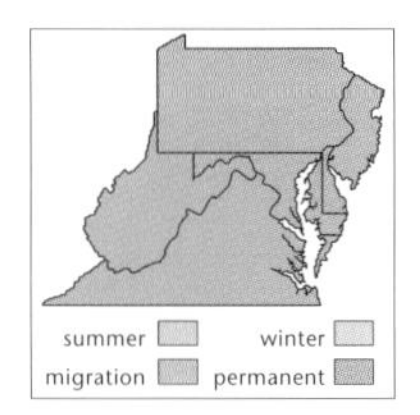

Voice A thin, gurgling "tseee."
Habitat Open coniferous and deciduous woodlands, bogs, swamps, and shrubby, overgrown fields; cemeteries, arboreta, residential areas where fruiting trees are found.
Abundance and Distribution Common summer resident* (Apr–Oct) nearly throughout; uncommon to rare along the Coastal Plain and Piedmont as a breeder; irregular winter visitor to fruiting trees throughout.
Where to Find National Arboretum, Washington, D.C.; Stokes State Forest, New Jersey; Allegheny National Forest, Pennsylvania.
Range Breeds across southern Canada and northern United States south to northern California, Kansas, and New York; winters in temperate United States south through Mexico and Central America to Panama and the Greater Antilles.

Wood-Warblers
(Family Parulidae)

This is a large group composed mostly of small species, many of which are migratory. Males of several species are brightly colored.

Blue-winged Warbler
Vermivora pinus
(L-5 W-8)

Greenish yellow above, yellow below; head yellow with black line through eye; bluish gray wings and tail with white wing bars and tail spots. *Female:* More greenish on head.

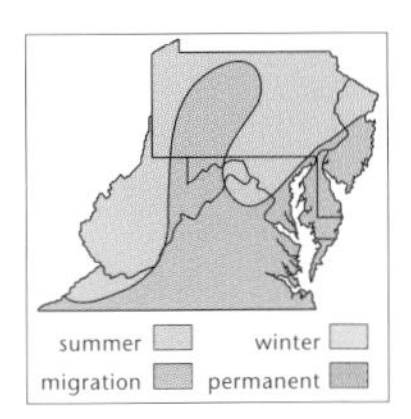

Voice Song—a dry, buzzy "beee bizzz"; call—"tsik."
Habitat Deciduous scrub, old fields.
Abundance and Distribution Uncommon to rare transient (Aug–Sep, May) throughout; uncommon to rare and local summer resident* (May–Sep) nearly throughout except Coastal Plain, Virginia Piedmont, and Appalachian highlands, where scarce or absent.
Where to Find FR 812 (Parkers Gap Road), Virginia; East Lynn Lake Wildlife Management Area, West Virginia; Erie National Wildlife Refuge, Pennsylvania.
Range Breeds in the eastern United States. Winters from southern Mexico to Panama. Interbreeds with Golden-winged Warbler to produce hybrid Brewster's (patterned like Blue-wing but with whitish underparts) and Lawrence's (patterned like Golden-wing but yellow below) warblers.

Golden-winged Warbler

Vermivora chrysoptera
(L-5 W-8)

Gray above, white below; golden crown and epaulets; black throat and ear patch; white tail spots. *Female and Immature:* Patterned like male but with gray throat and ear patch.

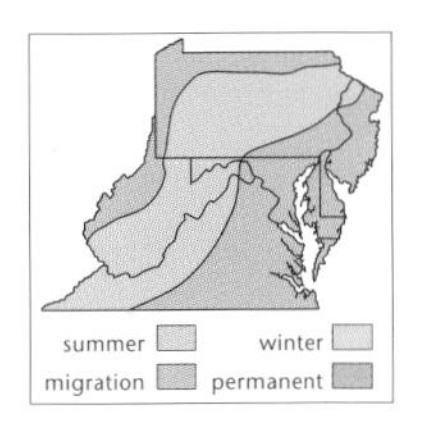

Voice Song—"bee biz biz biz"; call—"chip."
Habitat Deciduous scrub, old fields.
Abundance and Distribution Uncommon to rare transient (Aug–Sep, May) throughout; uncommon to rare and local summer resident* (May–Sep) in northwestern New Jersey, Pennsylvania (except Coastal Plain), West Virginia, western Maryland, and the Virginia mountains.

Where to Find Green Ridge State Forest, Maryland; R. D. Bailey Lake Wildlife Management Area, West Virginia; Clinch Valley College Campus, Virginia.
Range Breeds in the northeastern and north central United States. Winters from southern Mexico to Colombia and Venezuela. See Blue-winged Warbler for discussion of hybrids.

Tennessee Warbler

Vermivora peregrina
(L-5 W-8)

Olive above, white below, with gray cap, white eye stripe, and dark line through eye. *Immature:* Tinged with yellow.

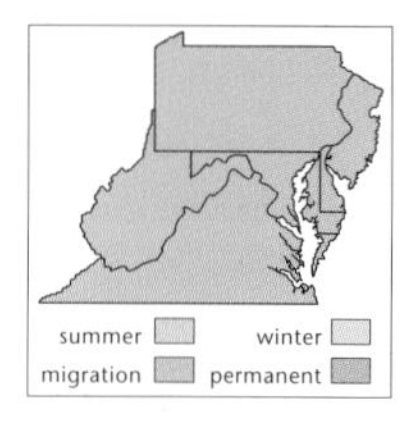

Voice Song—a series of rapid "tsip"s followed by a series of rapid "tsi"s. Call—a strong "tsip."
Similar Species The Tennessee Warbler lacks streaking on breast and has white undertail coverts. The Orange-crowned Warbler has faint streaking on breast and yellow undertail coverts. The Tennessee is greener above than the Orange-crown and has a more prominent eye stripe.
Habitat Breeds in coniferous forest; deciduous and mixed forest on migration.
Abundance and Distribution Common to uncommon fall transient (Sep) and uncommon to rare spring transient (May) nearly throughout, scarcer along the Coastal Plain and Piedmont.
Where to Find Jennings Environmental Education Center and Nature Reserve, Pennsylvania; High Knob Recreation Area, Virginia; Savage River State Forest, Maryland.
Range Breeds across boreal North America. Winters from southern Mexico to northern South America.

Orange-crowned Warbler

Vermivora celata

(L-5 W-8)

Greenish gray above, dingy yellow faintly streaked with gray below; grayish head with faint whitish eye stripe; orange crown visible on some birds at close range.

Voice Song—a weak, fading trill; call—a strong "cheet."
Similar Species The Orange-crowned Warbler has faint streaking on the breast and yellow undertail coverts, while the breast of the Tennessee Warbler is unstreaked and the undertail coverts are white; the Tennessee is greener above than the Orange-crown and has a more prominent eye stripe. Immatures of some Yellow Warblers are similar but have yellow (not gray) undertail lining.
Habitat Old fields, coastal scrub.
Abundance and Distribution Rare transient and winter visitor (Oct–Apr) along the immediate Maryland and Virginia coast.
Where to Find Presquile National Wildlife Refuge, Virginia; Blackwater National Wildlife Refuge, Maryland; Chincoteague National Wildlife Refuge, Virginia.
Range Breeds in western and northern North America. Winters from the southern United States south into Mexico, Belize, and Guatemala.

Nashville Warbler

Vermivora ruficapilla

(L-4 W-7)

Olive above, yellow below; gray head; white eyering; rufous cap. *Female:* Dingier; lacks reddish cap.

Voice Song—"Tsepit tsepit tsepit" followed by a trilled "tseeeeeeeeeeee"; call—"tsip."
Habitat Coniferous bogs; scrub; old fields, overgrown strip mines, clearcuts, and pine barrens.
Abundance and Distribution Uncommon to rare and local summer resident* (May–Sep) in northwestern New Jersey, Pennsylvania (except southeastern and extreme western portions), western Maryland, Virginia highlands (very rare), and highland bogs of Tucker and Grant counties in West Virginia; uncommon (inland) to rare (coast) transient (Sep–Oct, May) throughout.

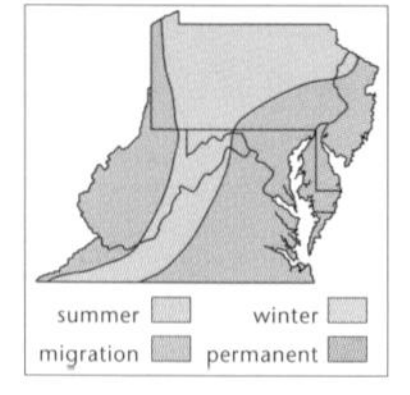

Where to Find Canaan Valley State Park, West Virginia; Allegheny National Forest, Pennsylvania; Swallow Falls State Park, Maryland.
Range Breeds across extreme north central and northeastern United States, south central and southeastern Canada, and northwestern United States. Winters from south Texas to Honduras.

Northern Parula
Parula americana
(L-5 W-8)

Bluish above; yellow throat and breast with black collar rimmed below with rust; white eyering, belly, and wing bars; greenish yellow on back. *Female and Immature:* Lack collar.

Voice Song—"brrrrrzzeeit," like running a thumbnail up a comb; also a slower, rising "zhe zhe zhe zeeeeit."
Habitat Swampy deciduous and mixed forest.

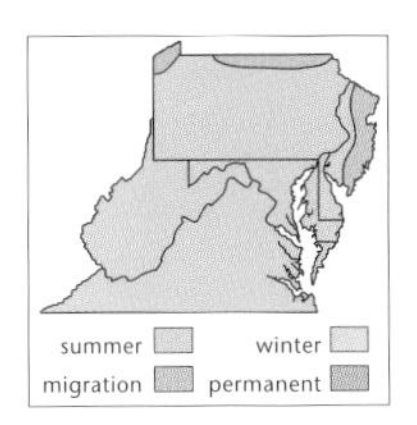

Abundance and Distribution Common to uncommon transient and summer resident* (Apr–Oct) nearly throughout, although scarce or absent as a breeder in New Jersey and much of extreme northern Pennsylvania.
Where to Find C&O Canal, Maryland; Great Dismal Swamp National Wildlife Refuge, Virginia; East Lynn Lake Wildlife Management Area, West Virginia.
Range Breeds in the eastern United States and southeastern Canada. Winters from southern Mexico to Panama, in West Indies.

Yellow Warbler
Dendroica petechia
(L-5 W-8)

Yellow throughout, somewhat dingier on the back; yellow tail spots. Male variably streaked below with reddish.

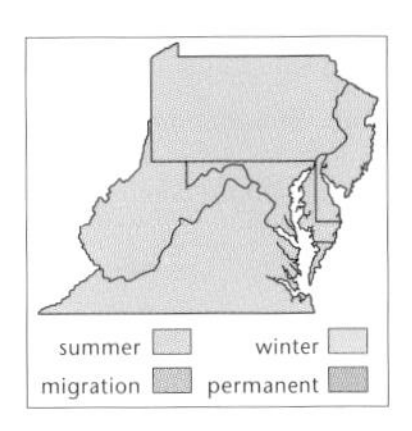

Voice Song—"tseet tseet tseet tsitsitsi tseet" ("Sweet, sweet, sweet, I'm so sweet"); call—"chip."
Similar Species Some immature Yellow Warblers are quite greenish, resembling immature Orange-crowned Warblers, but they have yellow (not gray) undertail lining.
Habitat Old fields, riparian thickets.
Abundance and Distribution Common transient and summer resident* (May–Sep) throughout.
Where to Find Bombay Hook National Wildlife Refuge, Delaware; Brigantine National Wildlife Refuge, New Jersey; John Heinz National Wildlife Refuge, Pennsylvania.
Range Breeds across most of North America. Winters from extreme southern United States through Mexico and Central America to northern and

central South America. Resident races in mangroves of West Indies, Central and South America.

Chestnut-sided Warbler

Dendroica pensylvanica

(L-5 W-8)

Greenish streaked with black above, white below with chestnut sides; yellow cap; white cheek and wing bars; black facial markings. *Winter:* Spring green above, white below with whitish yellow eye-ring, wing bars, and tail spots.

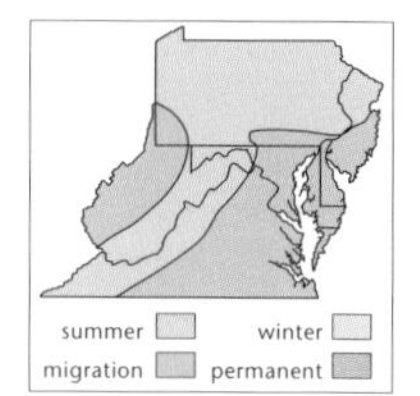

Voice Song—"pleased pleased pleased ta meetcha" and variations; call—"tsip."

Similar Species Bay-breasted is chestnut (male) or buffy (female) on breast, not white. Bright yellow-green back and white underparts separate the immature from other immature warblers.

Habitat Scrub, thickets, second growth, hedgerows.

Abundance and Distribution Common to uncommon summer resident* (May–Sep) in northern New Jersey, Pennsylvania (except southeast and southwest, where scarce and local), western Maryland, and the highlands of Virginia and eastern West Virginia.

Where to Find Dickey Ridge Visitor Center, Shenandoah National Park, Virginia; Canaan Valley State Park, West Virginia; Swallow Falls State Park, Maryland.

Range Breeds in northeastern United States and southeastern Canada. Winters from southern Mexico to Panama.

Magnolia Warbler

Dendroica magnolia

(L-5 W-8)

Black back, gray cap; yellow below broadly streaked with black; yellow rump; white wing and tail patches. *Female and Winter Male:* More brownish above than breeding male; breast streaking is faint and grayish.

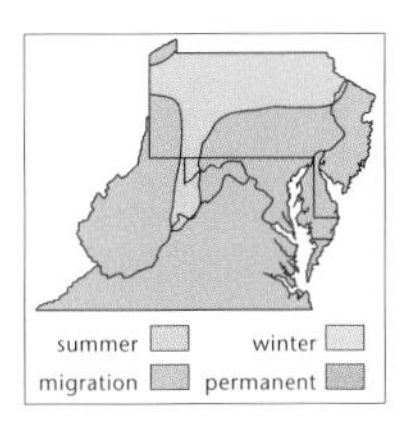

Habits Continually fans tail, showing white patches; often sallies for insects during foraging.
Voice Song—"tsweeta tsweeta tsweetee." Call is a hoarse, vireo-like "eeeeeh," very different from most other warblers.
Similar Species No other warbler has a white patch on the middle of the tail.
Habitat Thickets in coniferous forest, especially young spruce and hemlock, on the breeding ground; various woodlands and scrub at other times of the year.
Abundance and Distribution Common to uncommon and local summer resident* (May–Sep) in northern New Jersey, highlands and northern portions of Pennsylvania, western Maryland (Garrett County), Highland County, Virginia (Mount Rogers), and extreme highlands of the Allegheny Mountains in eastern West Virginia; common (fall) to uncommon (spring) transient (Sep, May) elsewhere.
Where to Find Canaan Valley State Park, West Virginia; Mount Rogers, Virginia; Swallow Falls State Park, Maryland.
Range Breeds across much of boreal Canada and the northeastern United States. Winters from central Mexico to Panama, in West Indies.

Cape May Warbler

Dendroica tigrina

(L-5 W-8)

Greenish with black streakings above, yellow breast with black streaks; yellow rump; chestnut ear patch; white undertail coverts, wing patch, and tail spots. *Female:* Dingier, with yellow ear patch.

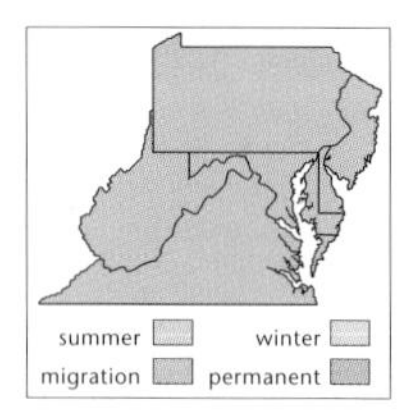

Voice Song—a weak, high-pitched "tsee tsee tsee tsee."

Similar Species Female Yellow-rumped Warbler lacks distinct streaking on throat and breast. Palm Warbler has yellow (not white) undertail coverts and wags tail.

Habitat Coniferous, mixed, and deciduous forest; riparian and oak woodland.

Abundance and Distribution Common to uncommon fall transient (May), uncommon to rare spring transient (Sep) throughout.

Where to Find Rock Creek Park, Washington, D.C.; John Heinz National Wildlife Refuge, Pennsylvania; Higbee Beach Wildlife Management Area, New Jersey.

Range Breeds in boreal regions of central and eastern Canada and northeastern United States. Winters in the West Indies.

Black-throated Blue Warbler

Dendroica caerulescens

(L-5 W-8)

Blue-black above; black face, throat, and sides; white breast, belly, and spot on primaries. Female is brownish above, dingy white below, with whitish eye stripe, dark ear patch, and (usually) white patch on primaries.

Voice Song—a rising "tsee tsee tsee tsuree"; call—"tsik."
Similar Species Female resembles Philadelphia Vireo and Tennessee Warbler, but dark cheek patch and white wing patch (when present) are distinctive.
Habitat Mixed woodlands, often with laurel or rhododendron understory, during the breeding season; various woodlands on migration and in winter.
Abundance and Distribution Common transient and summer resident* (May–Oct) in the highlands of Virginia, West Virginia, Pennsylvania, western Maryland, and northern New Jersey (scarce); common transient (Sep–Oct, May) elsewhere.
Where to Find FR 812 (Parkers Gap Rd), Virginia; Woodbine Recreation Area, West Virginia; Savage River State Forest, Maryland.
Range Breeds from southeastern Canada and northeastern United States south along the Appalachian chain to northern Georgia. Winters in the Caribbean basin.

Yellow-rumped Warbler

Dendroica coronata

(L-5 W-8)

Blue-gray above; black breast and sides; white belly; yellow cap, rump, and shoulder patch; white wing patch. Breeding males of the eastern and northern race have white throat; western race has yellow throat. *Female:* Brownish above, dingy below with faint streaking; yellow rump.

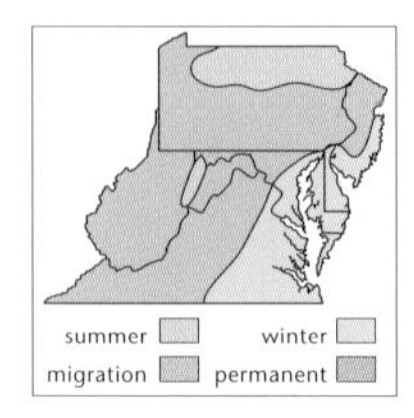

Voice Song—a weak trill, rising at the end "tsitsitsitsitsitsee." "Chit" call note often given in flight.
Habitat Breeds in coniferous and mixed forest; various woodlands and scrub on migration and in winter.
Abundance and Distribution Common to uncommon transient and uncommon to rare and local summer resident* (Apr–Sep) in the highlands of northern Pennsylvania, western Maryland (Garrett County), and mountains of eastern West Virginia; common transient (Sep–Oct, Apr–May) throughout; common winter resident (Sep–May) along the Coastal Plain and Piedmont.
Where to Find World's End State Park, Pennsylvania (breeding); Assawoman Wildlife Management Area, Delaware (migration, winter); National Arboretum, Washington, D.C. (migration, winter).
Range Breeds across northern boreal North America and in mountains of the west, south to southern Mexico. Winters from central and southern United States to Panama and the West Indies.

Black-throated Green Warbler

Dendroica virens

(L-5 W-8)

Green above; black bib; green crown; golden face; white belly. *Female:* Usually has some gray or black across breast.

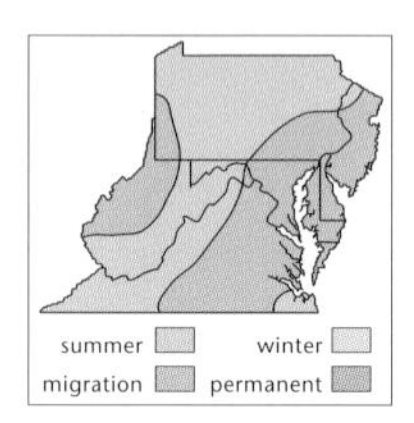

Voice Song—a lazy, insectlike "zee zee zee zoo zee" or "zoo zee zeezee zoo"; call—"chip."
Habitat Breeds in coniferous and mixed forest; found in broadleaf forest and scrub on migration and in winter.
Abundance and Distribution Common to uncommon and local summer resident* (May–Oct) in the

highlands of Virginia, West Virginia, Pennsylvania, western Maryland, northern New Jersey (scarce); also breeds in Virginia coastal lowlands of extreme southeastern corner; uncommon to rare transient (Sep–Oct, May) elsewhere.

Where to Find Great Dismal Swamp National Wildlife Refuge, Virginia; Swallow Falls State Park, Maryland; Canaan Valley State Park, West Virginia.

Range Breeds across central and southern Canada and north-central and northern United States, also south in the Appalachians to north Georgia and along the Coastal Plain to South Carolina. Winters from south Texas and south Florida through Mexico and Central America to Panama, in West Indies.

Blackburnian Warbler

Dendroica fusca

(L-5 W-8)

Black above, white below; orange bib; black-and-orange facial pattern. *Female:* Patterned like male, but black areas of male are grayish in female, orange areas are yellowish.

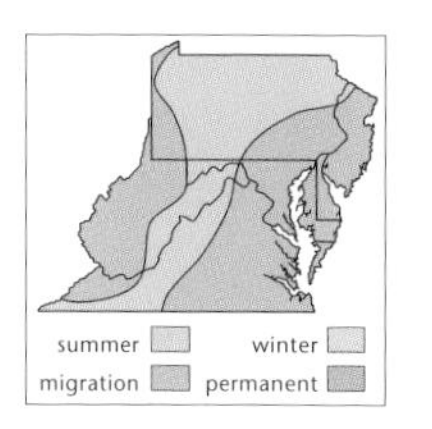

Voice Song—very high-pitched, rising "tsip tsip tsip tsip tsitsi tseeee"; call—"tsip."

Similar Species Female Black-throated Green is greenish or mottled black and green on the back. Female Blackburnian has pale striping on dark back.

Habitat Breeds in spruce-fir, mixed, and deciduous forest (in the Appalachians); found in various forest types on migration and in winter.

Abundance and Distribution Common to uncommon transient and uncommon to rare and local summer resident* (Apr–Sep) in the highlands of

northern Pennsylvania, western Maryland, and mountains of Virginia and eastern West Virginia; uncommon (inland) to rare (coast) transient (Sep, May) elsewhere, more numerous in fall.
Where to Find Woodbine Recreation Area, West Virginia; World's End State Park, Pennsylvania; Swallow Falls State Park, Maryland.
Range Breeds in the northeastern United States and southeastern Canada. Winters in Costa Rica, Panama, and northern South America.

Yellow-throated Warbler

Dendroica dominica
(L-5 W-8)

Dark gray above, white below with yellow bib; black mask; white eye stripe (yellow in some individuals); black streaks on sides.

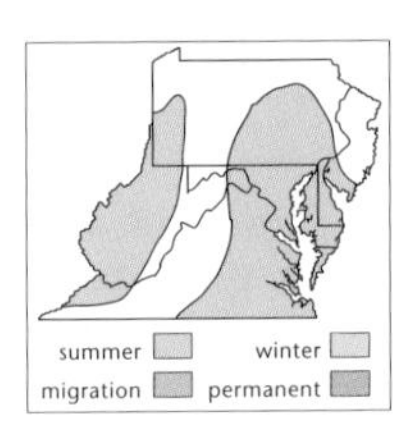

Habits Often forages by creeping along tree branches.
Voice Song—a strong, slurred series of notes ending with an upward inflection, "tew tew tew tew teweesee."
Habitat Conifers, cypress, sycamores; riparian woodland.
Abundance and Distribution Common to uncommon transient and summer resident* (Apr–Sep) in lowlands of southeastern (scarce and local) and southwestern Pennsylvania, western West Virginia, and the Piedmont and Coastal Plain of Maryland, Delaware, and Virginia.
Where to Find C&O Canal, Maryland; Great Dismal Swamp National Wildlife Refuge, Virginia; Dennis Creek Wildlife Management Area, New Jersey.

Range Breeds in the eastern United States. Winters from the Gulf coast of the United States south through Central America to Costa Rica, in Bahamas and Greater Antilles.

Pine Warbler

Dendroica pinus

(L-6 W-9)

Olive above, yellow below with faint, grayish streaking; whitish belly; yellow eye stripe; white wing bars; white tail spots.

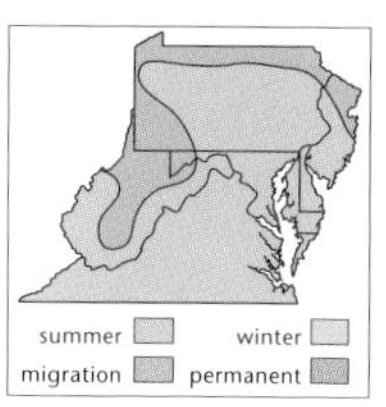

Voice Song—a series of chips similar to the Chipping Sparrow but slower, each chip distinct.
Similar Species Immature similar to immature Blackpoll and Bay-breasted but plain (not streaked) back. Could be confused with Yellow-throated Vireo, but the slimmer, smaller, more active Pine Warbler has yellow-green rump (not gray), white tail spots, and thin bill.
Habitat Pine forest.
Abundance and Distribution Common to uncommon summer resident* (Mar–Sep) in pines nearly throughout, although scarce or absent in northern and western Pennsylvania and northern New Jersey; uncommon to rare winter resident (Oct–Mar) along the Coastal Plain.
Where to Find National Arboretum, Washington, D.C.; Beech Fork State Park, West Virginia; Caledonia State Park, Pennsylvania.
Range Breeds in the eastern half of the United States and in south central and southeastern Canada, Bahamas, Hispaniola. Winters in southern portion of breeding range.

Prairie Warbler

Dendroica discolor

(L-5 W-7)

Greenish yellow above streaked with chestnut; yellow below with black markings on sides; yellow face with black eyeline and chin stripe; white wing bars. *Immature:* Similarly patterned but much dingier.

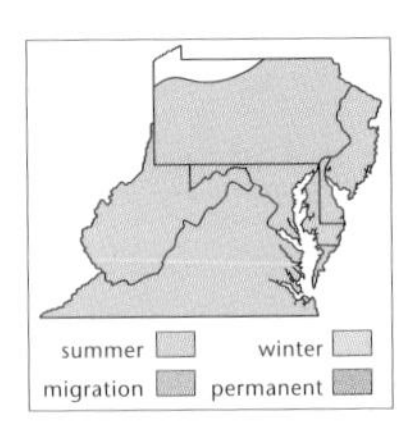

Habits Wags tail while foraging.
Voice Song—a rising series of buzzy "zee"s.
Similar Species Tail wagging habit separates this species from all but the immature Palm Warbler, which is brownish (not green) above and has grayish streaking on breast (not restricted to sides).
Habitat Scrubby coniferous and deciduous second growth; clearcuts; pine plantations.
Abundance and Distribution Common to uncommon summer resident* (May–Sep) in lowlands nearly throughout, although scarce and local or absent in extreme northern Pennsylvania.
Where to Find Higbee Beach Wildlife Management Area, New Jersey; Seneca Creek State Park, Maryland; George Washington Birthplace National Monument, Virginia.
Range Breeds in the eastern half of the United States. Winters in southern Florida and the Caribbean basin.

Palm Warbler

Dendroica palmarum

(L-5 W-7)

Olive with dark streaks above; yellowish or creamy below with brownish streaks; yellow undertail coverts; chestnut cap in breeding plumage; yellow eyestripe.

Habits Wags tail while foraging. Often forages low or on the ground.
Voice Song—a weak series of buzzy notes on the same pitch.
Similar Species The immature Palm Warbler is similar to the immature Pine Warbler but is brownish (not green) above and has grayish streaking on breast (not restricted to sides).

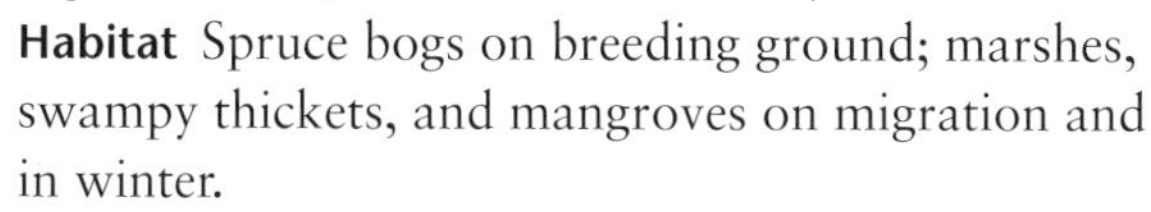

Habitat Spruce bogs on breeding ground; marshes, swampy thickets, and mangroves on migration and in winter.
Abundance and Distribution Common to uncommon transient (Sep–Oct, Apr) throughout, more numerous in fall; rare winter resident, mainly along the coast.
Where to Find Assawoman Wildlife Management Area, Delaware; Presque Isle State Park, Pennsylvania; Presquile National Wildlife Refuge, Virginia.
Range Breeds across central and eastern Canada and extreme north central and northeastern United States. Winters along the Atlantic and Gulf Coastal Plain and in the northern Caribbean Basin.

Bay-breasted Warbler

Dendroica castanea
(L-6 W-9)

Gray above with dark streaking; chestnut cap, throat, and sides; black mask; beige neck patch. *Female:* Olive above, beige below, often with some chestnut on sides.

Voice Song—a very high-pitched "wesee wesee wesee."

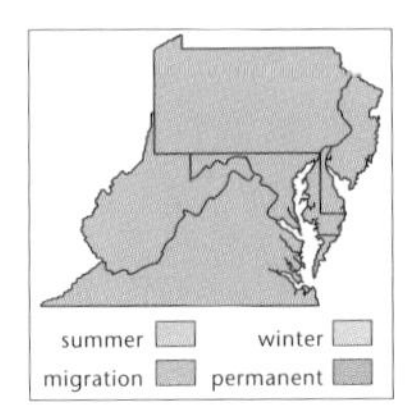

Similar Species Immature Blackpoll has faintly streaked (not plain) breast and back and pale (not dark) legs.
Habitat Breeds in coniferous forest; various woodlands on migration and in winter.
Abundance and Distribution Common to uncommon transient (Sep, May) throughout, more numerous in fall.
Where to Find Meadowside Nature Center, Maryland; Great Swamp National Wildlife Refuge, New Jersey; Beech Fork State Park, West Virginia.
Range Breeds across central and eastern Canada and extreme north central and northeastern United States. Winters in Panama, Colombia, and northwestern Venezuela.

Blackpoll Warbler

Dendroica striata

(L-6 W-9)

Grayish green streaked with black above, white below; black cap; white cheek; black chin stripe and streakings on side. *Female:* Olive cap streaked with black; faint white eye stripe; greenish cheek; whitish below with variable amounts of gray streaking.

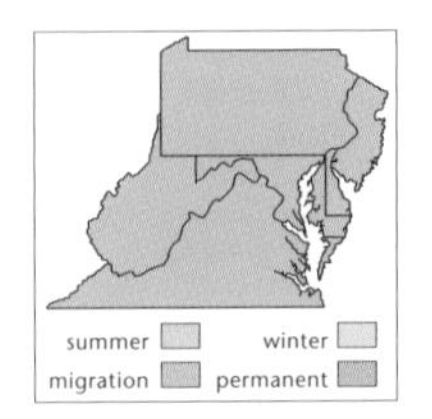

Voice Song—a weak series of "tsi"s, building in volume and then trailing off; call—a soft "chuk."
Similar Species Black-and-White Warbler resembles male, but crown has white median stripe (not solid black). See also Bay-breasted Warbler.
Habitat Breeds in coniferous forest; various forest and scrub sites on migration and in winter.
Abundance and Distribution Common transient (Sep–Oct, May) throughout; singing stragglers con-

tinue to pass through on migration well into June; more common in fall than in spring, except in Virginia, where it is scarcer in fall.
Where to Find Rock Creek Park, Washington, D.C.; Dickey Ridge Visitor Center, Shenandoah National Park, Virginia; Bowman's Hill Wildflower Preserve, Pennsylvania.
Range Breeds in northern boreal regions of North America. Winters in South America to northern Argentina.

Cerulean Warbler

Dendroica cerulea

(L-5 W-8)

A delicate bluish gray above, white below; black bar across chest and streakings down side; white wing bars. *Female:* Olive tinged with blue above; whitish below with grayish streakings; blue-gray crown; creamy eye stripe.

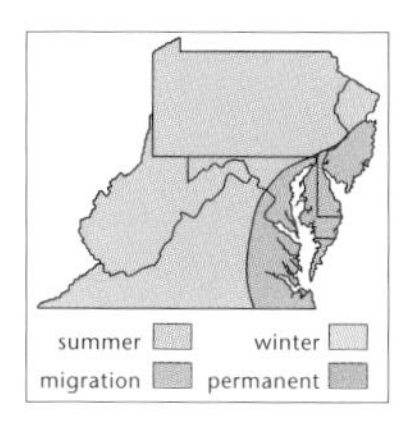

Voice Song—a series of 6–8 buzzy "tsee" notes, the last with an upward inflection; call—"tsip."
Similar Species Black throat band is distinctive for male. Female resembles female Blackpoll but has unstreaked back, prominent eyestripe, and dark (not yellowish) legs.
Habitat Deciduous forest, often tall oaks or sycamores.
Abundance and Distribution Uncommon summer resident* (May–Sep) in West Virginia, the northern Piedmont and mountains of Virginia, Pennsylvania (except extreme north, where scarce); rare and local summer resident* in the Coastal Plain.
Where to Find Allamuchy Mountain State Park, New Jersey; Catoctin Mountain Park, Maryland; Beech Fork State Park, West Virginia.

Range Breeds in the eastern United States except southeastern Coastal Plain. Winters in northern South America.

Black-and-white Warbler

Mniotilta varia

(L-5 W-9)

Boldly striped with black and white above and below. *Female:* Faint grayish streaking below.

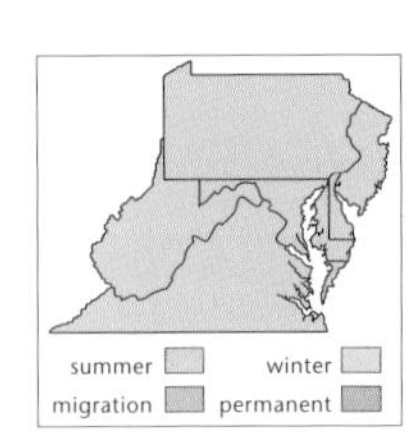

Habits Clambers up and down tree trunks and branches, nuthatch fashion.

Voice Song—a weak "wesee wesee wesee"; call—"pit."

Similar Species Only species with white median stripe in black cap and peculiar trunk-foraging behavior.

Habitat Deciduous and mixed forest.

Abundance and Distribution Common to uncommon transient and summer resident* (Apr–Sep) nearly throughout; scarcer as a summer resident on the Coastal Plain.

Where to Find Glover-Archbold Park, Washington, D.C. (migration); Dickey Ridge Visitor Center, Shenandoah National Park, Virginia; Woodbourne Sanctuary, Pennsylvania.

Range Breeds across Canada east of the mountains and in the eastern half of the United States. Winters from extreme southern United States, eastern Mexico, and Central America to northern South America; in West Indies.

American Redstart

Setophaga ruticilla

(L-5 W-9)

Black above and below with brilliant orange patches on tail, wings, and sides of breast; white undertail coverts. *Female:* Grayish brown above, whitish below with yellow patches on tail, wings, and sides of breast. *Immature Male:* Salmon-colored patches on sides of breast.

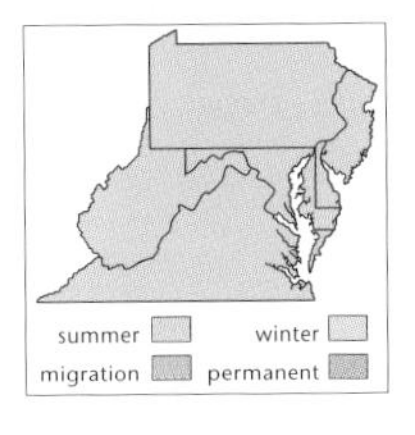

Habits This species catches much of its prey on the wing in brief sallies. Often fans tail and droops wings while foraging.

Voice One individual will often have 3 or 4 different songs, even alternating song types from one phrase to the next. Some common phrases are "tsee tsee tsee tsee tseet," "tsee tsee tsee tsee tsee-o," "teetsa teetsa teeetsa teetsa teet." Call—a strong "chip."

Habitat Deciduous and mixed forest.

Abundance and Distribution Common transient and summer resident* (May–Sep) throughout.

Where to Find Robert L. Graham (Nanticoke) Wildlife Area, Delaware; Sleepy Creek Wildlife Management Area, West Virginia; C&O Canal, Maryland.

Range Breeds across Canada south of the Arctic region and in the eastern half of the United States except the southeastern Coastal Plain. Winters from central Mexico south to northern South America, in West Indies.

Prothonotary Warbler

Protonotaria citrea

(L-6 W-9)

Golden orange head and breast; yellow-green back; blue-gray wings; white belly and undertail coverts. *Female:* Similar but yellow rather than orange.

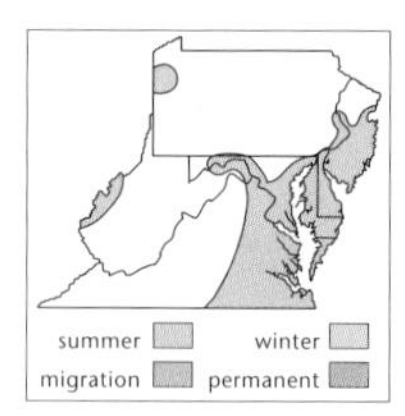

Voice Song—a loud, ringing "peeet weeet weeet weeet weeet weeet weeet"; call—"tsip."
Similar Species Female Wilson's Warbler has brownish green (not gray) wings and yellow (not white) undertail coverts. Yellow Warbler has yellow wing bars (none on Prothonotary).
Habitat Wooded swamps.
Abundance and Distribution Common summer resident* (May–Sep) along the Coastal Plain, uncommon to rare and local elsewhere; in West Virginia found mainly along the Ohio River and tributaries; in New Jersey mainly along the Delaware River and Bay and neighboring bottomlands; in inland Maryland, mainly along the Potomac; in Pennsylvania, mainly at Pymatuning State Park (nest boxes) and along the Delaware River.
Where to Find Myrtle Grove Wildlife Management Area, Maryland; Pymatuning State Park, Pennsylvania; Great Dismal Swamp National Wildlife Refuge, Virginia.
Range Breeds in the eastern half of the United States. Winters in southeastern Mexico and Caribbean basin.

Worm-eating Warbler

Helmitheros vermivorus

(L-6 W-9)

Olive above, buffy below; crown striped with black and buff.

Habits Forages in dead leaf clumps in trees or on the ground.

summer winter
migration permanent

Voice Song—an insectlike, buzzy trill; call—a strong "chip."

Habitat Deciduous forest, often on west-facing slopes.

Abundance and Distribution Uncommon transient and local summer resident* (May–Sep) nearly throughout; rare as a breeder on the Virginia Coastal Plain and in northern and western Pennsylvania.

Where to Find Cypress Swamp Conservation Area (North Pocomoke Swamp), Delaware (migration); Berwind Lake Wildlife Management Area, West Virginia; Delaware Water Gap National Recreation Area, Pennsylvania.

Range Breeds in the eastern United States. Winters from southern Mexico to Panama, in West Indies.

Swainson's Warbler

Limnothlypis swainsonii

(L-6 W-9)

Brown above, buff below with chestnut cap; whitish eye stripe.

Habits Feeds mainly on the ground.

Voice Song—a loud, clear "whee whee whitoo whee"; call—a ringing "chip."

Habitat Canebrakes, swampy thickets; also in rhododendron thickets in the southern Appalachians.

Abundance and Distribution Uncommon to rare and local summer resident* (May–Aug) in southern Virginia and southwestern West Virginia; scattered records elsewhere.
Where to Find Great Dismal Swamp National Wildlife Refuge, Virginia; Panther State Forest, West Virginia.
Range Breeds in the southeastern United States. Winters in the Bahamas, Greater Antilles, eastern Mexico and Yucatan.

Ovenbird

Seiurus aurocapillus
(L-6 W-10)

Olive above, white below with heavy, dark streaks; orange crown stripe bordered in black; white eyering.

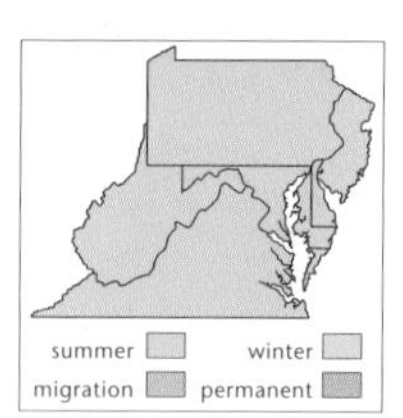

Habits Walks on forest floor, flicking leaves and duff while foraging for invertebrates.
Voice Song—a loud "teacher teacher teacher teacher."
Similar Species Northern and Louisiana waterthrushes have a prominent white or yellowish eyestripe. Ovenbirds do not bob while walking, as do waterthrushes.
Habitat Deciduous and mixed forest.
Abundance and Distribution Common transient and summer resident* (May–Sep) throughout.
Where to Find Glover-Archbold Park, Washington, D.C.; Beaver Swamp Wildlife Management Area, Cape May, New Jersey; Raccoon Creek State Park, Pennsylvania.
Range Central and eastern Canada and central and eastern United States south to east Kansas and north Georgia. Winters from southern Florida and

southern Mexico south through Central America to northern Venezuela, in West Indies.

Northern Waterthrush

Seiurus noveboracensis

(L-6 W-10)

Brown above, white or yellowish below with dark streaking on throat and breast; prominent creamy or yellowish eyestripe.

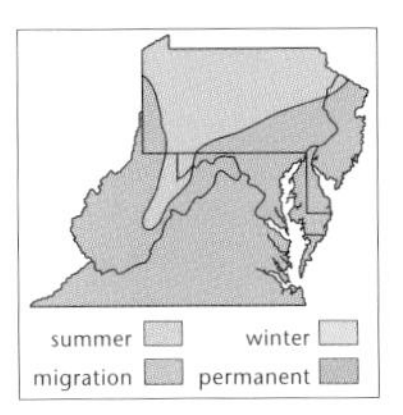

Habits Forages on the ground, bobbing as it walks, usually in boggy or wet areas.

Voice Song—"chi chi chi chewy chewy will will"; call—"chink."

Similar Species Louisiana Waterthrush has clear white throat (not streaked) and buffy flanks.

Habitat Swamps, bogs, swales, ponds, rivers, lakes; usually near stagnant water.

Abundance and Distribution Uncommon to rare and local summer resident* (May–Aug) in northern Pennsylvania and in southern Pennsylvania in the mountains, also the highlands of western Maryland, northern New Jersey (rare), and eastern West Virginia; uncommon transient (Aug–Sep, May) elsewhere.

Where to Find Erie National Wildlife Refuge, Pennsylvania (breeding); Canaan Valley State Park, West Virginia (breeding); Mason Neck National Wildlife Refuge, Virginia (migration).

Range Breeds in boreal North America south of the Arctic Circle. Winters from central Mexico south through Central America to northern South America, in Caribbean basin.

Louisiana Waterthrush

Seiurus motacilla

(L-6 W-10)

Brown above, creamy below with dark streaking on breast; prominent white eyestripe.

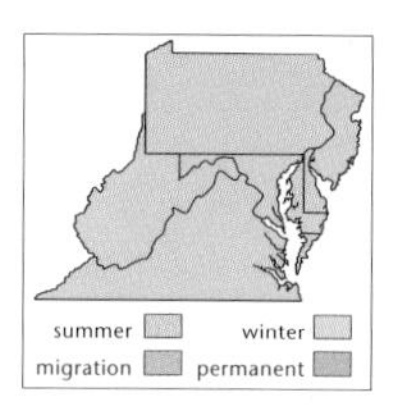

Habits Forages on the ground, bobbing as it walks, usually near running water.
Voice Song—"tsepit tsepit tsepit tsew titit ti ti."
Similar Species The Northern Waterthrush has streaking on the throat (not clear white as in Louisiana Waterthrush), and flanks are same color as breast and belly (not buffy as in the Louisiana).
Habitat Streams, rivers, swales, ponds.
Abundance and Distribution Common to uncommon summer resident* (Apr–Aug) in lowlands and mid-elevations along streams nearly throughout; rare and local in New Jersey, mainly in the north.
Where to Find C&O Canal, Maryland; Scotts Run Nature Preserve, Virginia; Ridley Creek State Park, Pennsylvania.
Range Breeds in the eastern United States. Winters from Mexico to northern South America, in West Indies.

Kentucky Warbler

Oporornis formosus

(L-5 W-8)

Green above, yellow below; yellow spectacles (forehead, eyestripe, eyering); black crown, lores, earpatch. Black more or less replaced by green in female.

Habits Forages by hopping (not walking) on the ground, picking insects from overhanging vegetation.

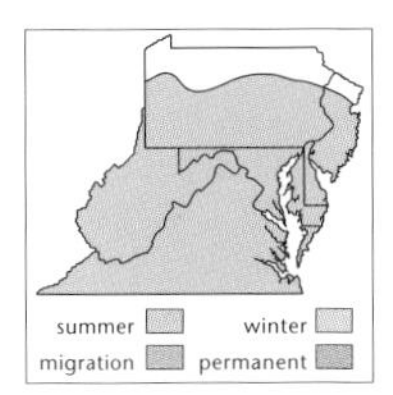

Voice Song—a clear, loud "choree choree choree choree choree." Call—a series of "chip"s, repeated as the bird hops, rising and falling in volume as the bird turns one way and then another.
Similar Species Male Common Yellowthroat has a black mask, no yellow spectacles.
Habitat Moist forest.
Abundance and Distribution Common to uncommon summer resident* (May–Aug) in lowlands and mid-elevations nearly throughout; rare and local in New Jersey and scarce or absent from northern Pennsylvania.
Where to Find Trap Pond State Park, Delaware; C&O Canal, Maryland; East Lynn Lake Wildlife Management Area, West Virginia.
Range Breeds in the eastern United States. Winters from southern Mexico to northern Colombia and northwestern Venezuela.

Connecticut Warbler

Oporornis agilis
(L-6 W-9)

Olive above, yellow below; gray hood; complete, white eyering. *Female:* Brownish yellow head.

Habits Forages low and on the ground.
Voice Song—a loud, clear "chipychip chipychip chipychipchipit"; call—"chink."
Similar Species Spring male has a complete white eyering (lacking in Mourning Warbler) and gray breast (blackish in Mourning). Females and immatures of these species are safely separable only in the hand on the basis of wing measurement minus tail measurement (Lanyon and Bull 1967).
Habitat Bogs, thickets.

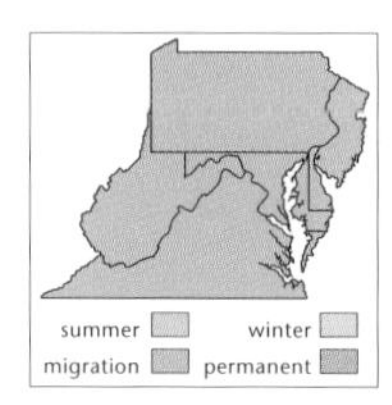

Abundance and Distribution Rare fall transient (Sep–Oct) throughout; scattered records in spring (May).
Where to Find Blackwater National Wildlife Refuge, Maryland; Ohiopyle State Park, Pennsylvania; Great Swamp National Wildlife Refuge, New Jersey.
Range Breeds in central Canada and extreme north central United States. Winters in northern South America.

Mourning Warbler
Oporornis philadelphia
(L-6 W-9)

Olive above, yellow below; gray hood with black on breast. *Female:* Brownish yellow head and partial eyering.

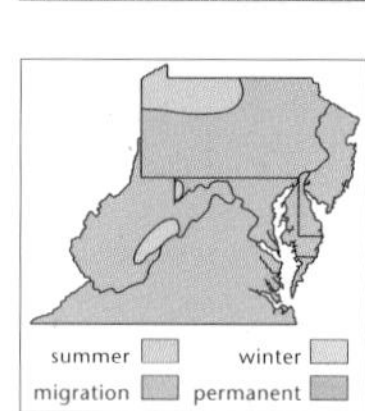

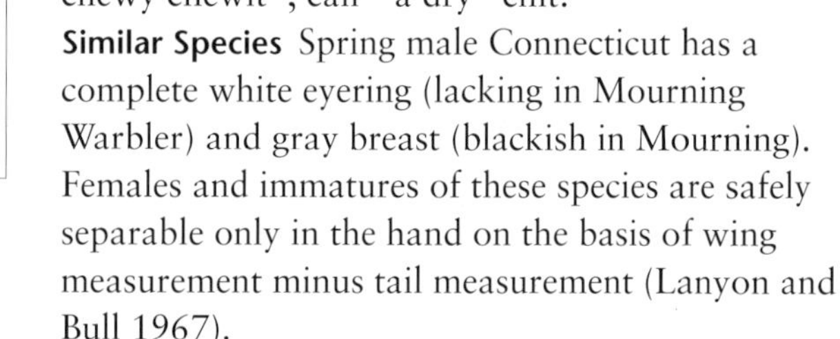

Habits Forages low and on the ground.
Voice Song—a loud, clear "chewy chewy chewy chewy chewit"; call—a dry "chit."
Similar Species Spring male Connecticut has a complete white eyering (lacking in Mourning Warbler) and gray breast (blackish in Mourning). Females and immatures of these species are safely separable only in the hand on the basis of wing measurement minus tail measurement (Lanyon and Bull 1967).
Habitat Dense thickets and tangles of second-growth woodlands.
Abundance and Distribution Uncommon to rare and local summer resident* (May–Sep) in northern Pennsylvania (mainly northwest); the Appalachians of eastern West Virginia and Garrett County, Maryland; and the highlands of Highland, Augusta, and Bath counties in Virginia; rare transient (Sep, May) elsewhere.

Where to Find Swallow Falls State Park, Maryland; Erie National Wildlife Refuge, Pennsylvania; Grayson Highlands State Park, Virginia.
Range Breeds in central and eastern Canada and extreme north central and northeastern United States. Winters from Nicaragua to northern South America.

Common Yellowthroat

Geothlypis trichas
(L-5 W-7)

Olive above, yellow below, with black mask. *Female:* Brownish above, bright yellow throat fading to whitish on belly; brownish on sides.

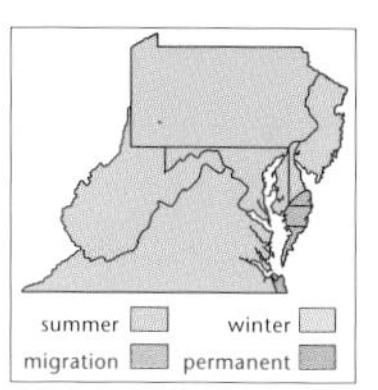

Habits Skulks in low, marshy vegetation.
Voice Song—"wichity wichity wichit"; call—a harsh "chuk."
Similar Species Lack of eyeline, eyering, and wing bars plus whitish belly separates female from other warblers.
Habitat Marshes, streams, estuaries, wet meadows, riparian areas; reed beds bordering rivers, ponds, and streams.
Abundance and Distribution Common transient and summer resident* (Apr–Sep) throughout; uncommon to rare winter resident (Oct–Apr), mainly along the southern Coastal Plain.
Where to Find Back Bay National Wildlife Refuge, Virginia (year-round); Prime Hook National Wildlife Refuge, Delaware; Cape May Migratory Bird Refuge (South Cape May Meadows), New Jersey.
Range Breeds from Canada south throughout the continent to the southern United States and in the highlands to southern Mexico. Winters from

the southern United States to Costa Rica, in Bahamas and Greater Antilles.

Hooded Warbler

Wilsonia citrina

(L-5 W-8)

Olive above, yellow below; black hood; yellow forehead and face; white tail spots. *Female:* Usually lacks hood but has greenish cap, yellow forehead and eyestripe.

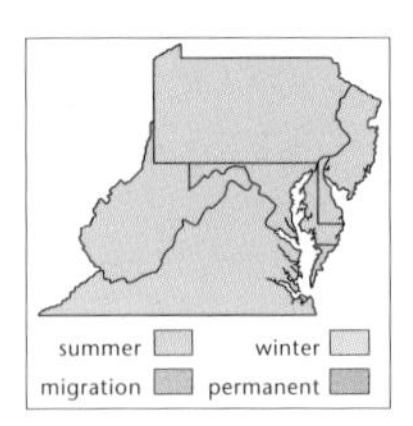

Habits Sallies from low perches for insects; often fans tail, exposing white spots.

Voice Song—"sweeta wee teeoo"; call—a clear musical "chip," usually given repeatedly for up to a minute.

Similar Species Female resembles female Wilson's Warbler, but Wilson's Warbler lacks the white tail marks of the Hooded Warbler and does not fan tail.

Habitat Dense thickets, tree falls in lowland deciduous, swamp, and riparian forest.

Abundance and Distribution Common to uncommon summer resident* (May–Sep) nearly throughout, although local in eastern Pennsylvania, New Jersey, and along the Coastal Plain.

Where to Find Jug Bay Natural Area, Maryland; Holly River State Park, West Virginia; Tioga/Hammond Lakes National Recreation Area, Pennsylvania.

Range Breeds in the eastern United States. Winters from southern Mexico to Panama.

Wilson's Warbler

Wilsonia pusilla

(L-5 W-7)

Olive above, yellow below with a black cap. *Female:* Often has only a partially black or completely greenish yellow crown.

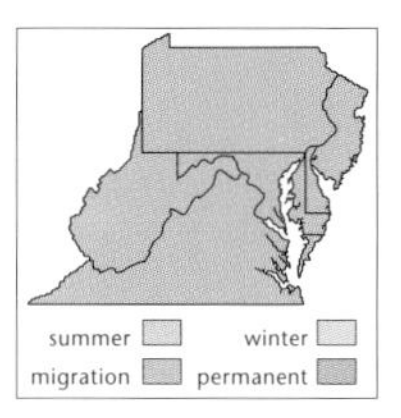

Habits Flycatches at mid- to upper canopy level, using short, sallying flights.

Voice Song—"Chee chee chee chee chipy-chipy-chipy-chipy" (almost a trill); call—a hoarse "ship."

Similar Species In some female Wilson's, the dark cap may be completely lacking. These birds can be distinguished from female Yellow Warblers by the brownish tail, lacking yellow spots. The larger, chunkier female Hooded Warbler can resemble the female Wilson's Warbler, but Wilson's Warbler lacks the white tail marks of the Hooded Warbler and does not fan tail.

Habitat Brushy thickets (willow, alder, aspen, dogwood).

Abundance and Distribution Uncommon to rare transient (Sep–Oct, May) nearly throughout, although scarcer along the southern Coastal Plain.

Where to Find John Heinz National Wildlife Refuge, Pennsylvania; Lily Pons Water Gardens, Maryland; Eastern Shore of Virginia National Wildlife Refuge, Virginia.

Range Breeds in boreal regions of northern and western North America; winters from southern California and Texas south to Panama.

Canada Warbler

Wilsonia canadensis

(L-5 W-8)

Slate gray above, yellow below with black "necklace" across breast; yellow lores and eyering. *Female:* Necklace is usually fainter.

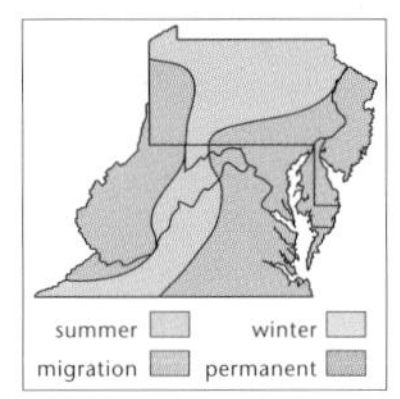

Voice Song—a high, thin, slurred series of notes, "tsi tsi tsi tsewy tsi."

Habitat A variety of coniferous and mixed forests and thickets.

Abundance and Distribution Common to uncommon and local summer resident* (May–Aug) in northern Pennsylvania and in southern Pennsylvania in the mountains, also the highlands of Virginia, western Maryland, northern New Jersey, and the Appalachians of eastern West Virginia; uncommon to rare transient (Aug–Sep, May) elsewhere.

Where to Find Swallow Falls State Park, Maryland; Canaan Valley State Park, West Virginia; High Knob Recreation Area, Virginia.

Range Breeds in eastern and central Canada from Labrador to northeastern British Columbia and in boreal United States from Minnesota to New England; south in the Appalachians to northern Georgia. Winters in northern South America.

Yellow-breasted Chat

Icteria virens

(L-7 W-10)

A nearly thrush-sized warbler; brown above; yellow throat and breast; white belly and undertail coverts; white eyering and supraloral stripe; lores black or grayish.

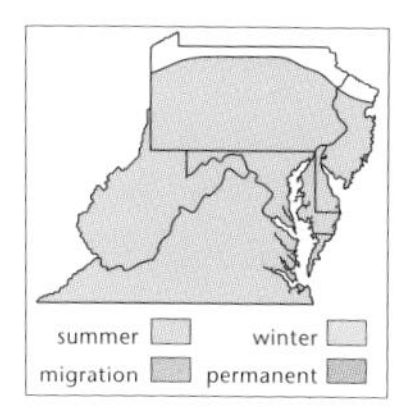

Habits Very shy; has flight song.
Voice A varied series of clear whistles and harsh, scolding "chak"s and "jeer"s.
Habitat Dense thickets, brushy pastures, forest undergrowth.
Abundance and Distribution Common but local summer resident* (May–Sep) nearly throughout; scarce in northern Pennsylvania and in highlands; rare winter visitor along the Coastal Plain.
Where to Find Bombay Hook National Wildlife Refuge, Delaware; Raccoon Creek State Park, Pennsylvania; Higbee Beach Wildlife Management Area, New Jersey.
Range Breeds in scattered regions nearly throughout the United States and southern Canada south to central Mexico. Winters from central Mexico to Panama.

Tanagers
(Family Thraupidae)

Most tanagers are neotropical in distribution. Only 4 species occur in the United States, 3 of which are found regularly in the Mid-Atlantic. Males are predominantly red (Hepatic Tanager, Summer Tanager, Scarlet Tanager) or yellow and red (Western Tanager). Females are brownish or greenish.

Summer Tanager
Piranga rubra
(L-8 W-13)

Red. *Female and Immature Male:* Tawny brown above, more yellowish below; second-year males and some females are blotched with red.

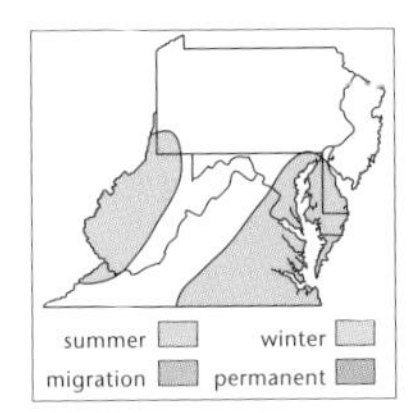

Voice Song—slurred, robinlike phrases; call—"pit-a-chuk."
Similar Species The female Baltimore Oriole is somewhat similar but has white wing bars and a long, pointed oriole bill. The female Summer Tanager lacks wing bars and has a relatively blunt, tanager bill. The female Scarlet Tanager is olive green above, not tawny as is the female Summer Tanager.
Habitat Open deciduous and mixed woodlands.
Abundance and Distribution Common to uncommon and local summer resident* (May–Sep) in western West Virginia, in extreme southwestern Pennsylvania (Green County), and along the Coastal Plain of Delaware, Maryland, and Virginia; uncommon in the Piedmont and valleys and mid-elevations of the mountains of Maryland and Virginia; rare or absent elsewhere.
Where to Find Redden State Forest, Delaware; Presquile National Wildlife Refuge, Virginia; Blackwater National Wildlife Refuge, Maryland.
Range Breeds across eastern and southern portions of the United States and northern Mexico; winters from southern Mexico through Central America to northern South America.

Scarlet Tanager

Piranga olivacea
(L-8 W-13)

Red with black wings. *Winter male:* Greenish above with black wings and tail, splotched with red during molt. *Female and Immature Male:* Olive above, yellowish below.

Voice Song—hoarse, loud, robinlike phrases; call—"chik-burr."

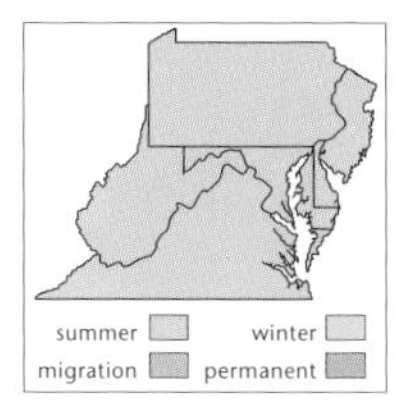

Similar Species Female is distinguished from other female tanagers by greenish (rather than brownish or grayish) cast to plumage and whitish (not greenish or yellowish) wing linings.

Habitat Deciduous and mixed forest; oak and riparian woodland.

Abundance and Distribution Common transient and summer resident* (May–Sep) throughout, more numerous as a breeder inland than along the Coastal Plain.

Where to Find Highpoint State Park, New Jersey; Cook Forest State Park, Pennsylvania; Dolly Sods Scenic Area, West Virginia.

Range Breeds in northeastern United States and southeastern Canada; winters in northern South America.

Sparrows and Towhees

(Family Emberizidae)

Small to medium-sized birds with conical bills, generally cryptically colored in various shades and patterns of browns and grays.

Eastern Towhee

Pipilo erythrophthalmus

(L-8 W-11)

Red eye; black head, breast, and back; rufous sides; white belly; tail black and rounded with white corners. *Female:* Patterned similarly but brown instead of black.

Habits This bird spends most of its time on the ground, using backward kick-hops to scatter duff and expose seeds and invertebrate prey.

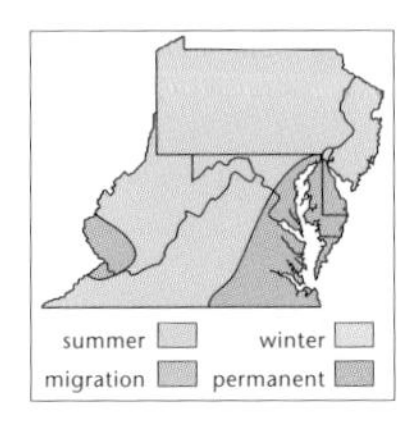

Voice Song—"drink your teeee" and variations; call—"chewink," "shrrinnk."
Habitat Undergrowth and thickets of deciduous and mixed woodlands.
Abundance and Distribution Common transient and summer resident* (Mar–Oct) throughout; common to uncommon winter resident (Nov–Feb) in southwestern West Virginia and the Coastal Plain and Piedmont of Delaware, Maryland, and Virginia; rare (mainly at feeders) or absent elsewhere in the region in winter.
Where to Find Rock Creek Park, Washington, D.C.; Jug Bay Natural Area, Maryland; Hemlock Springs Overlook, Shenandoah National Park, Virginia.
Range Breeds from extreme southern Canada across the United States (except most of Texas) and in the highlands of Mexico and Guatemala; winters from central and southern United States to Mexico and Guatemala.

Bachman's Sparrow

Aimophila aestivalis

(L-6 W-8)

Grayish streaked with brown above; pale gray below; dark malar stripe; gray cheek; central crown stripe bordered by brown.

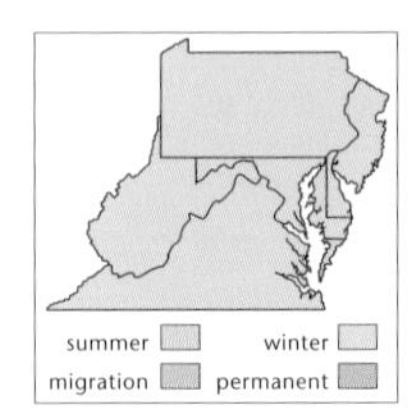

Habits Secretive.
Voice Song—"tseee chi chi chi," "tsoooo chew chew chew," and other variations on a similar theme.
Similar Species The Field Sparrow is rustier with a pinkish (not dark) bill; Grasshopper Sparrow is yellowish buff with creamy central crown stripe (not gray).

Habitat Open pine woods and savanna; brushy, overgrown fields.
Abundance and Distribution Rare and local resident* of southeastern Virginia.
Where to Find Sussex County, Virginia; Southampton County, Virginia.
Range Southeastern United States.

American Tree Sparrow

Spizella arborea

(L-6 W-9)

Streaked brownish above; dingy white below with dark breast spot; rufous crown; 2 white wing bars.

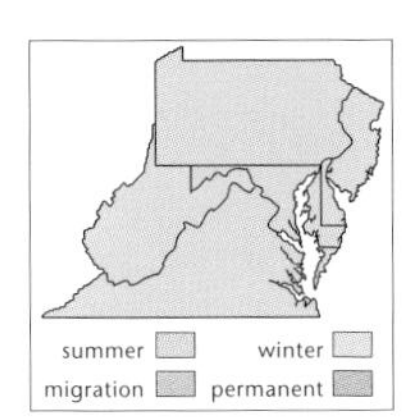

Voice Song—a rapid, high-pitched series of notes, "tse tse tse tsetl tse" and similar variations; call—"tsetl-de."
Similar Species This is the largest rufous-capped sparrow and the only one with a dark spot on an unstreaked breast.
Habitat Open, shrubby areas; weedy fields; overgrown pastures; grasslands.
Abundance and Distribution Common to uncommon transient and winter resident (Nov–Mar) nearly throughout, scarcer along the Coastal Plain.
Where to Find Huntley Meadows County Park, Virginia; Ohiopyle State Park, Pennsylvania; Brigantine National Wildlife Refuge, New Jersey.
Range Breeds in bog, tundra, and willow thickets of northern North America; winters in southern Canada, northern and central United States.

Chipping Sparrow

Spizella passerina

(L-5 W-8)

A small sparrow, streaked rusty brown above, dingy white below, with 2 white wing bars; rufous cap (somewhat streaked in winter); white eyebrow; black eyeline. *Immature:* Streaked crown; gray or buffy eyebrow; brown cheek patch.

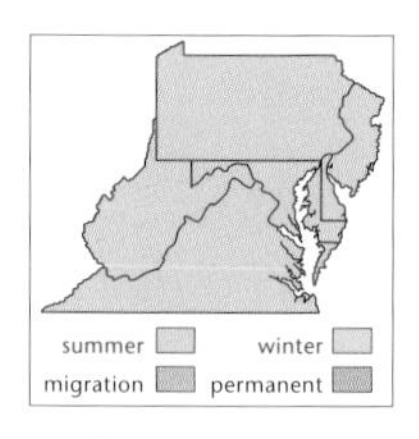

Voice Song—a rapid, metallic trill; call—"tsip."
Similar Species Adult facial pattern is distinctive; immature is similar to Clay-colored Sparrow but has grayish rather than buffy rump.
Habitat Open pine forests, woodlands, orchards, parks, suburbs, and cemeteries with scattered coniferous trees.
Abundance and Distribution Common summer resident* (Apr–Oct) throughout; rare winter visitor, mainly along the Coastal Plain.
Where to Find National Arboretum, Washington, D.C.; Jug Bay Natural Area, Maryland; Sleepy Creek Wildlife Management Area, West Virginia.
Range Breeds over most of North America south of the tundra, south through Mexico and Central America to Nicaragua; winters along coast and in southern portions of the breeding range.

Clay-colored Sparrow

Spizella pallida

(L-5 W-8)

Streaked brown above, buffy below; streaked crown with central gray stripe; grayish eyebrow; buffy cheek patch outlined in dark brown; gray nape; 2 white wing bars.

Voice Song—a buzzy "zee zee zee," with the number and speed of the zees varying; call—"sip."
Similar Species The immature Clay-colored Sparrow has a buffy rump rather than grayish rump as in the immature Chipping Sparrow.
Habitat Grasslands, brushy pastures.
Abundance and Distribution Rare fall transient (Sep–Oct) in coastal Maryland and Virginia; scattered breeding records in western Pennsylvania.
Where to Find Assateague Island National Seashore, Maryland; Chincoteague National Wildlife Refuge, Virginia.
Range Breeds from central Canada to the north central United States; winters from Texas south to Guatemala.

Field Sparrow
Spizella pusilla
(L-6 W-8)

Pink bill; streaked brown above, buff below; crown with gray central stripe bordered by rusty stripes; 2 white wing bars.

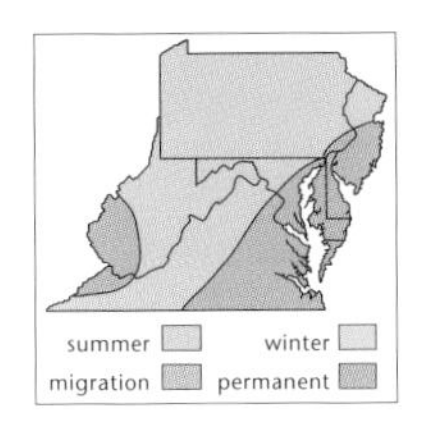

Voice Song—a series of "tew"s, beginning slowly and accelerating to a trill; call—"tsee."
Similar Species No other plain-breasted sparrow has a pink bill.
Habitat Old fields, brushy pastures.
Abundance and Distribution Common summer resident* (Mar–Nov) throughout; common to uncommon winter resident (Dec–Feb) in southwestern West Virginia, Piedmont, and Coastal Plain; scarce or absent in winter in Pennsylvania, northern New Jersey, and highland areas.
Where to Find Meadowside Nature Center, Maryland; Yellow Creek State Park, Pennsylvania; East

Lynn Lake Wildlife Management Area, West Virginia.

Range Breeds across eastern half of the United States and in southeastern Canada; winters in the southern half of the breeding range south to Florida and northeastern Mexico.

Vesper Sparrow

Pooecetes gramineus

(L-6 W-10)

Grayish streaked with brown above; white with brown streaks below; rusty shoulder patch; white outer tail feathers.

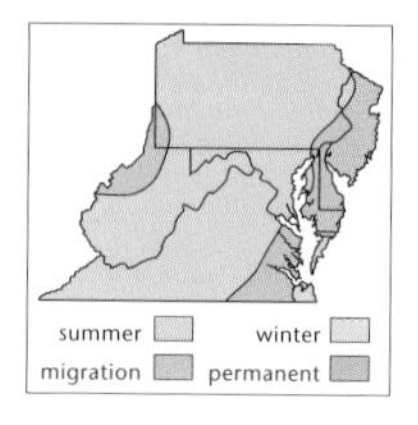

Voice Song—"chew chew chee chi-chi-chi tititit"; call—"chip."

Similar Species Savannah Sparrow has yellow lores and brown (not white) outer tail feathers.

Habitat Grasslands, pastures, scrub, agricultural fields.

Abundance and Distribution Common to uncommon transient and summer resident* (Mar–Oct), rare and local along the Coastal Plain and Piedmont as a breeder; uncommon to rare in winter (Nov–Feb), scarcer inland.

Where to Find Assawoman Wildlife Management Area, Delaware (winter); Sky Meadows State Park, Virginia (breeding); Buzzard Swamp Wildlife Management Area, Pennsylvania.

Range Breeds across much of northern North America to the central United States; winters in the southern United States and Mexico.

Lark Sparrow

Chondestes grammacus

(L-7 W-11)

Streaked brown above; dingy below; distinctive chestnut, white, and black face pattern; white throat; black breast spot; white corners on black, rounded tail.

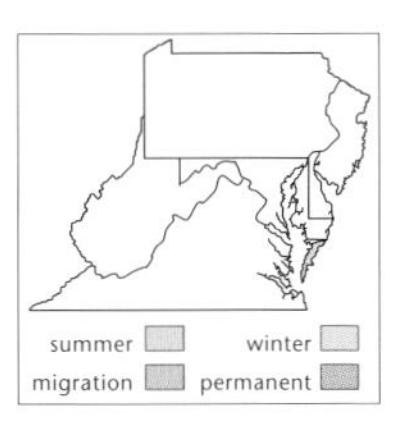

Voice Song—towhee-like “drink-your-teee,” often followed by various trills; call—“tseek.”
Habitat Grasslands, pastures, coastal prairie and dunes, agricultural fields.
Abundance and Distribution Rare fall transient (Sep), mainly along the Virginia coast; recent breeding records from Wayne County, West Virginia; scattered records from other places and seasons.
Where to Find Chincoteague National Wildlife Refuge, Virginia.
Range Breeds in the Canadian prairie states and across most of the United States except eastern, forested regions, south into northern Mexico; winters from the southern United States to southern Mexico.

Savannah Sparrow

Passerculus sandwichensis

(L-6 W-9)

Buff striped with brown above; whitish variously streaked with brown below; yellow or yellowish lores; whitish or yellowish eyebrow; often with dark, central breast spot. Plumage is highly variable in amount of yellow on face and streaking on breast according to subspecies, several of which winter in the region.

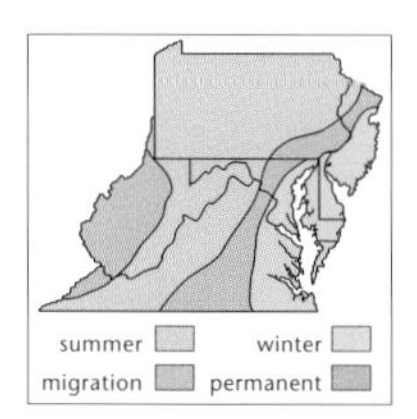

Habits Often in flocks.
Voice Song—a high-pitched, insectlike "tseet tsitit tsee tsoo"; call—"tsee."
Similar Species Savannah Sparrow has yellow lores and brown (not white) outer tail feathers; Vesper Sparrow has white outer tail feathers and lacks yellow lores.
Habitat Grasslands, pastures, agricultural fields, coastal marshes.
Abundance and Distribution Common to uncommon transient and summer resident* (Mar–Nov) in Pennsylvania, the northern panhandle of West Virginia, and highlands of eastern West Virginia, Virginia, and western Maryland; rare and local as a breeder in New Jersey salt marshes; common to uncommon transient and winter resident (Sep–Apr), mainly along the coast.
Where to Find Chincoteague National Wildlife Refuge, Virginia (year-round); Bear Mountain Road (CR 601), Highland County, Virginia (breeding); Prince Gallitzin State Park, Pennsylvania (breeding).
Range Breeds throughout the northern half of North America south to the central United States; also breeds in the central highlands of Mexico and Guatemala. Winters in the coastal and southern United States south to Honduras.

Grasshopper Sparrow

Ammodramus savannarum

(L-5 W-8)

A stubby, short-tailed bird; streaked brown above; creamy buff below; buffy crown stripe; yellow at bend of wing; yellowish or buffy lores and eyebrow.

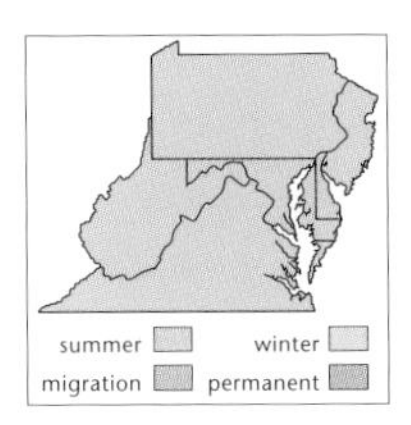

Voice Song—an insectlike "tsi-pi-ti-zzzzzzzzz"; call—a weak "kitik."
Habitat Overgrazed pasture (prefers short stubble).
Abundance and Distribution Common to uncommon and local transient and summer resident* (Apr–Sep) throughout; rare to casual in winter (Oct–Mar).
Where to Find Seneca Creek State Park, Maryland; Bombay Hook National Wildlife Refuge, Delaware; Prince Gallitzin State Park, Pennsylvania.
Range Breeds across northern and central United States; southern Mexico to northwestern South America; in Bahamas and Cuba. Winters in the southern United States, Mexico, and elsewhere within its tropical breeding range.

Henslow's Sparrow
Ammodramus henslowii
(L-5 W-7)

Streaked rusty brown above with rusty wings; buffy breast with dark streaks; whitish belly; grayish green head with dark brown crown and malar stripes. *Immature:* Breast streaking is faint or absent.

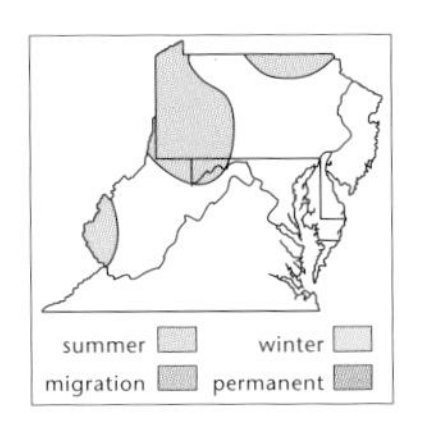

Voice Song—a high-pitched, repeated "tse-ik"; call—"tsip."
Similar Species Henslow's Sparrow has rusty wings (not grayish brown) and lacks the yellow lores of Savannah Sparrow and the yellow bend of wing of Grasshopper Sparrow.
Habitat Wet meadows, sedge marshes, weedy fields.
Abundance and Distribution Uncommon to rare and local transient and summer resident* (Apr–Sep) in scattered localities throughout the

region. The principal known concentration is in western Pennsylvania, with lesser numbers in western Maryland and western West Virginia. Formerly more numerous in the region. However, the bird is difficult to locate, and numbers may be higher than suspected.

Where to Find Prince Gallitzin State Park, Pennsylvania; Kerr Reservoir, Virginia.

Range Breeds in the northeastern and north central United States and southeastern Canada; winters in the southeastern United States.

Le Conte's Sparrow

Ammodramus leconteii

(L-5 W-7)

Streaked brown above; whitish below with dark streaking on sides; whitish central crown stripe bordered by dark brown stripes; buffy yellow eyebrow stripe; gray cheek patch outlined by buffy yellow.

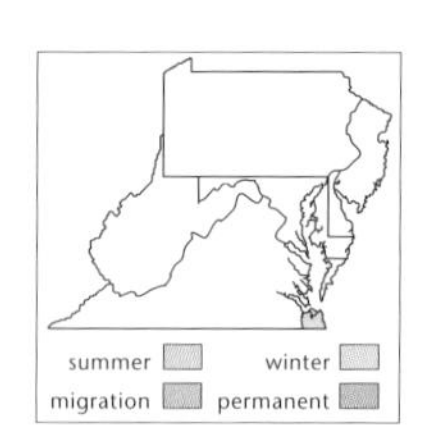

Voice Song—a high-pitched, insectlike buzz, "tsi tsi tzzzzz"; call—"tseak."

Similar Species Le Conte's Sparrow has unstreaked breast like the Grasshopper Sparrow and sharp-tailed sparrows, but the Grasshopper Sparrow has yellow at the bend of the wing and lacks streaking on flanks; sharp-tailed sparrows have gray nape (Le Conte's is buffy).

Habitat Tall grasslands, wet meadows, salt marsh, rank fields.

Abundance and Distribution Rare winter visitor (Oct–Apr) to extreme southeastern Virginia.

Where to Find Eastern Shore of Virginia National Wildlife Refuge, Virginia.

Range Breeds in central Canada (British Columbia to Quebec) and extreme north central United

States (Montana to Michigan); winters in southeastern United States.

Nelson's Sharp-tailed Sparrow

Ammodramus nelsoni

(L-6 W-8)

Streaked dark brown above; buffy throat and buffy-orange breast; belly whitish; gray crown stripe bordered by dark brown crown stripes; buffy orange eyebrow; gray nape.

Voice Song—a hoarse, buzzy "chur-chur-aaaaa zee-zurr zee-zurr," often uses a flight song as well; call—"tsuk."

Similar Species Nelson's Sharp-tailed Sparrow (winter resident in region) has buffy throat and tawny, unstreaked breast and lacks gray cheek patch; Salt Marsh Sharp-tailed Sparrow (permanent resident in region) has white throat, buffy, streaked breast, and gray cheek patch. Both species of Sharp-tailed Sparrows have a gray nape (Le Conte's is buffy).

Habitat Wet grasslands, coastal and inland marshes.

Abundance and Distribution Uncommon and local winter resident (Oct–Apr) along the Coastal Plain.

Where to Find Bombay Hook National Wildlife Refuge, Delaware; Assateague Island National Seashore, Maryland; Chincoteague National Wildlife Refuge, Virginia.

Range Breeds in central Canada and northeastern and north central United States (North Dakota,

Minnesota, Maine); winters along the coast of southeastern United States from Virginia to Texas and northeastern Mexico.

Saltmarsh Sharp-tailed Sparrow

Ammodramus caudacutus

(L-6 W-8)

Streaked dark brown above; buffy breast and flanks with faint brown streaks; belly whitish; gray crown stripe bordered by dark brown crown stripes; buffy orange eyebrow; gray cheek outlined by buffy orange stripes above and below; white throat; gray nape.

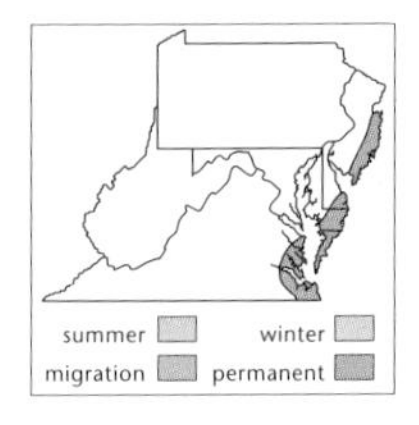

Voice Song—a "whisper song" involving a prolonged series of soft, wheezy twitters and trills.
Similar Species Salt Marsh Sharp-tailed Sparrow (permanent resident in region) has white throat, buffy, streaked breast, and gray cheek patch. Nelson's Sharp-tailed Sparrow (winter resident in region) has buffy throat and tawny, unstreaked breast and lacks gray cheek patch. Sharp-tailed Sparrow has gray nape (Le Conte's is buffy).
Habitat Coastal salt marsh.
Abundance and Distribution Common to uncommon and local resident* of salt marshes along the immediate coast.
Where to Find Brigantine National Wildlife Refuge, New Jersey; Blackwater National Wildlife Refuge, Maryland; Eastern Shore of Virginia National Wildlife Refuge, Virginia
Range Breeds along the Atlantic Coast of the

United States from Maine to North Carolina; winters along the Atlantic Coast from New York to the central east coast of Florida.

Seaside Sparrow

Ammodramus maritimus

(L-6 W-8)

Yellow lores; white throat; grayish streaked with dark brown above; buffy breast and whitish belly streaked with brown; a stocky bird with longish bill and short tail.

Voice Song—"brrrt zee zzurr zee," reminiscent of Red-winged Blackbird; call—"kak."
Similar Species Le Conte's and sharp-tailed sparrows lack yellow lores and clear, white throat of Seaside Sparrow.
Habitat Salt marshes.
Abundance and Distribution Common resident* along the immediate coast, somewhat scarcer in winter.
Where to Find Bombay Hook National Wildlife Refuge, Delaware; Assateague Island National Seashore, Maryland; Chincoteague National Wildlife Refuge, Virginia.
Range Resident along the coast of the eastern United States from Maine to Texas.

Fox Sparrow

Passerella iliaca

(L-7 W-11)

Hefty, for a sparrow—nearly thrush-sized; streaked dark or rusty brown above; whitish below with heavy dark

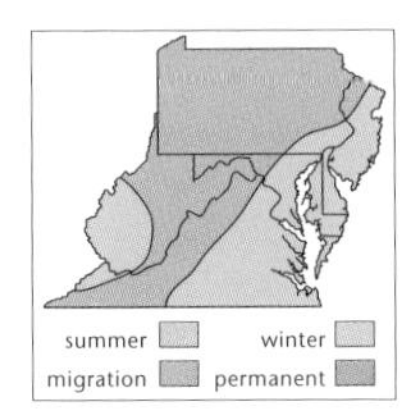

or rusty streakings that often coalesce as a blotch on the breast; rusty rump and tail.

Habits Forages in towhee fashion, jump-kicking its way through forest duff.
Voice Song—whistled, with long (3–4 seconds) varied phrases; call—"tshek."
Similar Species Hermit Thrush has long, thrush bill (not short, conical bill of sparrow) and unstreaked back.
Habitat Thickets and undergrowth of deciduous, mixed, and coniferous woodlands; hedgerows; scrub, brush piles.
Abundance and Distribution Common to uncommon transient (Nov, Mar) and uncommon to rare winter resident (Dec–Feb) along the Coastal Plain, in the Piedmont and southwestern West Virginia, or at feeders.
Where to Find McKee-Beshers Wildlife Management Area (Hughes Hollow), Maryland; Colonial Parkway, Virginia; Bombay Hook National Wildlife Refuge, Delaware.
Range Breeds across the northern tier of North America and in the mountains of the west; winters in coastal and southern United States.

Song Sparrow
Melospiza melodia
(L-6 W-9)

Streaked brown above; whitish below with heavy brown streaks and central breast spot; gray eyebrow; dark whisker and postorbital stripe.

Voice Song—"chik sik-i-sik choree k-sik-i-sik" with many variations; call—a nasal "chink."

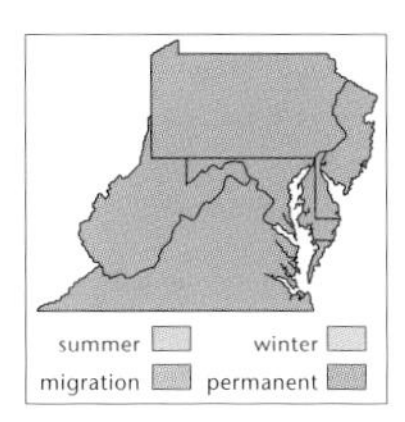

Similar Species Fox Sparrow head is brownish and lacks light/dark striping of Song Sparrow.
Habitat Swamps; inland and coastal marshes; riparian thickets (reeds, sedges); wet meadows; brushy fields.
Abundance and Distribution Common resident* nearly throughout; scarce in highlands; withdraws to dense stream-border thickets and marshes in winter.
Where to Find Beaver Swamp Wildlife Management Area, Cape May, New Jersey; Lewis Wetzel Wildlife Management Area, West Virginia; John Heinz National Wildlife Refuge, Pennsylvania.
Range Breeds across temperate and boreal North America; winters in temperate breeding range, southern United States and northern Mexico; resident population in central Mexico.

Lincoln's Sparrow

Melospiza lincolnii

(L-6 W-8)

Streaked brown above; patterned gray-and-brown face; white throat and belly; distinctive finely streaked, buffy breast band.

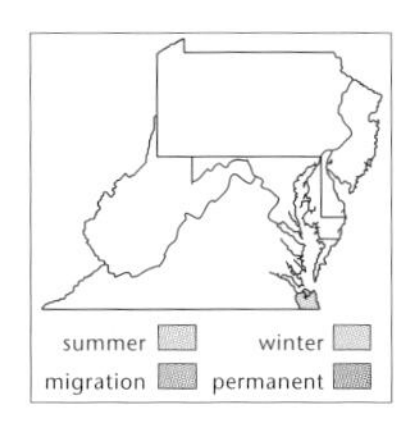

Voice Song—a series of brief trills at different pitches; call—"shuk."
Similar Species Finely streaked, buffy breast band is unique.
Habitat Brushy fields, hedgerows.
Abundance and Distribution Rare transient and winter resident (Oct–May) in southeastern Virginia.
Where to Find Back Bay National Wildlife Refuge, Virginia.

Range Breeds across the northern tier of North America and in the mountains of the west; winters in the coastal and southern United States south to Honduras.

Swamp Sparrow

Melospiza georgiana

(L-6 W-8)

Rusty crown with central grayish stripe; gray face; streaked brown above with rusty wings; whitish throat but otherwise grayish below with tawny flanks, faintly streaked.

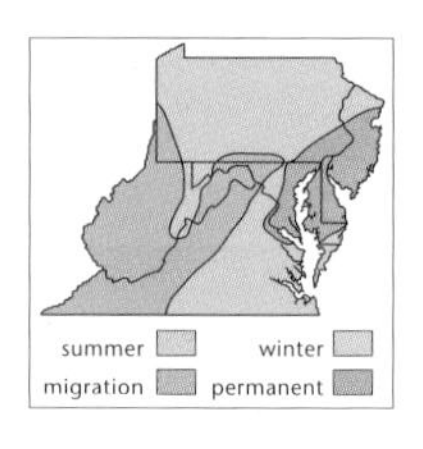

Voice Song—long "chipy-chipy-chipy" trills at various pitches and speeds; call—"chip."
Similar Species Only the Swamp Sparrow, among rusty-crowned sparrows of the region, lacks white wing bars.
Habitat Northern peat bogs (summer); coastal and inland marshes, wet grasslands, brushy pastures (winter).
Abundance and Distribution Uncommon and local summer resident* (May–Sep) in much of Pennsylvania and New Jersey; local breeding populations in highlands and coastal marshes of Maryland and Delaware, Highland County (Virginia), and the Appalachian highlands of West Virginia; common transient and winter resident (Sep–Apr) along the Piedmont and Coastal Plain; transient (Sep–Oct, Mar–Apr) elsewhere.
Where to Find Pymatuning State Park, Pennsylvania (breeding); Assawoman Wildlife Management Area, Delaware (winter); Great Swamp National Wildlife Refuge, New Jersey.
Range Breeds in central and eastern Canada and north central and northeastern United States; win-

ters in eastern and south central United States south to Mexico.

White-throated Sparrow

Zonotrichia albicollis

(L-7 W-9)

White throat; alternating black and white (or black and buff) crown stipes; yellow lores; streaked brown above; grayish below.

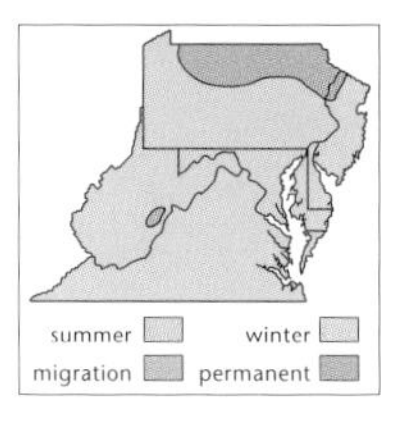

Voice Song—a thin, wavering whistle, often heard in March in thickets, "poor sam peabody peabody"; call—"seet."

Habitat Coniferous bogs (breeding); woodland thickets, brushy fields, feeders (migration, winter).

Abundance and Distribution Uncommon and local summer resident* (Apr–Sep) in northern Pennsylvania; also breeding records from the highlands of eastern West Virginia and northwestern New Jersey; common transient and winter resident (Oct–Apr) nearly throughout, scarce or absent from highlands in winter.

Where to Find National Arboretum, Washington, D.C. (migration, winter); R. B. Winter State Park, Pennsylvania (breeding); C&O Canal, Maryland (migration, winter).

Range Breeds across most of boreal Canada and northeastern and north central United States; winters in eastern and southern United States and northern Mexico.

White-crowned Sparrow

Zonotrichia leucophrys

(L-7 W-10)

Black-and-white striped crown; gray neck, breast, and belly; streaked gray and brown back; pinkish bill. *Immature:* Crown stripes are brown and gray.

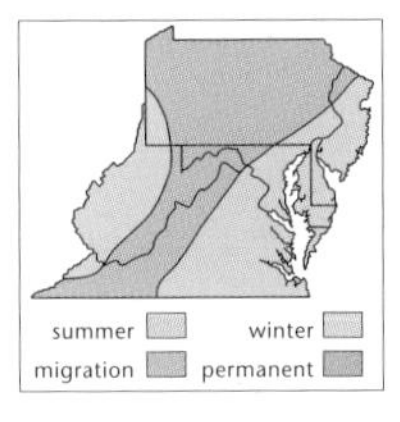

Voice Song—"tsee tsee tsee zzeech-i chi-i-i"; call—"chip."

Similar Species The White-throated Sparrow is also a large, plain-breasted, striped-crowned sparrow, but it has a white throat and yellow lores.

Habitat Thickets in coniferous and deciduous woodlands, brushy fields.

Abundance and Distribution Common to uncommon transient (Oct–Nov, Mar–Apr) throughout; uncommon to rare and local and winter resident (Nov–Mar) in southwestern Pennsylvania, western West Virginia, the Piedmont, and Coastal Plain, often at feeders and multiflora rose hedges.

Where to Find Lily Pons Water Gardens, Maryland; Lucketts, Hwy 661, Virginia; Nockamixon State Park, Pennsylvania.

Range Breeds in northern and western North America; winters across most of the United States south to central Mexico.

Dark-eyed Junco

Junco hyemalis

(L-6 W-10)

Entirely dark gray except for white belly and outer tail feathers; pinkish bill. *Female:* Similar but brownish rather than gray.

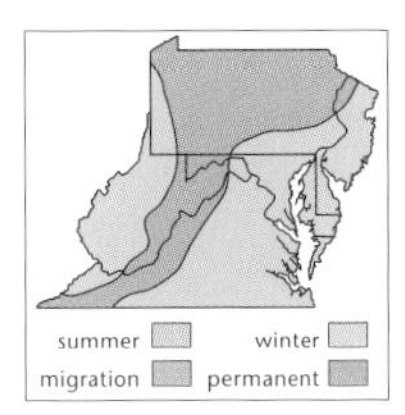

Voice Song—a trill, given at different speeds and pitches; call—"tsik."
Habitat Breeds in coniferous and mixed forests; open mixed woodlands; grasslands; agricultural fields, feeders (migration, winter).
Abundance and Distribution Common to uncommon resident* in northern Pennsylvania and in southern Pennsylvania in the mountains, also the highlands (above 3,000 feet) of western Maryland, northern New Jersey (rare), eastern West Virginia, and Virginia; common transient and winter resident (Oct–Apr) elsewhere.
Where to Find Dickey Ridge Visitor Center, Shenandoah National Park, Virginia (breeding); Swallow Falls State Park, Maryland (breeding); Canaan Valley State Park, West Virginia (breeding).
Range Breeds across northern North America and in the eastern and western United States, south in the mountains; winters from southern Canada south through the United States to northern Mexico.

Lapland Longspur

Calcarius lapponicus

(L-6 W-11)

Black head and breast with white or buff face pattern; rusty nape; streaked brown above; white belly; tail is all dark except for outermost tail feathers. *Winter Male and Female:* Brownish crown; buffy eyebrow, nape, and throat with darker brown mottlings on breast and flanks; buffy cheek outlined by darker brown.

Voice Song—short phrases of squeaky, slurred notes, given in flight; call—"tseeu."
Habitat Plowed fields, overgrazed pasture.

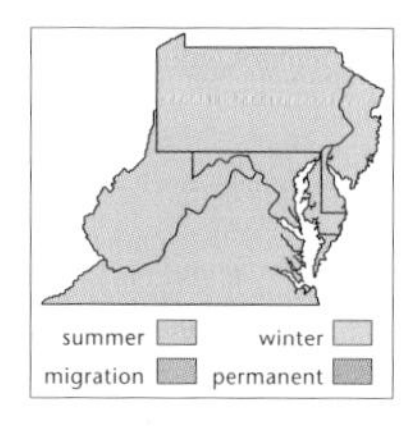

Abundance and Distribution Rare to casual transient and winter visitor (Nov–Feb) throughout.
Where to Find Bombay Hook National Wildlife Refuge, Delaware; Chincoteague National Wildlife Refuge, Virginia; John Heinz National Wildlife Refuge, Pennsylvania.
Range Breeds in tundra of extreme northern North America and Eurasia; winters in temperate regions of the Old and New World.

Snow Bunting

Plectrophenax nivalis
(L-7 W-11)

White head, rump, and underparts; black back; wings white with black primaries. *Winter Male and Female:* White is tinged with buff, and dark back is mottled with white.

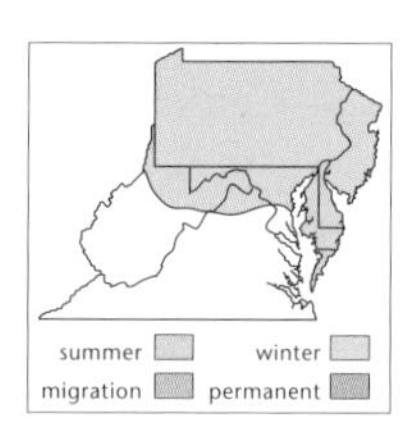

Voice Song—repeated musical, whistled "tsee tsee chewee" and similar phrases; flight call—"tseoo."
Habitat Lake shores, beaches, fields, pastures, agricultural areas (especially recently manured fields).
Abundance and Distribution Uncommon to rare and irregular winter visitor (Nov–Feb) from Pennsylvania and New Jersey to northern Virginia and along the coast; casual elsewhere.
Where to Find Prince Gallitzin State Park, Pennsylvania; Chincoteague National Wildlife Refuge, Virginia; Cape May Migratory Bird Refuge, New Jersey.
Range Breeds circumpolar in tundra; winters in southern boreal and northern temperate regions of Old and New World.

Cardinals, New World Grosbeaks, and Buntings

(Family Cardinalidae)

Small to medium-sized birds with conical bills; adult males generally are brightly colored, while females and immatures are cryptically colored.

Northern Cardinal

Cardinalis cardinalis

(L-9 W-12)

Red with crest; black face patch at base of red bill. *Female and Immature Male:* Crested like male but greenish brown, paler below, bill brownish or reddish.

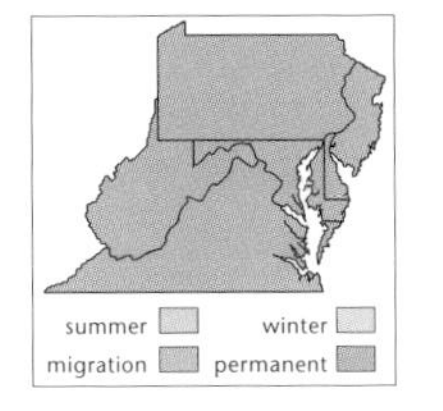

Voice Song—loud, ringing whistle, "whit-chew," repeated; call—a sharp "peak."
Habitat Thickets and tangles of open deciduous forest and second growth, hedgerows, residential parklands.
Abundance and Distribution Common resident* throughout.
Where to Find National Arboretum, Washington, D.C.; John Heinz National Wildlife Refuge, Pennsylvania; East Lynn Lake Wildlife Management Area, West Virginia.
Range Eastern United States and southeastern Canada; southwestern United States, Mexico, Guatemala, and Belize.

Rose-breasted Grosbeak

Pheucticus ludovicianus

(L-8 W-13)

Black head, back, wings, and tail; brownish in winter; red breast; white belly, rump, wing patches, and tail spots. *Female:* Mottled brown above, buffy below heavily streaked with dark brown; white eyebrow and wing bars. *First Year Male:* Like female but shows rose tints on breast and underwing coverts. *Second Year Male:* Patterned much like adult male but splotched brown and black on head and back.

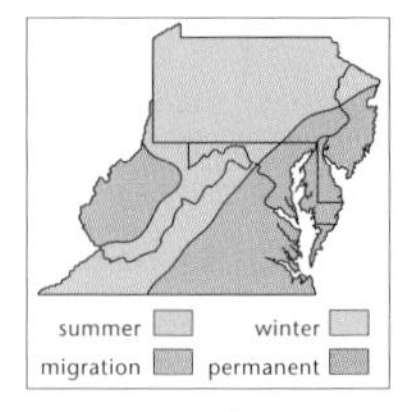

Voice Song—rapid, robinlike phrases; call—a sharp "keek" or "kik."
Habitat Deciduous forest and parklands.
Abundance and Distribution Uncommon and local summer resident* (May–Sep) in Pennsylvania, northern New Jersey, and the highlands (above 3,000 feet) of western Maryland, eastern West Virginia, and Virginia; common to uncommon transient (Sep, May) elsewhere.
Where to Find Stokes State Forest, New Jersey; Swallow Falls State Park, Maryland; Erie National Wildlife Refuge, Pennsylvania.
Range Breeds in northeastern United States, central and southeastern Canada; winters from southern Mexico south through Central America to northern South America and western Cuba.

Blue Grosbeak

Guiraca caerulea

(L-7 W-11)

Dark blue with 2 rusty wing bars. *Female and Immature Male:* Brownish above, paler below, with

tawny wing bars; often has a blush of blue on shoulder or rump.

Habits Flicks and fans tail.
Voice Song—a rapid series of up-and-down warbles, some notes harsh, some slurred; call—"chink."
Similar Species Male and female Indigo Buntings resemble corresponding sexes of Blue Grosbeak but are smaller and lack tawny wing bars and massive grosbeak bill.

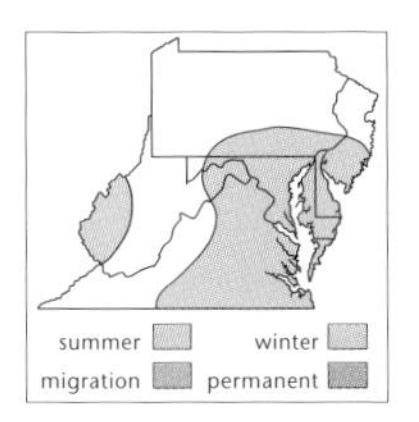

Habitat Thickets, scrub, brushy pastures, hedgerows.
Abundance and Distribution Common to uncommon summer resident* (May–Sep) in the Coastal Plain and Piedmont from southern New Jersey south, in valleys of the mountain region of Virginia, and in western West Virginia.
Where to Find Jug Bay Natural Area, Maryland; George Washington Birthplace National Monument, Virginia; Beech Fork State Park, West Virginia.
Range Breeds from central and southern United States through Mexico and Central America to Costa Rica; winters from northern Mexico to Panama; rarely Cuba.

Indigo Bunting

Passerina cyanea

(L-6 W-9)

Indigo blue. *Female and Immature Male:* Brown above, paler below with faint streaking on breast. *Winter Adult and Second Year Males:* Bluish with variable amounts of brown on back and wings.

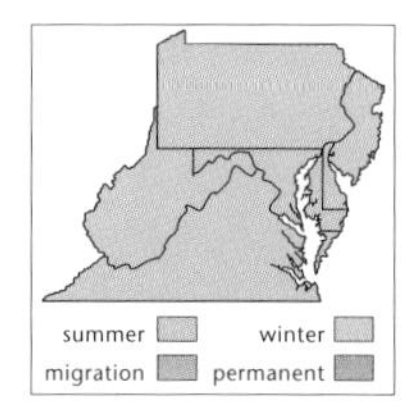

Voice Song—warbled phrases of paired or triplet notes; call—"tsink."
Similar Species Male and female Indigo Buntings resemble corresponding sexes of Blue Grosbeak but are smaller and lack tawny wing bars and massive grosbeak bill.
Habitat Thickets, hedgerows, brushy fields.
Abundance and Distribution Common transient and summer resident* (May–Sep) nearly throughout.
Where to Find Trap Pond State Park, Delaware; Locust Lake State Park, Pennsylvania; Stokes State Forest, New Jersey.
Range Breeds from extreme southeastern and south central Canada south through eastern and southwestern United States; winters from central Mexico to Panama and the West Indies.

Dickcissel

Spiza americana
(L-6 W-9)

Patterned like a miniature meadowlark—black bib (gray in winter) and yellow breast; streaked brown above; grayish head with creamy eyebrow; rusty red wing patch. *Female and Immature Male:* Patterned like male but paler yellow below and without black bib.

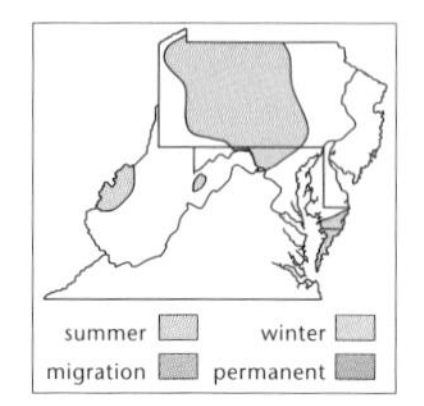

Voice Song—a dry "chik sizzzle," also a "brrrzeet" given in flight.
Similar Species Pale females and immatures resemble House Sparrow female but usually have traces of yellow on pale white (not dirty white) breast, a clear, whitish or yellowish eyebrow, and some chestnut on shoulder.

Habitat Grasslands, agricultural fields.
Abundance and Distribution Rare, irregular, and local summer resident* (May–Sep) at scattered localities that change from year to year.
Where to Find Mount Zion Strip Mines, Pennsylvania; Lucketts, Hwy 661, Virginia; Presquile National Wildlife Refuge, Virginia.
Range Breeds across eastern and central United States and south central Canada; winters mainly in northern South America.

Blackbirds and Orioles
(Family Icteridae)

Most migratory icterids are dimorphic, with large, strikingly colored males and smaller, more cryptically colored females.

Bobolink
Dolichonyx oryzivorus
(L-7 W-12)

Black with creamy nape; white rump and shoulder patch. *Winter Male and Female:* Streaked brown and yellow-buff above; yellow-buff below; crown with central buff stripe bordered by dark brown stripes; buff eyebrow.

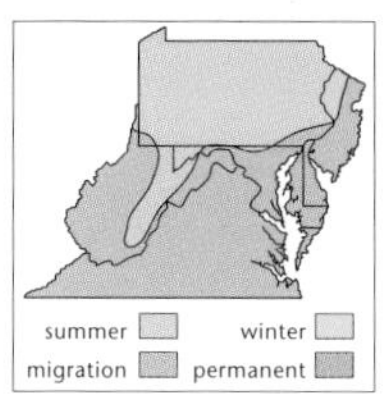

Voice Song—twittering, bubbling series of squeaks, "cherk"s, "ink"s, along with a few "bob-o-link"s and other, similar phrases thrown in, given in flight; call—"tink."
Similar Species Blackbird size and bill separate this bird from sparrows and buntings.

Habitat Grasslands, grain and hay fields, brushy pastures.
Abundance and Distribution Uncommon to rare and local summer resident,* nesting in hay fields of Pennsylvania, northwestern New Jersey, western Maryland, the eastern panhandle of West Virginia, and Highland County, Virginia; common fall (Aug–Sep) and uncommon spring (May) transient throughout. This species has declined significantly as a breeding bird in the region in recent years.
Where to Find Pymatuning Wildlife Management Area, Pennsylvania; farm fields in Garrett County, Maryland; farm fields in Tucker County, West Virginia.
Range Breeds in the northern United States and southern Canada; winters in South America.

Red-winged Blackbird

Agelaius phoeniceus

(L-8 W-14)

Black with red epaulets bordered in orange. *Female and Immature Male:* Dark brown above, whitish below heavily streaked with dark brown; whitish eyebrow and malar stripes. *Second Year Male:* Intermediate between female and male—black blotching, some orange on epaulet.

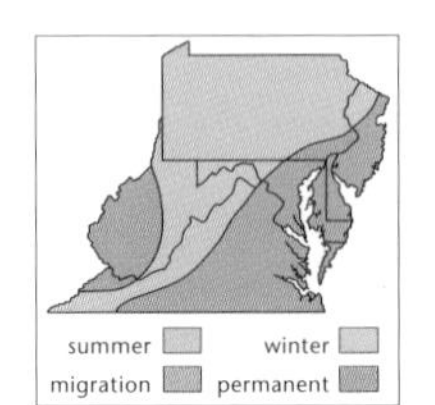

Voice Song—"konk-ka-ree"; call—a harsh "shek."
Similar Species Blackbird size and bill separate females from female Purple and House finches, which are also heavily streaked.
Habitat Inland and coastal marshes (breeding); brushy fields; tall grasslands; grain and hay fields; grain storage areas (winter).
Abundance and Distribution Common transient

and summer resident* (Mar–Oct) throughout; common winter resident (Nov–Feb) in coastal marshes; also winters elsewhere in lowlands of the region, generally often in large flocks.
Where to Find Lily Pons Water Gardens, Maryland; Cape May Migratory Bird Refuge (South Cape May Meadows), New Jersey; Brandywine Creek State Park, Delaware.
Range Breeds nearly throughout North America from the Arctic Circle south to Costa Rica; winters from temperate portions of breeding range south; resident populations in Bahamas and Cuba.

Eastern Meadowlark
Sturnella magna
(L-10 W-15)

Streaked brown and white above; yellow below with a black or brownish V on the breast; crown striped with buff and dark brown; tail is dark in center, white on outer edges.

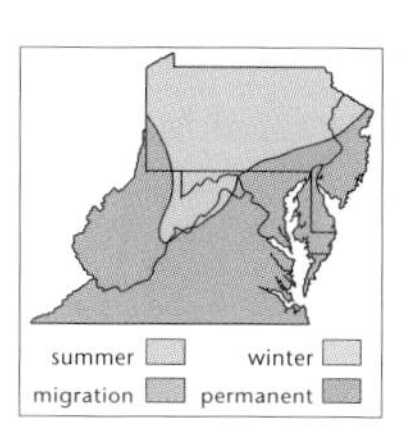

Voice Song—a plaintive "see-ur see-ur," the second phrase at a lower pitch than the first; call—a rattling, harsh "ka-kak-kak-kak-kak."
Similar Species Western Meadowlark is paler on the back; malar stripe is white in Eastern, yellow in Western. There are 2 breeding records for the Western Meadowlark in Pennsylvania and scattered observations of singing birds elsewhere in the region.
Habitat Grasslands; grain and hay fields; overgrown pastures.
Abundance and Distribution Common transient and summer resident* (Mar–Oct) throughout; common to uncommon or rare winter resident (Nov–Feb), seen mainly in agricultural fields and

marshes in lowlands. This species has declined as a breeding bird in the region in recent years.
Where to Find Lucketts, Hwy 661, Virginia; Mud Level Rd, Shippensburg, Pennsylvania; Eastern Neck Wildlife Refuge, Maryland.
Range Breeds from southeastern Canada and eastern and southern United States west to Arizona, in Mexico through Central and northern South America, in Cuba; winters through most of breeding range except northern portions.

Yellow-headed Blackbird

Xanthocephalus xanthocephalus
(L-10 W-16)

Black body with yellow head and breast; white wing patch. *Female and Immature Male:* Brown body; yellowish breast and throat; yellowish eyebrow.

Voice Song—"like a buzz saw biting a hard log" (Edwards in Oberholser 1974, 808); call—a croak.
Habitat Marshes, brushy pastures, agricultural fields. Favorite sites for these and several other blackbird species are cattle feedlots and grain elevators in winter.
Abundance and Distribution Rare transient, mainly in fall (Aug–Oct), and winter visitor along the immediate coast; scarcer inland.
Where to Find Hog Island Waterfowl Management Area, Virginia; Chincoteague National Wildlife Refuge, Virginia.
Range Breeds in south central and southwestern Canada, north central and northwestern United States; winters from southern California, Arizona, and New Mexico south to central Mexico.

Rusty Blackbird

Euphagus carolinus

(L-9 W-14)

Entirely black with creamy yellow eye. *Breeding Female:* Grayer. *Winter Male:* Black is tinged with rust; often shows a buffy eyebrow. *Winter Female:* Rusty above, buffy below, with prominent buffy eyebrow.

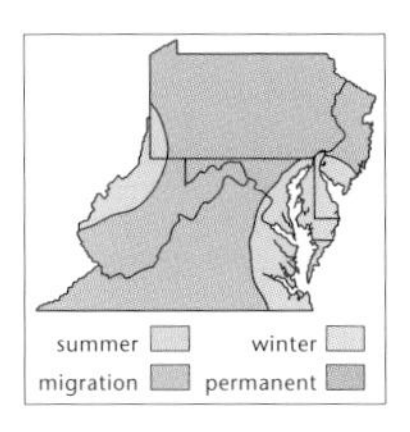

Voice Song—"curtl seee" repeated; call—"chik."
Similar Species Prominent creamy eye separates this bird from female Brewer's Blackbird. The few fall male Brewer's that show some rust are black (not rusty as in Rusty Blackbird) along the trailing edge of the secondaries.
Habitat Deciduous and coniferous forests, swamps, wooded edges of marshes.
Abundance and Distribution Common to uncommon transient (Oct–Nov, Mar–Apr) throughout; uncommon to rare and local winter visitor (Dec–Feb) mainly along the coast from Cape May south and in bottomland swamps and marshes.
Where to Find Dyke Marsh, Virginia; Allegheny National Forest, Pennsylvania; Dennis Creek Wildlife Management Area, New Jersey.
Range Breeds in boreal coniferous forest and bogs across the northern tier of North America; winters in the eastern United States.

Brewer's Blackbird

Euphagus cyanocephalus

(L-9 W-15)

Entirely black with purplish gloss on head (in proper light); yellow eye; some fall males are tinged rusty.

Female: Dark brown above with dark brown eye, slightly paler below.

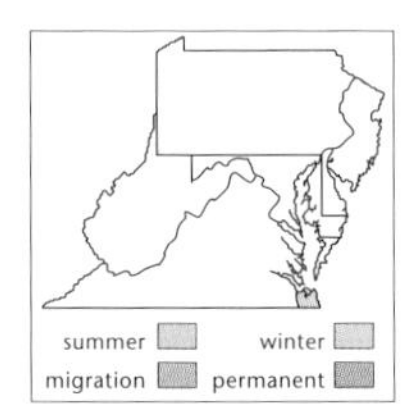

Voice Song—"chik-a-chik-a-perzee chik-a-chik-a-perzee"; call—"chik."
Similar Species The male Brewer's Blackbird is similar in color pattern to the Common Grackle (black with purplish gloss on head) but is much smaller (about the size of a Red-winged Blackbird) than the grackle (Blue Jay size) and has a relatively short, square-tipped tail, not a long, graduated tail like the grackle. The female Brewer's Blackbird is solid brown below (not buffy with faint gray streaking, as is the smaller female Brown-headed Cowbird).
Habitat Grasslands, pastures, agricultural fields; feedlots, grain elevators.
Abundance and Distribution Rare fall and winter visitor (Nov–Mar) mainly along the immediate coast of southern Virginia; scarcer elsewhere in the region.
Where to Find Back Bay National Wildlife Refuge, Virginia.
Range Breeds in the western and central United States and Canada; winters in the breeding range from southwestern Canada and the western United States southward into the southern United States to central Mexico.

Common Grackle

Quiscalus quiscula
(L-12 W-17)

Entirely black with purple gloss on head in proper light; tail long and rounded; cream-colored eye. *Female:* Dull black; whitish eye; tail not as long as male's.

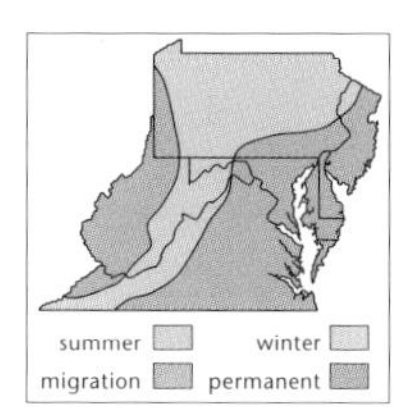

Voice Song—a repeated squeak, like a rusty hinge, with "chek"s interspersed; call—"chek."
Similar Species Cowbirds and blackbirds have short, square tails (not long and rounded as in grackles). The female Boat-tailed Grackle has a longer tail and is buffy rather than black.
Habitat Open woodlands, urban areas, agricultural fields, pastures; feeders; grain storage areas.
Abundance and Distribution Common summer resident* (Mar–Oct) throughout; common although local winter resident (Nov–Feb) in western West Virginia, Piedmont, Coastal Plain, and at scattered lowland localities elsewhere in the region, often concentrated in large flocks at roosts and feeding sites.
Where to Find The Mall, Washington, D.C.; parks in Philadelphia, Pennsylvania; parks in Newark, New Jersey.
Range Breeds east of the Rockies in Canada and United States; winters in the southern half of the breeding range.

Boat-tailed Grackle
Quiscalus major
(Male L-17 W-23; Female L-13 W-18)

Entirely black with purplish gloss on head (in proper light); tail wedge-shaped and longer than body; creamy eye. Female: Brown above, paler below; buffy eyebrow; brown eye; wedge-shaped tail not as long as male's.

Voice Song—a series of harsh "eeek"s, "aaahhk"s, and similar squawks.
Similar Species The female Boat-tailed Grackle has a longer tail than the female Common Grackle and is buffy rather than black.

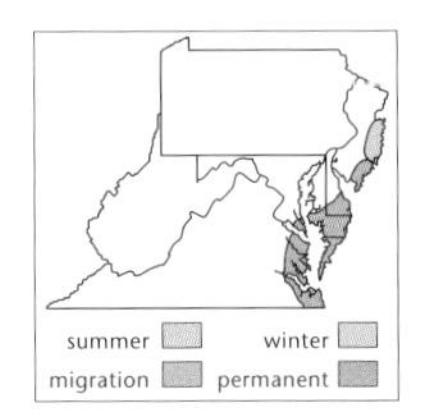

Habitat Coastal marshes.
Abundance and Distribution Common but local summer resident* (Apr–Oct) along the immediate coast; winter resident (Nov–Mar) from Cape May southward, although in lower numbers.
Where to Find South Cape May Meadows, New Jersey; Chincoteague National Wildlife Refuge, Virginia; Point Lookout State Park, Maryland.
Range Coastal eastern United States from New York to Texas.

Brown-headed Cowbird
Molothrus ater
(L-7 W-13)

Black body; brown head. *Female:* Brown above, paler below with grayish streakings.

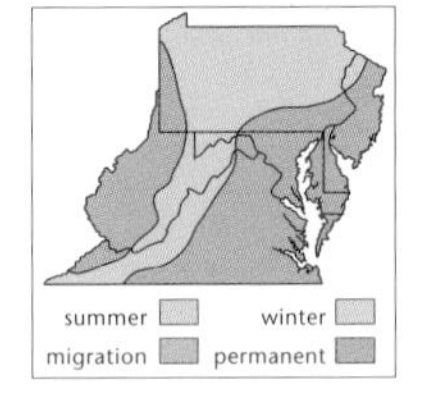

Habits A social parasite, laying its eggs in other birds' nests.
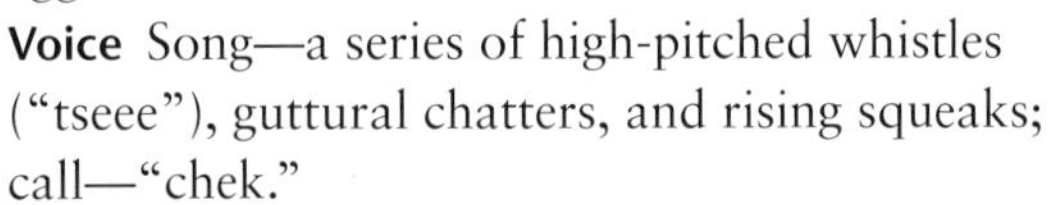
Voice Song—a series of high-pitched whistles ("tseee"), guttural chatters, and rising squeaks; call—"chek."
Similar Species The female Brown-headed Cowbird is pale below with faint grayish streaks; the larger female Brewer's Blackbird is solid brown below.
Habitat Pastures, agricultural fields, feedlots, grain elevators, scrub, open woodlands.
Abundance and Distribution Common transient and summer resident* (Apr–Oct) throughout; common although local winter resident (Nov–Mar) in western West Virginia, Piedmont, Coastal Plain, and at scattered lowland localities elsewhere in the region, often in large flocks.
Where to Find National Zoological Park, Washington, D.C.; Brandywine Creek State Park, Delaware; Beech Fork State Park, West Virginia.

Range Breeds across most of North America from south of the Arctic to central Mexico; winters in the southern half of its breeding range.

Orchard Oriole

Icterus spurius

(L-7 W-10)

Black hood, wings, and tail; chestnut belly, wing patch, lower back, and rump. *Female:* Greenish yellow above, yellow below, with 2 white wing bars; blue-gray legs. *Second Year Male:* Like female but with black throat and breast.

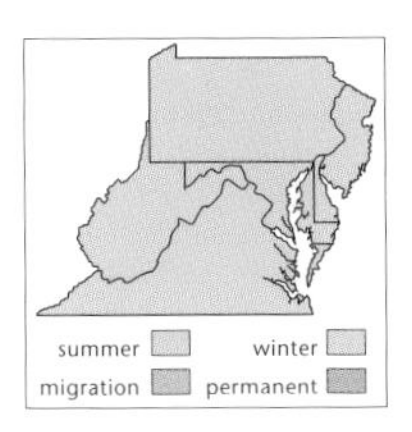

Voice Song—"tsee tso tsee tsoo tewit tewit tewit tseerr" and similar wandering twitters; call—"kuk."

Similar Species The female Orchard Oriole has an unstreaked greenish yellow back; the larger female Baltimore Oriole is mottled tawny brown and black on the back.

Habitat Riparian woodlands, orchards, brushy pastures, scrub.

Abundance and Distribution Common to uncommon summer resident* (May–Aug) at lowlands and mid-elevations from southern Pennsylvania and New Jersey south throughout the region; uncommon to rare and local in northern areas and highlands.

Where to Find Seneca Creek State Park, Maryland; Keystone State Park, Pennsylvania; Huntley Meadows Park, Virginia.

Range Breeds across the eastern and central United States south into central Mexico; winters from central Mexico south to northern South America.

Baltimore Oriole

Icterus galbula

(L-8 W-12)

Black hood, back, and wings; tail black at base and center, but outer terminal portions orange; orange belly, rump, and shoulder patch. *Female and Immature:* Orange-brown above, yellow-orange below with varying amounts of black on face and throat; white wing bars.

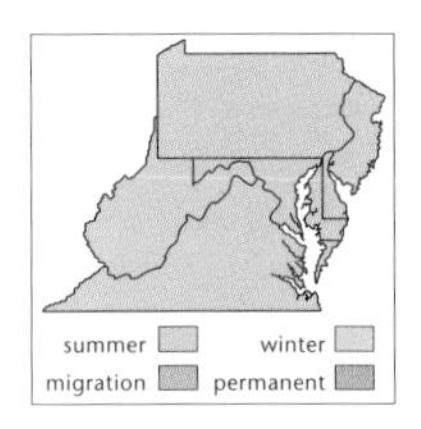

Voice Song—slurred rapid series of easily imitated whistles; call—a chatter, also "wheweee."
Similar Species The female Baltimore Oriole has an orangish belly; the smaller, female Orchard Oriole is yellowish or greenish yellow on the belly.
Habitat Riparian woodland (sycamore, cottonwood, willow), orchards, open deciduous woodlands, hedgerows, second growth.
Abundance and Distribution Common transient (Aug–Sep, May) throughout; common to uncommon summer resident* (May–Aug) nearly throughout except along Coastal Plain, where scarce and local; rare along coast in winter (feeders).
Where to Find C&O Canal, Maryland; Bluestone Wildlife Management Area, West Virginia; Great Swamp National Wildlife Refuge, New Jersey.
Range Breeds in the eastern United States and southeastern Canada; winters from central Mexico to northern South America, also in Greater Antilles.

Old World Finches
(Family Fringillidae)

Like many of the emberizids (buntings, New World sparrows), finches are small to medium-sized birds with thick, conical bills for eating seeds and fruits.

Pine Grosbeak
Pinicola enucleator
(L-9 W-14)

Rosy head, breast, and rump; gray flanks and belly; black back, wings, and tail; white wing bars; heavy, black grosbeak bill. *Female:* Gray body; head tinged with yellow; whitish shading below eye; white wing bars.

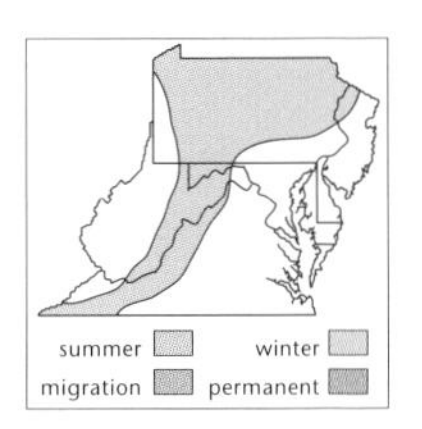

Voice Song—a weak twitter, rising and falling; call—a faint "che chu."
Habitat Breeds in coniferous forest, often bordering streams or tarns; found in hemlocks, pines, sumac, mountain ash, and similar fruiting trees in winter, often in small flocks.
Abundance and Distribution Rare to casual and irregular winter visitor (Nov–Mar) throughout, more numerous in northern portions and highlands of the region.
Where to Find Cook Forest State Park, Pennsylvania; Buzzard Swamp Wildlife Management Area, Pennsylvania.
Range Breeds in boreal forest of Old and New World, south to New Mexico in the Rockies; winters in southern portions of the breeding range south into north temperate regions of the Old and New World.

Purple Finch

Carpodacus purpureus

(L-6 W-10)

Rosy head and breast; whitish belly; brown above suffused with rose; rose rump; tail is notched; undertail coverts white. *Female and Immature Male:* Brown above; white below heavily streaked with brown; brown head with white eyebrow and malar stripes; undertail coverts white.

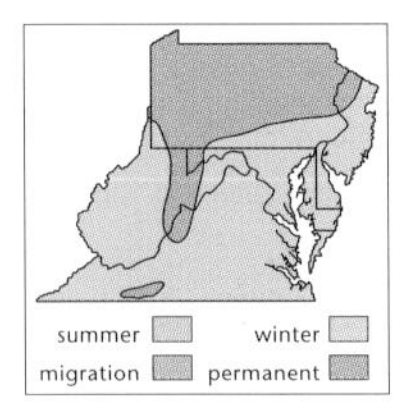

Voice Song—a rapid, tumbling series of slurred notes and trills; call—"chik."

Similar Species The "chit" call note of the Purple Finch is distinct from the nasal "wink" call note of the House Finch. The male Purple Finch is rosier overall and lacks brown streaking on the flanks of the male House Finch. The female Purple Finch has a pronounced dark jaw stripe and white eyeline, which the female House Finch lacks.

Habitat Coniferous forest and parklands; feeders in winter.

Abundance and Distribution Common to uncommon summer resident* (May–Aug) in northern Pennsylvania and in southern Pennsylvania in the mountains, the highlands (above 3,000 feet) of western Maryland, northwestern New Jersey (rare), eastern West Virginia, and Virginia (Highland Co., Mt. Rogers area); common to uncommon and local transient and winter resident (Oct–Apr) elsewhere.

Where to Find Woodbine Recreation Area, West Virginia (breeding); Erie National Wildlife Refuge (resident); National Arboretum, Washington, D.C. (winter).

Range Breeds across Canada, northern United States, and western mountains of United States south into Baja California; winters throughout much of the United States except the Great Plains and western deserts.

House Finch

Carpodacus mexicanus

(L-6 W-10)

Brown above with red brow stripe; brown cheeks; rosy breast; whitish streaked with brown below. *Female and Immature:* Brown above; buffy head and underparts finely streaked with brown; buffy preorbital stripe in some.

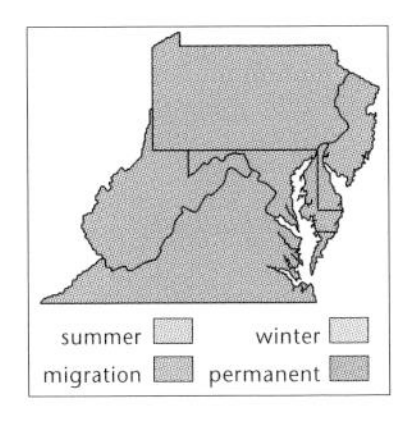

Voice Song—a long series of squeaks and warbles; call—a nasal "wink."

Similar Species The nasal "wink" call note of the House Finch is easily distinguishable from the sharp "chik" of the Purple Finch. The House Finch has a square tail and prominent streaking on flanks and undertail coverts; Purple Finch has a notched tail and white (unstreaked) undertail coverts. The male Purple Finch is rosier overall and lacks brown streaking on the flanks of the male House Finch. The female Purple Finch has a pronounced dark jaw stripe and white eyeline, which the female House Finch lacks.

Habitat Scrub, old fields; agricultural, residential, and urban areas; usually nests in scrubby conifers, including ornamentals; common at feeders.

Abundance and Distribution Common to uncommon resident* nearly throughout; some southward migration in winter.

Where to Find National Arboretum, Washington, D.C.; Virginia State Arboretum, White Post, Virginia; John Heinz National Wildlife Refuge, Pennsylvania.

Range Resident from southwestern Canada throughout much of the United States except Great Plains, south to southern Mexico.

Red Crossbill

Loxia curvirostra

(L-7 W-11)

Red with dark wings and tail; crossed bill. *Female and Immature:* Yellowish with dark wings and tail.

Voice Song—2-note "tsoo tee" repeated 3 or 4 times followed by a trill; call—repeated "kip" notes in flight.

Similar Species The "chri-chri" call notes of the White-winged Crossbill are distinct from the "kip" call note of the Red Crossbill; White-winged Crossbill has 2 white wing bars, which the Red Crossbill lacks.

Habitat A variety of coniferous species during nonbreeding periods.

Abundance and Distribution Rare and irregular winter visitor (Oct–Apr); travels in small flocks widely during the nonbreeding period, wherever pine nut crops are available; much more numerous in some years than others; has bred* in the region.

Where to Find Hoffman Sanctuary, New Jersey; Mountain Lake, Virginia; Allegheny National Forest, Pennsylvania.

Range Resident in boreal regions of the Old and New World, south in the mountains of the west through Mexico and the highlands of Central America to Nicaragua; winters in breeding range and irregularly south in north temperate regions of the world.

White-winged Crossbill

Loxia leucoptera

(L-7 W-11)

Rosy red body with dark band across back; black tail and wings with broad white wing bars; crossed bill. *Female and Immature:* Brown tinged with yellow above; yellowish below; dark wings with white wing bars.

Voice Song—trilled phrases at various speeds and pitches; call—a raspy "chri-chri-chri-chri."
Similar Species The "chri-chri" call notes of the White-winged Crossbill are distinct from the "kip" call note of the Red Crossbill; White-winged Crossbill has two white wing bars, which the Red Crossbill lacks.
Habitat Coniferous and mixed forest.
Abundance and Distribution Rare and irregular winter visitor (Oct–Apr); travels in small flocks widely during the nonbreeding period, wherever pine nut crops are available; more numerous in some years than others.
Where to Find Buzzard Swamp Wildlife Management Area, Pennsylvania; Mount Rogers, Virginia; Pine Barrens, Burlington County, New Jersey.
Range Resident in boreal northern tier of North America ranging south in winter into northern temperate zone.

Common Redpoll

Carduelis flammea

(L-5 W-9)

Front half of crown red; back of head and back streaked brown and white; black throat; white below tinged with rose on breast and flanks; brown streaks on flanks. *Female:* Little or no rose on breast.

Voice Song—a series of chips, churrs, and buzzy trills with whiney "tsewee" calls thrown in.
Similar Species Hoary Redpoll is much paler with faint or no streaking below.
Habitat Agricultural fields, brushy pastures, feeders.
Abundance and Distribution Rare and irregular winter visitor (Nov–Mar) in northern portions and highland meadows of the region; casual elsewhere. Travels in small flocks widely during the nonbreeding period; more numerous in some years than others.
Where to Find Jennings Environmental Education Center and Nature Reserve, Pennsylvania; Long Branch Nature Center, Virginia; Brigantine National Wildlife Refuge, New Jersey.
Range Breeds in the high Arctic of both Old and New World; winters in boreal and northern temperate regions.

Pine Siskin

Carduelis pinus
(L-5 W-9)

Streaked brown above; whitish below with brown streaks; yellow wing patch and at base of tail.

Voice Song—a sequence of "chipy chipy" notes interspersed with raspy, rising "zeeeech" calls; call—a nasal "schreee."
Similar Species The heavily streaked body and yellow patches on wings and base of tail distinguish this bird from other small finches.
Habitat A wide variety of coniferous and mixed forests; feeders in winter.

Abundance and Distribution Uncommon to rare and local summer resident* (Apr–Sep) in spruce forests of northern and highland Pennsylvania; has bred* in other parts of the region (New Jersey, West Virginia, Virginia) but only sporadically; common to uncommon or rare and irregular winter visitor (Nov–Mar), scarcer and less regular in southern parts of the region. Travels in small flocks widely during the nonbreeding period; more numerous in some years than others.
Where to Find National Arboretum, Washington, D.C. (winter); Allegheny National Forest, Pennsylvania (summer); Mountain Lake, Virginia (summer some years).
Range Breeds in boreal regions of northern North America and in western mountains south through United States and highlands of Mexico to Veracruz; winters in all except the extreme northern portions of the breeding range and in most of the temperate United States.

American Goldfinch

Carduelis tristis

(L-5 W-9)

Yellow body; black cap, wings, and tail; white at base of tail and wing bar; yellow shoulder patch. *Female and Winter Male:* Brownish above; yellowish or buff breast; whitish belly; dark wings with white wing bars.

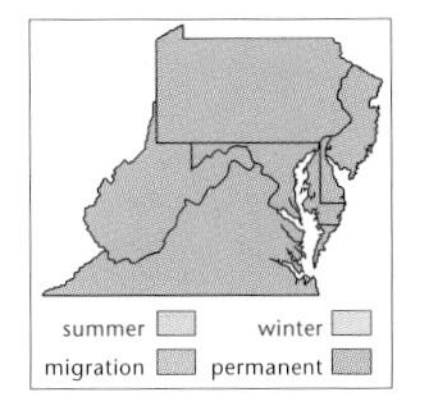

Habits Usually in flocks; dipping-soaring flight, like a roller coaster; almost always giving characteristic flight call—"ker-chik ker-chik-chik-chik."
Voice Song—a sequence of trills, whiney "tsoowee"s, and "ker-chik"s; flight call—a charac-

teristic and unmistakable "ker-chik ker-chik-chik-chik."

Similar Species The American Goldfinch female is the only small bird in the region with a conical bill that is brownish above and whitish below.

Habitat Grasslands, brushy pastures, old fields; feeders in winter and spring.

Abundance and Distribution Common summer resident* (Apr–Oct) throughout; common to uncommon winter resident (Nov–Mar) in lowlands and mid-elevations, often in flocks at feeders and in brushy fields; numbers vary widely in winter, depending on food availability and weather.

Where to Find Meadowside Nature Center, Maryland; Brandywine Creek State Park, Delaware; Beech Fork State Park, West Virginia.

Range Breeds across southern Canada, northern and central United States to southern California and northern Baja California in the west; winters in central and southern United States and northern Mexico.

Evening Grosbeak

Coccothraustes vespertinus

(L-8 W-13)

A chubby bird with heavy, yellowish or whitish bill; yellow body; black crown and brownish head with yellow forehead and eyebrow; black tail and wings with white wing patch. *Female:* Grayish above, buffy below; dark malar stripe; white wing patch.

Habits Usually in flocks.

Voice Calls—a sharp "peak" and a hoarse "peer."

Habitat Coniferous and mixed forest; often at feeders in winter.

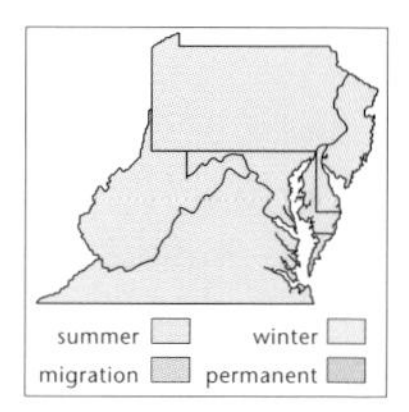

Abundance and Distribution Common to uncommon or rare and highly irregular winter visitor (Nov–Mar) throughout, often in flocks in conifers and at feeders; numbers vary widely in winter, depending on food availability and weather.
Where to Find Meadowside Nature Center feeding station, Maryland; R. B. Winter State Park, Pennsylvania; James River Park, Virginia.
Range Breeds in boreal portions of central and southern Canada, northern and western United States, south in western mountains to western and central Mexico; winters in breeding range and in temperate and southern United States.

Old World Sparrows
(Family Passeridae)

This small group, formerly considered to be members of the same family as weaver finches (Ploceidae), is native to the Old World.

House Sparrow
Passer domesticus
(L-6 W-10)

A chunky, heavy-billed bird; brown above with heavy, dark brown streaks; dingy gray below; gray cap; chestnut nape; black lores, chin, and bib. *Female:* Streaked buff and brown above; dingy gray below; pale buff postorbital stripe.

Voice Song—"chip cheap chip chip chi-chi-chi chip . . . "; call—"cheap."
Similar Species Pale female and immature Dickcissels resemble House Sparrow female but usually

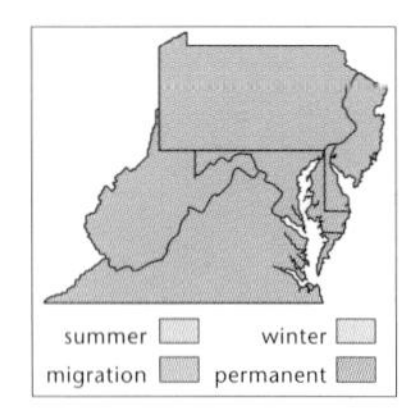

have traces of yellow on pale white (not dirty white) breast, a clear, whitish or yellowish eyebrow, and some chestnut on shoulder.

Habitat Urban areas, pastures, agricultural fields, feed lots, farms, grain elevators.

Abundance and Distribution Common resident* throughout.

Where to Find The Mall, Washington, D.C.; Baltimore, Maryland; Philadelphia, Pennsylvania; Newark, New Jersey.

Range Resident in boreal, temperate, and subtropical regions of Old and New World; currently expanding into tropical regions. Introduced into the Western Hemisphere in 1850.

Appendix

Hypothetical, Casual, Accidental, and Extinct Species

Western Grebe	*Aechmophorus occidentalis*
Clark's Grebe	*Aechmophorus clarkii*
Yellow-nosed Albatross	*Thalassarche chlororhynchos*
Kermadec Petrel	*Pterodroma neglecta*
Black-capped Petrel	*Pterodroma hasitata*
White-faced Storm-Petrel	*Pelagodroma marina*
Band-rumped Storm-Petrel	*Oceanodroma castro*
White-tailed Tropicbird	*Phaethon lepturus*
Brown Booby	*Sula leucogaster*
Magnificent Frigatebird	*Fregata magnificens*
Reddish Egret	*Egretta rufescens*
White-faced Ibis	*Plegadis chihi*
Roseate Spoonbill	*Ajaia ajaja*
Greater Flamingo	*Phoenicopterus ruber*
Lesser White-fronted Goose	*Anser erythropus*
Ross's Goose	*Chen rossii*
Barnacle Goose	*Branta leucopsis*
Trumpeter Swan	*Cygnus buccinator*
Cinnamon Teal	*Anas cyanoptera*
White-cheeked Pintail	*Anas bahamensis*
Tufted Duck	*Aythya fuligula*
Barrow's Goldeneye	*Bucephala islandica*
Masked Duck	*Nomonyx dominicus*
American Swallow-tailed Kite	*Elanoides forficatus*
Mississippi Kite	*Ictinia mississippiensis*
White-tailed Eagle	*Haliaeetus albicilla*
Swainson's Hawk	*Buteo swainsoni*
Ferruginous Hawk	*Buteo regalis*
Eurasian Kestrel	*Falco tinnunculus*
Gyrfalcon	*Falco rusticolus*
Greater Prairie-Chicken (Heath Hen)	*Tympanuchus cupido* (EXTIRPATED)
Corn Crake	*Crex crex*
Paint-billed Crake	*Neocrex erythrops*
Spotted Rail	*Pardirallus maculatus*
Limpkin	*Aramus guarauna*
Sandhill Crane	*Grus canadensis*
Whooping Crane	*Grus americana*
Northern Lapwing	*Vanellus vanellus*
Mongolian Plover	*Charadrius mongolus*

Mountain Plover	*Charadrius montanus*
Spotted Redshank	*Tringa erythropus*
Eskimo Curlew	*Numenius borealis* (EXTINCT)
Black-tailed Godwit	*Limosa limosa*
Bar-tailed Godwit	*Limosa lapponica*
Temminck's Stint	*Calidris temminckii*
Sharp-tailed Sandpiper	*Calidris acuminata*
Curlew Sandpiper	*Calidris ferruginea*
Spoonbill Sandpiper	*Eurynorhynchus pygmeus*
Great Snipe	*Gallinago media*
Eurasian Woodcock	*Scolopax rusticola*
South Polar Skua	*Stercorarius maccormicki*
Mew Gull	*Larus canus*
California Gull	*Larus californicus*
Yellow-legged Gull	*Larus cachinnans*
Thayer's Gull	*Larus thayeri*
Sabine's Gull	*Xema sabini*
Ivory Gull	*Pagophila eburnea*
Elegant Tern	*Sterna elegans*
White-winged Tern	*Chlidonias leucopterus*
Whiskered Tern	*Chlidonias hybridus*
Brown Noddy	*Anous stolidus*
Common Murre	*Uria aalge*
Thick-billed Murre	*Uria lomvia*
Black Guillemot	*Cepphus grylle*
Atlantic Puffin	*Fratercula arctica*
White-winged Dove	*Zenaida asiatica*
Passenger Pigeon	*Ectopistes migratorius* (EXTINCT)
Monk Parakeet	*Myiopsitta monachus*
Carolina Parakeet	*Conuropsis carolinensis* (EXTINCT)
Groove-billed Ani	*Crotophaga sulcirostris*
Northern Hawk Owl	*Surnia ulula*
Burrowing Owl	*Athene cunicularia*
Boreal Owl	*Aegolius funereus*
Black-chinned Hummingbird	*Archilochus alexandri*
Calliope Hummingbird	*Stellula calliope*
Rufous Hummingbird	*Selasphorus rufus*
Three-toed Woodpecker	*Picoides tridactylus*
Black-backed Woodpecker	*Picoides arcticus*
Ivory-billed Woodpecker	*Campehilus principalis*
Western Wood-Pewee	*Contopus sordidulus*
Pacific-Slope Flycatcher	*Empidonax oberholseri*
Say's Phoebe	*Sayornis saya*
Vermilion Flycatcher	*Pyrocephalus rubinus*

Ash-throated Flycatcher	*Myiarchus cinerascens*
Great Kiskadee	*Pitangus sulphuratus*
Cassin's Kingbird	*Tyrannus vociferans*
Gray Kingbird	*Tyrannus dominicensis*
Fork-tailed Flycatcher	*Tyrannus forficatus*
Scissor-tailed Flycatcher	*Tyrannus savana*
Bell's Vireo	*Vireo bellii*
Gray Jay	*Perisoreus canadensis*
Black-billed Magpie	*Pica hudsonia*
Violet-green Swallow	*Tachycineta thalassina*
Cave Swallow	*Petrochelidon fulva*
Boreal Chickadee	*Poecile hudsonica*
Rock Wren	*Salpinctes obsoletus*
Northern Wheatear	*Oenanthe oenanthe*
Mountain Bluebird	*Sialia currucoides*
Varied Thrush	*Ixoreus naevius*
Sage Thrasher	*Oreoscoptes montanus*
Sprague's Pipit	*Anthus spragueii*
Bohemian Waxwing	*Bombycilla garrulus*
Bachman's Warbler	*Vermivora bachmanii* (EXTINCT)
Black-throated Gray Warbler	*Dendroica nigrescens*
Townsend's Warbler	*Dendroica townsendi*
Sutton's Warbler	*Dendroica potomac (Hybrid? Dendroica dominica × Parula americana)*
Kirtland's Warbler	*Dendroica kirtlandii*
Western Tanager	*Piranga ludoviciana*
Green-tailed Towhee	*Pipilo chlorurus*
Spotted Towhee	*Pipilo maculatus*
Canyon Towhee	*Pipilo fuscus*
Black-throated Sparrow	*Amphispiza bilineata*
Harris's Sparrow	*Zonotrichia querula*
Golden-crowned Sparrow	*Zonotrichia atricapilla*
Baird's Sparrow	*Ammodramus bairdii*
Smith's Longspur	*Calcarius pictus*
Chestnut-collared Longspur	*Calcarius ornatus*
Black-headed Grosbeak	*Pheucticus melanocephalus*
Lazuli Bunting	*Passerina amoena*
Painted Bunting	*Passerina ciris*
Western Meadowlark	*Sturnella neglecta*
Bullock's Oriole	*Icterus bullockii*
Brambling	*Fringilla montifringilla*
Hoary Redpoll	*Carduelis hornemanni*

Credits

Page numbers are listed after each photographer or illustrator (t, top; b, bottom). A second number in parentheses after a credit indicates that the photograph also appears in the Quick Guide to the Most Common Birds.

Vernon Eugene Grove Jr.: 107, 109b, 116, 118, 120, 123 (7), 124, 125, 126, 127, 128, 129tb, 130, 131, 133, 135 (6), 137t, 144, 146 (7), 148, 149, 167, 173, 175 (6), 181, 182, 183, 186, 189, 190, 192, 193t, 194, 196, 199, 201, 205, 206tb, 208t (6), 219b, 221, 224, 238 (6), 240b, 245, 252b (7), 256, 257b, 258, 259tb, 260, 262, 268b, 279b, 282, 283, 292, 298 (7), 302b, 317, 323 (6), 364, 365b, 366, 378, 381t, 385 (7), 386, 387b, 390, 401 (6)

David F. Parmelee: 105, 110, 114, 155, 159b, 171, 178t (6), 212, 213t, 222, 225, 229t, 237b, 250

John H. Rappole: 134, 140, 142, 143t, 145, 147, 150, 151tb, 152, 161, 162, 163, 166, 172, 177, 180, 188, 193b, 195, 197 (6), 200, 203b, 204, 209, 211, 215b, 218, 229b (6), 235, 240, 241, 242, 246t, 248t, 252t (6), 265, 266, 269, 270 (6), 271t, 272, 275t, 276 (7), 280, 290b (6), 291, 293, 294, 295t, 299 (7), 300, 301, 302t, 306, 325, 339, 359, 361, 372, 376b, 379 (6), 389, 395, 399

Barth Schorre: 117tb (6), 137, 138 (7), 165, 191, 198, 203t, 207, 234, 239, 246b, 249, 253, 255, 257t, 268t, 273 (7), 277, 285, 286tb, 287, 289, 305, 314b, 315, 318, 319tb (6, 7), 321tb, 322, 326, 327, 328, 330tb, 331, 332, 333, 335, 336, 337, 338, 340, 341tb, 342, 343, 344, 345 (7), 346, 347t, 348, 349, 350b, 352, 353, 354, 355tb, 357, 358 (7), 362tb, 365t, 375, 376t, 380, 381b (6), 382, 384 (7), 388, 391, 392 (6), 398

VIREO (Academy of Natural Sciences): S. Bahrt, 393; G. Bailey, 369; R. & N. Bowers, 367, 373; J. Cameron, 360; John Cancalosi, 251; A. & S. Carey, 179; R. J. Chandler, 219b; H. Clarke, 159t, 324; R. Crossley, 227b; A. Cruickshank, 168; H. Cruickshank, 122; R. Curtis, 213b, 284, 310, 350t, 394; Rob Curtis, 351; R. & S. Day, 202; J. H. Dick, 156b; J. Dunning, 312; S. Faccio, 316; J.

Fuhrman, 230, 232, 295b (7), 296, 311; B. Gadsby, 143b; W. Greene, 261, 278, 334b, 400; J. S. Greenlaw, 370; C. E. Harold, 185b; J. Heidecker, 267 (7), 279t, 374; B. Henry, 304, 383, 396, 397tb; K. T. Karlson, 109t; J. Kinney, 112t; S. J. Lang, 187, 281; G. Lasley, 112b, 368; P. La Tourret, 297; G. LeBaron, 247; S. Maka, 271b; H. O. Miller, 248b; A. Morris, 106, 139, 154, 160, 164, 208b, 214, 217, 220, 223, 231, 233, 236, 237t, 243, 244, 264, 275b, 309; A. & E. Morris, 119, 132, 136, 371b; M. Patrikeev, 254, 387t; R. L. Pitman, 111, 113b, 115; Michel Robert, 184; F. Rowher, 157; J. Schumacher, 314t (7); R. & A. Simpson, 185t, 263; B. Small, 274, 329; Brian E. Small, 108, 158, 216, 288, 308, 334t, 377; M. Stubblefield, 210; D. Tipling, 113t, 153; F. Truslow, 303; T. J. Ulrich, 156t; T. Vezo, 121, 215t, 290t, 307b; Doug Wechsler, 371t; J. Wedge, 307t; B. K. Wheeler, 169, 176, 178b; C. Witt, 228; J. R. Woodward, 363; D. & M. Zimmerman, 170

Kevin Winker: 227t, 347b

References

American Ornithologists' Union. 1998. *Check-list of North American Birds*. Seventh Edition. Lawrence, Kans.: American Ornithologists' Union.

———. 2000. Forty-second supplement to the American Ornithologists' Union *Check-list of North American Birds*. *Auk* 117:847–858.

Bailey, H. H. 1913. *The Birds of Virginia*. Lynchburg, Va.: J. P. Bell Co.

Bartram, W. 1983. *Travels of William Bartram*, edited by M. Van Doren. New York: Dover Publications.

Boyle, W. J., Jr. 1986. *A Guide to Bird Finding in New Jersey*. New Brunswick, N.J.: Rutgers University Press.

Brauning, D. W. (ed.). 1992. *Atlas of Breeding Birds of Pennsylvania*. Pittsburgh: University of Pittsburgh Press.

Brett, J. J. 1991. *The Mountain and the Migration: A Guide to Hawk Mountain*. Ithaca: Cornell University Press.

Broun, M. 1999. *Hawks Aloft: The Story of Hawk Mountain*. Mechanicsburg, Pa.: Stackpole Books.

Buckelew, A. R, Jr., and G. A. Hall. 1994. *The West Virginia Breeding Bird Atlas*. Pittsburgh: University of Pittsburgh Press.

Bull, J., and J. Farrand Jr. 1977. *The Audubon Society Field Guide to North American Birds: Eastern Region*. New York: Alfred A. Knopf.

Clark, W. S., and B. K. Wheeler. 1987. *A Field Guide to Hawks of North America*. Boston: Houghton Mifflin Co.

Clement, P., A. Harris, and J. Davis. 1993. *Finches and Sparrows: An Identification Guide*. Princeton: Princeton University Press.

Cooke, W. W. 1915. *Bird Migration. USDA Bulletin* 185:1–47.

Coues, E., and D. W. Prentiss. 1883. *Avifauna Columbiana: Being a List of Birds Ascertained to Inhabit the District of Columbia, with the Times of Arrival and Departure of Such as Are Non-residents, and Brief Notices of Habits, etc.* Bull. U.S. National Mus., No. 26. Washington, D.C.: Smithsonian Institution.

Curson, J., D. Quinn, and D. Beadle. 1994. *New World Warblers.* London: A&C Black.

DeGraaf, R. M., and J. H. Rappole. 1995. *Neotropical Migratory Birds: Natural History, Distribution, and Population Change.* Ithaca, N.Y.: Cornell University Press.

DeGraaf, R. M., V. E. Scott, R. H. Hamre, L. Ernst, and S. H. Anderson. 1991. *Forest and Rangeland Birds of the United States: Natural History and Habitat Use.* Agriculture Handbook 688. Washington, D.C.: Forest Service, U.S. Department of Agriculture.

Dunne, P., and C. Sutton. 1989. *Hawks in Flight: Flight Identification of North American Migrant Raptors.* Boston: Houghton Mifflin.

Ford, P. 1995. *Birder's Guide to Pennsylvania.* Houston: Gulf Publishing Co.

Grant, P. J. 1982. *Gulls: A Guide to Identification.* Stoke on Trent, U.K.: Poyser, Calton.

Greenlaw, J. S. 1993. Behavioral and morphological diversification in Sharp-tailed Sparrows (*Ammodramus caudacutus*) of the Atlantic Coast. *Auk* 110:286–303.

Hall, G. A. 1983. *West Virginia Birds: Distribution and Ecology.* Special Publication 7. Pittsburgh: Carnegie Museum of Natural History.

Handley, C. S. 1976. *Birds of the Great Kanawha Valley (1770–1975).* Parsons, W.V.: McClain Printing Co.

Harrison, G. H. 1976. *Roger Tory Peterson's Dozen Birding Hot Spots.* New York: Simon and Schuster.

Harrison, P. 1991. *Seabirds: An Identification Guide.* Boston: Houghton Mifflin Co.

Hayman, P., J. Marchant, and T. Prater. 1986. *Shorebirds: An Identification Guide to the Waders of the World.* Boston: Houghton Mifflin Co.

Hines, R. W. 1985. *Ducks at a Distance: A Waterfowl Identification Guide.* Washington, D.C.: U.S. Fish and Wildlife Service.

Hunt, C. B. 1967. *Physiography of the United States.* San Francisco: W. H. Freeman and Co.

———. 1974. *Natural Regions of the United States and Canada.* San Francisco: W. H. Freeman and Co.

Johnsgard, P. A. 1979. *A Guide to North American Waterfowl.* Bloomington: Indiana University Press.

Johnston, D. W. 1997. *A Birder's Guide to Virginia*. Colorado Springs: American Birding Association.

———. 2000. *Mountain Lake Region and Its Bird Life*. Martinsville: Virginia Museum of Natural History.

Johnston, I. H. 1923. *Birds of West Virginia: Their Economic Value and Aesthetic Beauty*. Charleston, W.V.: State Department of Agriculture.

Kain, T. 1987. Virginia's birdlife: An annotated checklist. *Virginia Avifauna*, no. 3. Virginia Society of Ornithology.

King, J. 1953. *Telling Trees*. New York: William Sloane Associates.

Küchler, A. W. 1975. *Potential Natural Vegetation of the Coterminous United States*. Washington, D.C.: American Geographical Society.

Lanyon, W. E., and J. Bull. 1967. Identification of Connecticut, Mourning, and MacGillivray's warblers. *Bird-Banding* 38:187–194.

Leck, C. F. 1975. *Birds of New Jersey: Their Habits and Habitats*. New Brunswick, N.J.: Rutgers University Press.

———. 1984. *The Status and Distribution of New Jersey's Birds*. New Brunswick, N.J.: Rutgers University Press.

Lewis, T. 1746. *The Fairfax Line: Thomas Lewis's Journal of 1746; with Footnotes and an Index by John W. Wayland, Ph.D.* Printed by Henkel Press in 1925, New Market, Va.

Maryland Ornithological Society. 1986. *Field Checklist of Maryland and D.C.* Baltimore: Maryland Ornithological Society.

Maryland Ornithological Society, Montgomery County Chapter. 1995. *The Birds of Montgomery County*. Potomac, Md.: Maryland Ornithological Society.

Maynard, L. W. 1902. *Birds of Washington and Vicinity: Including Adjacent Parts of Maryland and Virginia*. Washington, D.C.: Woodward and Lothrop,

McWilliams, G. M., and D. W. Brauning. 2000. *The Birds of Pennsylvania*. Ithaca, N.Y.: Cornell University Press.

National Geographic Society. 1987. *Field Guide to the Birds of North America*. 2d Ed. Washington, D.C.: National Geographic Society.

National Park Service. 1994. *Birds of Shenandoah National Park*. Luray, Va.: Shenandoah Natural History Association.

Oberholser, H. 1974. *The Bird Life of Texas.* Austin: University of Texas Press.

Peterson, R. T. 1980. *A Field Guide to the Birds: A Completely New Guide to All the Birds of the Eastern and Central United States.* 4th Ed. Boston: Houghton Mifflin Co.

Phillips, A. R. 1975. Semipalmated Sandpiper identification, migration, summer and winter ranges. *American Birds* 29:799–806.

———. 1986. *The known birds of North and Middle America. Part I: Hirundinidae to Mimidae; Certhiidae.* Denver: Privately published.

———. 1991. *The known birds of North and Middle America. Part II: Bombycillidae; Sylviidae to Sturnidae; Vireonidae.* Denver: Privately published.

Phillips, A. R., M. A. Howe, and W. E. Lanyon. 1966. Identification of the flycatchers of eastern North America with special emphasis on the genus *Empidonax. Bird-Banding* 37:153–171.

Rising, J. D. 1996. *A Guide to the Identification and Natural History of the Sparrows of the United States and Canada.* New York: Academic Press.

Rives, W. C. 1889–1890. *A Catalogue of the Birds of the Virginias.* Proceedings of the Newport Natural History Society, Document 7. Newport, R.I.: Newport Natural History Society.

Robbins, C. S. (ed.). 1996. *Atlas of the Breeding Birds of Maryland and the District of Columbia.* Pittsburgh: University of Pittspurgh Press.

Robbins, C. S., and D. Bystrak. 1977. *Field List of the Birds of Maryland.* 2nd Ed. Baltimore: Maryland Ornithological Society.

Robbins, C. S., B. Brunn, and H. S. Zimm. 1983. *Birds of North America: A Guide to Identification.* Rev. Ed. New York: Golden Press.

Shriner, C. A. 1896. *The Birds of New Jersey.* Trenton, N.J.: Fish and Game Commission of the State of New Jersey.

Sibley, D. 1993. *The Birds of Cape May.* Cape May Point, N.J.: Cape May Bird Observatory.

Simpson, M. B., Jr. 1992. *Birds of the Blue Ridge Mountains: A Guide for the Blue Ridge Parkway, Great Smoky*

Mountains, Shenandoah National Park, and Neighboring Areas. Chapel Hill, N.C.: University of North Carolina Press.

Stewart, R. E., and C. S. Robbins. 1958. *Birds of Maryland and the District of Columbia.* North American Fauna, no. 62. Washington, D.C.: U.S. Fish and Wildlife Service.

Stokes, D. W., and L. Q. Stokes. 1996. *Stokes Field Guide to Birds: Eastern Region.* Boston: Little, Brown and Co.

Stone, W. 1937. *Bird Studies at Old Cape May,* vols. 1 and 2. Philadelphia: Delaware Ornithological Club.

Talbot, W. 1672. *The discoveries of John Lederer, in three several marches from Virginia, to the west of Carolina, and other parts of the continent: begun in March 1669, and ended in September 1670, together with a general map of the whole territory which he traversed.* Printed by F. C. for Samuel Heyrick at Grays-Inne Gate in Holborn, London. Reprinted by George P. Humphrey, Rochester, N.Y., in 1902.

Todd, W. E. C. 1940. *Birds of Western Pennsylvania.* Pittsburgh: University of Pittsburgh Press.

Warren, B. H. 1890. *Report on the Birds of Pennsylvania.* 2nd Ed. Harrisburg, Pa.: State Board of Agriculture.

Wayland, J. W. 1989. *Twenty-five Chapters on the Shenandoah Valley.* 4th Ed. Harrisonburg, Va: C. J. Carrier Co.

Wilds, C. 1992. *Finding Birds in the National Capital Area.* 2nd Ed. Washington, D.C.: Smithsonian Institution Press.

Wood, M. 1979. *Birds of Pennsylvania.* University Park, Pa.: Pennsylvania State University.

Index

Numbers in **boldface** refer to species accounts; numbers in *italics* refer to illustrations.